SELECTED SOLUTIONS

for

PHYSICS

Third Edition Revised

DAVID HALLIDAY

ROBERT RESNICK

EDWARD DERRINGH

Prepared by

EDWARD DERRINGH

Wentworth Institute of Technology
Boston, Massachusetts

JOHN WILEY & SONS NEW YORK
CHICHESTER BRISBANE TORONTO SINGAPORE

ISBN 0 471 09712 8
Printed in the United States of America

10 9

CONTENTS

Chapter 1 1
Chapter 2 5
Chapter 3 14
Chapter 4 22
Chapter 5 32
Chapter 6 42
Chapter 7 53
Chapter 8 62
Chapter 9 74
Chapter 10 83
Chapter 11 94
Chapter 12 99
Chapter 13 110
Chapter 14 119
Chapter 15 126
Chapter 16 139
Chapter 17 153
Chapter 18 162
Chapter 19 169
Chapter 20 178
Chapter 21 185
Chapter 22 191
Chapter 23 197
Chapter 24 207
Chapter 25 213

Chapter 26 219
Chapter 27 225
Chapter 28 233
Chapter 29 243
Chapter 30 253
Chapter 31 263
Chapter 32 269
Chapter 33 280
Chapter 34 288
Chapter 35 299
Chapter 36 306
Chapter 37 312
Chapter 38 316
Chapter 39 321
Chapter 40 328
Chapter 41 333
Chapter 42 339
Chapter 43 345
Chapter 44 351
Chapter 45 361
Chapter 46 369
Chapter 47 374
Chapter 48 382
Chapter 49 387
Chapter 50 396

CHAPTER 1

1-6

(a) 1 light-year = (186,000 mi/s)(3.156×10^7 s) = 5.870×10^{12} mi so that

$$1 \text{ mi} = \frac{1}{5.870 \times 10^{12}} = 1.7 \times 10^{-13} \text{ ly}.$$

Therefore,

$$1 \text{ AU} = 92.9 \times 10^6 \text{ mi} = (92.9 \times 10^6 \text{ mi})(1.7 \times 10^{-13} \text{ ly/mi}),$$

$$1 \text{ AU} = 1.58 \times 10^{-5} \text{ ly} \quad \underline{\text{Ans}}.$$

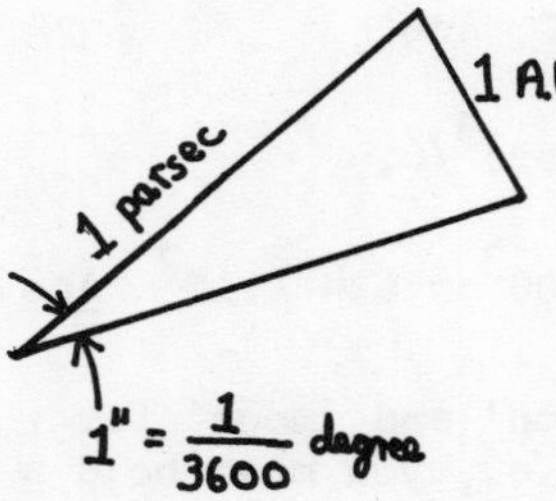

Since 1 radian = 206265", then by definition,

$$206265 \text{ AU} = 1 \text{ pc};$$

hence,

$$1 \text{ AU} = \frac{1}{206265} \text{ pc},$$

$$1 \text{ AU} = 4.85 \times 10^{-6} \text{ pc} \quad \underline{\text{Ans}}.$$

(b) From (a), 1 ly = 5.87×10^{12} mi Ans.
Since 1 AU = 4.85×10^{-6} pc = 15.8×10^{-6} ly, it follows that

$$1 \text{ pc} = \frac{15.8}{4.85} = 3.26 \text{ ly}$$

and therefore

$$1 \text{ pc} = (3.26)(5.87 \times 10^{12} \text{ mi}) = 1.91 \times 10^{13} \text{ mi} \quad \underline{\text{Ans}}.$$

1-9

The apparent angular diameters θ of the sun and moon in the sky are virtually identical. Thus the situation at eclipse is as shown in the sketch. Let the earth-moon distance be r, the earth-sun distance R, and call d the diameter of the moon and D that of the sun.

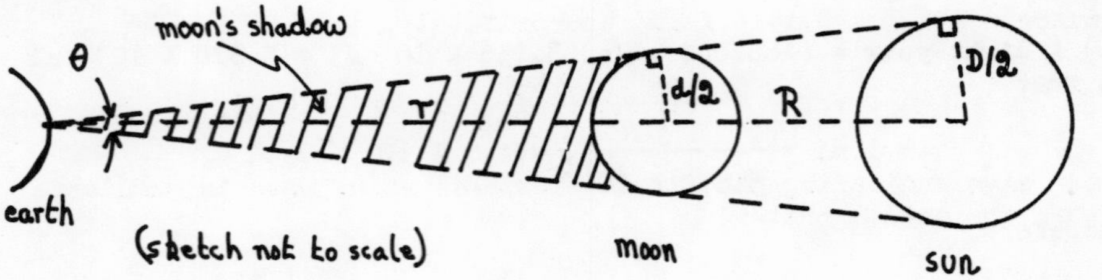

(a) From the sketch,

$$\sin(\tfrac{1}{2}\theta) = \tfrac{1}{2}d/r = \tfrac{1}{2}D/R,$$

$$D/d = R/r = (400\ r)/r = 400 \quad \text{Ans.}$$

(b) Since the volume of a sphere = $\pi(\text{diameter})^3/6$,

$$\text{volume of sun/volume of moon} = D^3/d^3 = 400^3 = 6.4 \times 10^7 \quad \underline{\text{Ans}}.$$

(c) In the figure, substitute 'dime' for'moon' and 'moon' for 'sun'. You will find that θ = 0.0092 rad (i.e., you must hold a dime about 190 cm from your eye to just eclipse the full moon; do NOT try this with the sun). The average earth-moon distance is r = 380,000 km. Since θ is very small, sinθ ≈ θ and

$$d = r\theta = (3.8 \times 10^5\ \text{km})(0.0092) = 3500\ \text{km} \quad \underline{\text{Ans}}.$$

1-12

(a) Since 1 g = 10^{-3} kg and 1 ℓ = 1000 cm^3, it follows that $1\ g/cm^3 = (10^{-3}\ kg)/(10^{-3}\ \ell) = 1\ kg/\ell$ Ans.

(b) 1.0 liter of water contains 1.0 kg of water, from (a). Since 10 h = (10)(3600 s) = 36,000 s, the mass flow rate = mass/time is

$$1.0\ \text{kg}/36{,}000\ \text{s} = 2.78 \times 10^{-5}\ \text{kg/s} \quad \underline{\text{Ans}}.$$

1-13

The year used in civil affairs is the tropical year and this contains 365.2422 mean solar days. Using the old definition of the second (1 mean solar day = 86,400 seconds),

$$1 \text{ year} = (365.2422)(86{,}400 \text{ s}) = 3.1557 \times 10^7 \text{ s}.$$

Since $\pi \times 10^7 = 3.1416 \times 10^7$ approximately,

$$\% \text{ error} = \frac{3.1416 - 3.1557}{3.1557} \times 100 = -0.447\%,$$

the sign indicating that the approximate value used is smaller than the actual number.

1-16

Let f = average rotation rate of the earth during the year. From the graph, the deviation in f at midsummer is +60 and the deviation in spring is -90 (the values depend on the exact dates chosen for comparison), leading to a net deviation of 150 parts in 10^{10} of the average rate; that is

$$\text{net deviation in rate} = 150 \times 10^{-10} f.$$

The "total equivalent" rate is $f + 150 \times 10^{-10} f$, corresponding to a rotation period equal to

$$(f + 150 \times 10^{-10} f)^{-1} \approx f^{-1}(1 - 150 \times 10^{-10})$$
$$= (86{,}400 \text{ s})(1 - 150 \times 10^{-10}).$$

Hence, the difference in period between midsummer and spring is

$$(86{,}400)(150 \times 10^{-10}) \text{ s} = 1.3 \times 10^{-3} \text{ s} \quad \underline{\text{Ans}},$$

the period in midsummer being shorter.

<u>1-18</u>

The last day of the twenty centuries is longer than the first day by

$$(20 \text{ centuries})(10^{-3} \text{ s/century})$$

which is 0.020 s. The average day during the twenty centuries is (0 + 0.020)/2 = 0.010 s longer than the first day, since the increase occurs uniformly. The cumulative effect is

$$(\text{average difference})(\text{number of days})$$

$$= (0.010 \text{ s/average day})(365.25 \text{ days} \times 2000),$$

$$= 7305 \text{ s} = 2.0 \text{ h} \quad \underline{\text{Ans}}.$$

CHAPTER 2

2-3

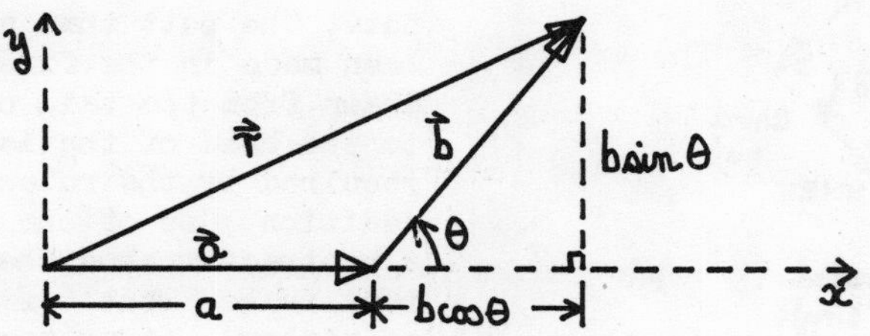

Place $\vec{a}$ along the x-axis. The components of $\vec{b}$ are

$$b_x = b\cos\theta,\ b_y = b\sin\theta.$$

Hence, applying the Pythagorean theorem,

$$r^2 = (a + b\cos\theta)^2 + (b\sin\theta)^2 = a^2 + 2ab\cos\theta + b^2,$$

since $\sin^2\theta + \cos^2\theta = 1$. The angle θ between $\vec{a}$ and $\vec{b}$ ranges from 0° to 360°. For 0°, cosθ = 1, its largest possible value; thus,

$$r_{max}^2 = a^2 + 2ab + b^2;\ r_{max} = a + b.$$

The smallest possible value of cosθ is for cos180° = -1 and therefore

$$r_{min}^2 = a^2 - 2ab + b^2 = (a - b)^2 = (b - a)^2,$$

$$r_{min} = |a - b|.$$

That is, $r_{min} = a - b$, or $b - a$, whichever is positive.

2-5

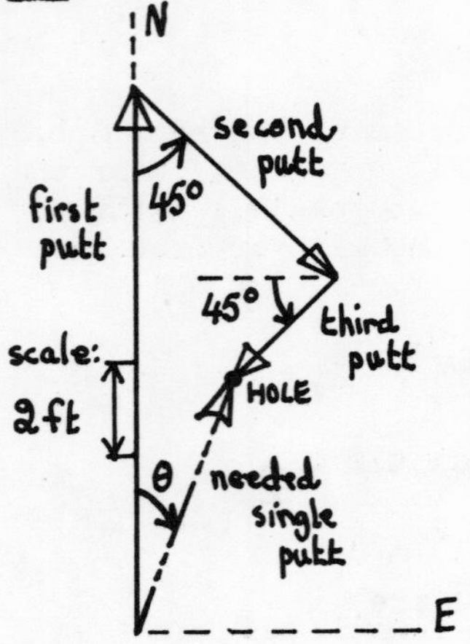

The displacements of the ball due to the successive putts are shown in the sketch, with the scale used shown on the first putt. The putt that should have been made in the first place is drawn from the tail of the first to the head of the last, as required by the rules of vector addition. Use of the scale and a protractor show the desired putt to be about 6 ft (length of vector), at an angle θ of about 20 degrees east of north.

2-10

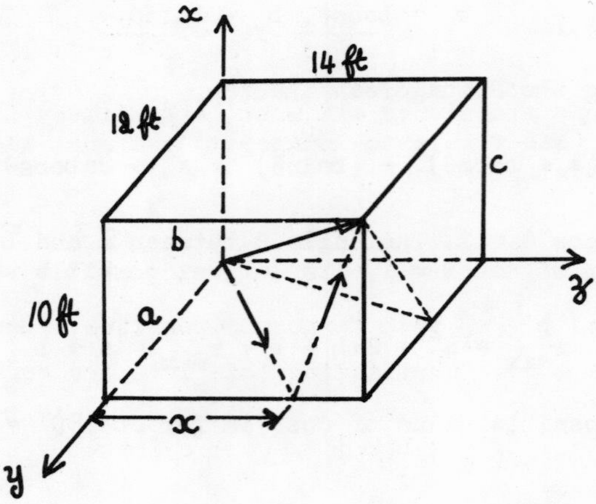

(a) From the figure, $\vec{D} = 10\vec{i} + 12\vec{j} + 14\vec{k}$ and therefore

$$D = (10^2 + 12^2 + 14^2)^{\frac{1}{2}} = 21.0 \text{ ft} \quad \underline{\text{Ans}}.$$

(b) Answering the questions in turn:

No: a straight line is the shortest distance;
Yes: fly need not fly in a straight line;
Yes: it could fly in a straight line (unlikely).

(c) For the room as oriented above and choice of corners shown,

$$\vec{D} = 10\vec{i} + 12\vec{j} + 14\vec{k} \quad \underline{\text{Ans}}.$$

(d) For this part, call the lengths of the sides of the room a, b, c as indicated (do not set a = 12 ft, b = 14 ft, c = 10 ft), for the fly has a choice of walls to walk on first; two possible paths to the opposite corner are shown; focus on the one with arrows. The total length L of the path is

$$L = [a^2 + x^2]^{\frac{1}{2}} + [(b - x)^2 + c^2]^{\frac{1}{2}}.$$

To find the shortest path set $dL/dx = 0$; this gives

$$(c^2 - a^2)x^2 + 2a^2bx - a^2b^2 = 0.$$

The two solutions follow from the quadratic formula and are

$$x_1 = \frac{ab}{a - c}\ ;\ x_2 = \frac{ab}{a + c}.$$

Now x represents a distance and must be positive. Likewise, $b - x$ is a distance (see figure to locate it) and must also be positive; these latter distances are

$$b - x_1 = \frac{bc}{c - a}\ ;\ b - x_2 = \frac{bc}{c + a}.$$

Clearly, x_1 and $b - x_1$ both cannot be positive. Therefore, choose $x_2 = x = ab/(a + c)$. Substituting this into the equation for L gives

$$L_{min} = L(x_2) = [(a + c)^2 + b^2]^{\frac{1}{2}},$$

For the smallest possible L_{min}, select a, b, c so that 2ac is the smallest possible: a = 10 ft, c = 12 ft or a = 12 ft, c = 10 ft. In either case, b = 14 ft and

$$L_{min} = (12^2 + 14^2 + 10^2 + 240)^{\frac{1}{2}} = 26.1 \text{ ft} \quad \underline{\text{Ans}}.$$

<u>2-12</u>

Orient the coordinate axes so that one of the vectors lies along either the x or y-axis; for example, let $\vec{a}$ lie along the x-axis.

Then,

$$\vec{a} = a\vec{i} \; ; \; \vec{b} = (b\cos\theta)\vec{i} + (b\sin\theta)\vec{j}$$

and therefore

$$\vec{a} + \vec{b} = (a + b\cos\theta)\vec{i} + (b\sin\theta)\vec{j};$$

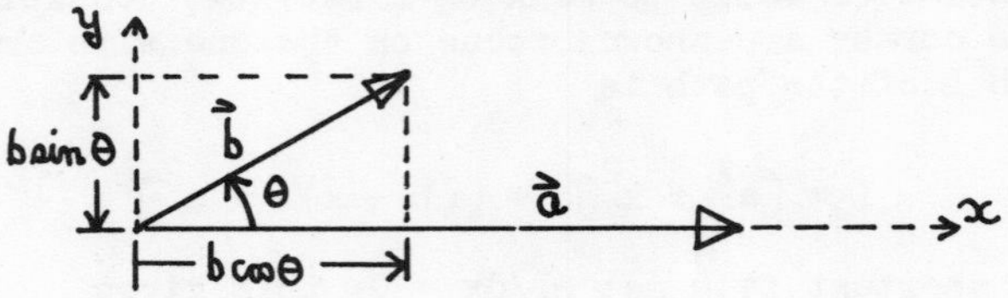

hence,

$$|\vec{a} + \vec{b}| = [(a + b\cos\theta)^2 + (b\sin\theta)^2]^{\frac{1}{2}} = [a^2 + b^2 + 2ab\cos\theta]^{\frac{1}{2}} \quad \underline{\text{Ans}}.$$

2-17

(a) The displacement vector $\vec{D}$ joins Washington and Manila along the straight line connecting them, the line passing through the earth, and points toward Manila.

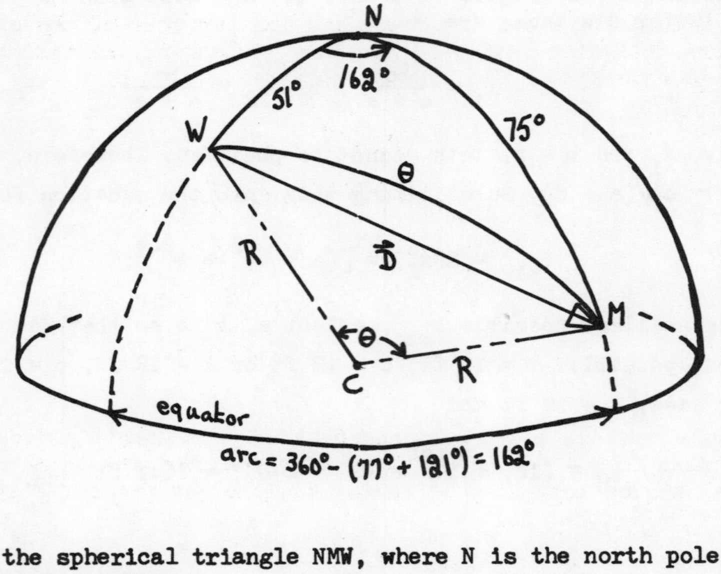

(b) In the spherical triangle NMW, where N is the north pole, the law of cosines gives

$$\cos\theta = \cos 51^\circ \cos 75^\circ + \sin 51^\circ \sin 75^\circ \cos 162^\circ = -0.55105;$$
$$\theta = 123.4^\circ.$$

If R = 6378 km, the radius of the earth, it is clear from the plane triangle WCM, where C is the center of the earth, that

$$D = 2R\sin\tfrac{1}{2}\theta = 11{,}230 \text{ km} \quad \underline{\text{Ans}}.$$

2-19

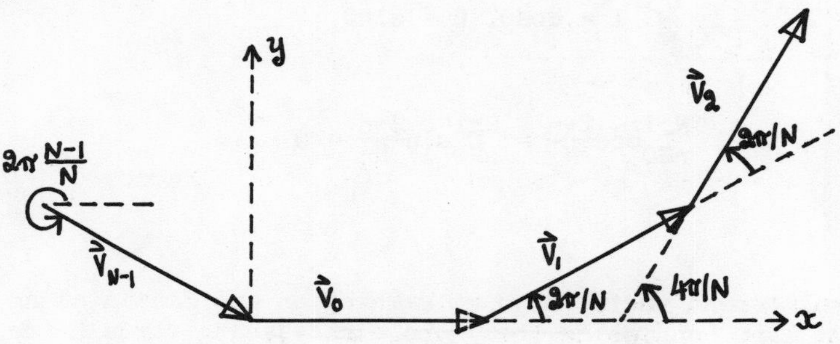

In the figure are shown the first and last vectors of the chain of N vectors. Aligning the coordinate axes as shown, and taking the x and y-components of each vector gives

$$\vec{v}_0 = \vec{i},$$
$$\vec{v}_1 = \cos\frac{2\pi}{N}\vec{i} + \sin\frac{2\pi}{N}\vec{j},$$
$$\vec{v}_2 = \cos\frac{4\pi}{N}\vec{i} + \sin\frac{4\pi}{N}\vec{j},$$
$$\vec{v}_n = \cos\frac{2\pi n}{N}\vec{i} + \sin\frac{2\pi n}{N}\vec{j},$$
$$\vec{v}_{N-1} = \cos\frac{2\pi(N-1)}{N}\vec{i} + \sin\frac{2\pi(N-1)}{N}\vec{j},$$

all of the vectors being considered as of unit length. Evidently the vectors form a polygon of N equal sides, for the total of the external angles between each vector and the one preceding it is 2π rad = 360°; that is,

$$\sum_{n=0}^{N-1}\vec{v}_n = 0\vec{i} + 0\vec{j}.$$

Adding the vector components given above, this becomes

$$\left[1 + \cos\frac{2\pi}{N} + \cos\frac{4\pi}{N} + \ldots + \cos\frac{2\pi(N-1)}{N}\right]\vec{i}$$
$$+ \left[0 + \sin\frac{2\pi}{N} + \sin\frac{4\pi}{N} + \ldots + \sin\frac{2\pi(N-1)}{N}\right]\vec{j} = 0.$$

The terms in the square brackets must vanish identically. Making the substitutions

$$1 = \cos 0, \; 0 = \sin 0,$$

yields

$$\sum_{n=0}^{N-1}\cos\frac{2\pi n}{N} = \sum_{n=0}^{N-1}\sin\frac{2\pi n}{N} = 0.$$

2-29

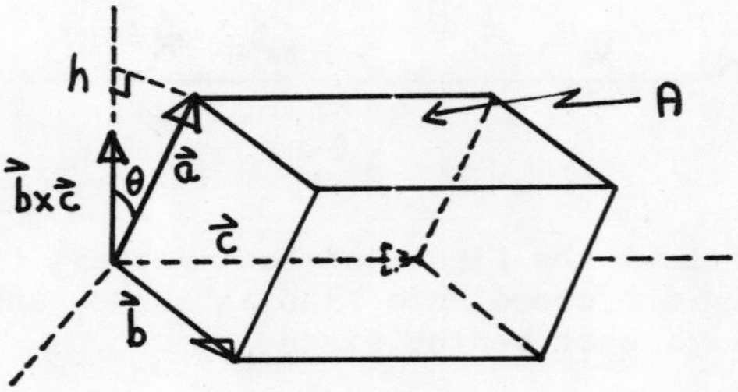

Since the area of a parallelogram is the base length times the height,

$$|\vec{b} \times \vec{c}| = \text{area of base} = \text{area of top} = A,$$

and therefore

$$\vec{a}\cdot(\vec{b} \times \vec{c}) = a|\vec{b} \times \vec{c}|\cos(\vec{a},\vec{b} \times \vec{c}) = aA\cos\theta = (a\cos\theta)A = hA,$$

which is the volume of the parallepiped.

2-35

(a) From the text diagram, $\vec{b} = a\vec{i} + a\vec{j}$, $\vec{c} = a\vec{j} + a\vec{k}$. Hence,

$$\vec{d} = \vec{b} \times \vec{c} = a^2(\vec{i} + \vec{j}) \times (\vec{j} + \vec{k}),$$

$$\vec{d} = a^2(\vec{i} \times \vec{j} + \vec{i} \times \vec{k} + \vec{j} \times \vec{j} + \vec{j} \times \vec{k}),$$

$$\vec{d} = a^2(\vec{k} - \vec{j} + 0 + \vec{i}) = a^2\vec{i} - a^2\vec{j} + a^2\vec{k} \quad \underline{\text{Ans}}.$$

(b) $\vec{b}\cdot\vec{c} = a.0 + a.a + 0.a = a^2$ Ans.
$\vec{d}\cdot\vec{c} = \vec{d}\cdot\vec{b} = 0$ since $\vec{d} = \vec{b} \times \vec{c}$ is perpendicular to both $\vec{b}$ and $\vec{c}$.

(c) Again, from the text figure, $\vec{e} = a\vec{i} + a\vec{j} + a\vec{k}$; from (a), $\vec{b} = a\vec{i} + a\vec{j}$. Then,

$$\vec{b}\cdot\vec{e} = a^2 + a^2 + 0 = 2a^2 = eb\cos\angle(e,b),$$

$$2a^2 = (a\sqrt{3})(a\sqrt{2})\cos\theta,$$

$$\theta = \cos^{-1}(\sqrt{6}/3) = 35°16' \quad \text{Ans.}$$

2-37

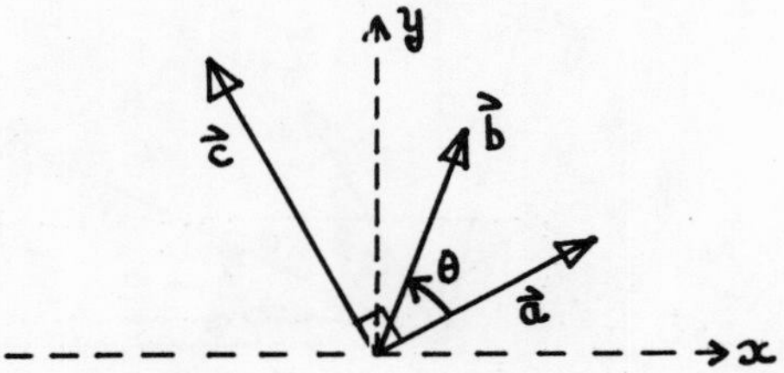

(a) The magnitudes of the vectors are a = 3.5777, b = 4.5277. Using the formulae for scalar product,

$$\vec{a}\cdot\vec{b} = ab\cos\theta = (3.5777)(4.5277)\cos\theta = 16.1988\cos\theta,$$

$$\vec{a}\cdot\vec{b} = (3.2)(0.5) + (1.6)(4.5) = 8.8,$$

$$8.8 = 16.1988\cos\theta,$$

$$\theta = 57° \quad \underline{\text{Ans}}.$$

(b) In this part, the vector $\vec{b}$ has no role. Write $\vec{c} = c_x\vec{i} + c_y\vec{j}$. Since c = 5,

$$25 = c_x^2 + c_y^2.$$

Also, since $\vec{c}$ is perpendicular to $\vec{a}$, $\vec{c}\cdot\vec{a} = 0$, so that

$$3.2c_x + 1.6c_y = 0.$$

This last equation gives $c_y = -2c_x$. Substituting this into the first equation yields

$$25 = c_x^2 + 4c_x^2 = 5c_x^2,$$

$$c_x = \pm\sqrt{5} = \pm 2.236;\ c_y = \mp 4.472 \quad \underline{\text{Ans}}.$$

The vector $\vec{c}$ with components given by the lower set of signs is shown in the figure; the other possibility is antiparallel to this.

2-39

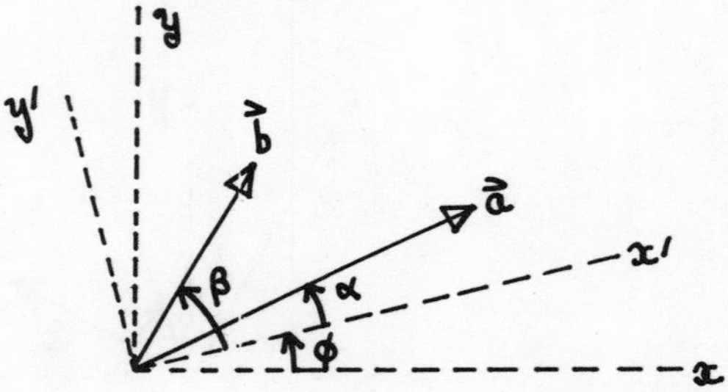

The vectors $\vec{a}$ and $\vec{b}$ expressed in the two coordinate systems are

$$\vec{a} = (a\cos\alpha)\vec{i}' + (a\sin\alpha)\vec{j}' = a\cos(\alpha + \phi)\vec{i} + a\sin(\alpha + \phi)\vec{j};$$
$$\vec{b} = (b\cos\beta)\vec{i}' + (b\sin\beta)\vec{j}' = b\cos(\beta + \phi)\vec{i} + b\sin(\beta + \phi)\vec{j}.$$

Similar relations hold between the unit vectors:

$$\vec{i}' = \cos\phi\vec{i} + \sin\phi\vec{j};$$
$$\vec{j}' = -\sin\phi\vec{i} + \cos\phi\vec{j}.$$

Therefore,

$$(\vec{a} + \vec{b})' = (a\cos\alpha + b\cos\beta)\vec{i}' + (a\sin\alpha + b\sin\beta)\vec{j}'$$
$$= (a\cos\alpha + b\cos\beta)(\cos\phi\vec{i} + \sin\phi\vec{j})$$
$$+ (a\sin\alpha + b\sin\beta)(-\sin\phi\vec{i} + \cos\phi\vec{j})$$

$$= [a(\cos\alpha\cos\phi - \sin\alpha\sin\phi) + b(\cos\beta\cos\phi - \sin\beta\sin\phi)]\vec{i}$$

$$+ [a(\cos\alpha\sin\phi + \sin\alpha\cos\phi) + b(\cos\beta\sin\phi + \sin\beta\cos\phi)]\vec{j}$$

$$= [a\cos(\alpha + \phi) + b\cos(\beta + \phi)]\vec{i}$$

$$+ [a\sin(\alpha + \phi) + b\sin(\beta + \phi)]\vec{j}$$

$$= \vec{a} + \vec{b} \text{ in unprimed system.}$$

CHAPTER 3

3-4

Let $\vec{D}$ be the train's displacement during the total travel time T. Then, by definition, the average velocity v is

$$\bar{\vec{v}} = \vec{D}/T.$$

Since 60 km/h = 1 km/min, the successive distances travelled are 40 km, 20 km and 50 km. A diagram showing the displacement vectors is easily constructed and from this the total displacement $\vec{D}$ is found to be

$$\vec{D} = [40\vec{i} + (20\sqrt{2}/2\vec{i} + 20\sqrt{2}/2\vec{j}) - 50\vec{i}] \text{ km},$$

$$\vec{D} = 4.142\vec{i} + 14.142\vec{j}.$$

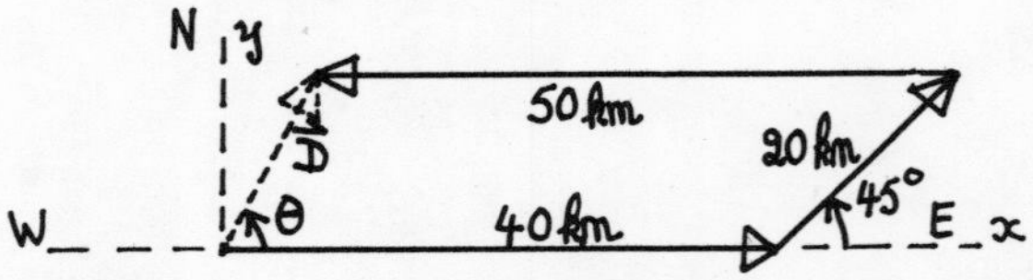

The total time T of travel is, of course, 40 min + 20 min + 50 min = 110 min = 11/6 h. Since the magnitude of $\vec{D}$ is

$$D = (4.142^2 + 14.142^2)^{\frac{1}{2}} = 14.736 \text{ km},$$

then

$$\bar{v} = \frac{14.736 \text{ km}}{11/6 \text{ h}} = 8.0 \text{ km/h},$$

and the direction of the average velocity is at an angle θ north of east where

$$\theta = \tan^{-1}(14.142/4.142) = 73.7° \quad \underline{\text{Ans}}.$$

3-18

The measured stopping distance s and the assumed maximum possible deceleration a are

$$s = 19.2 \text{ ft}; \; a = 32 \text{ ft/s}^2.$$

Then, the car's greatest possible initial speed v was

$$v = (2as)^{\frac{1}{2}} = [2(32\ ft/s^2)(19.2\ ft)]^{\frac{1}{2}},$$

$$v = 35\ ft/s,$$

or v = 24 mi/h. The car's speed could not have been greater than this for, if it was, it would have taken more than 19.2 ft to bring the car to a halt, the deceleration being the maximum possible, by assumption.

3-26

Let the vehicle be moving at a speed v_0 when, at t = 0, the driver slams on the brakes a distance x from the barrier, which it strikes 4.0 s later at a speed v_f. Then, if a is the acceleration, and since 35 mi/h = 51.33 ft/s,

(a)

$$x = \tfrac{1}{2}at^2 + v_0t + x_0,$$

$$110 = \tfrac{1}{2}a(4)^2 + (51.33)4 + 0,$$

$$a = -11.9\ ft/s^2 \quad \underline{Ans}.$$

(b)

$$v_f = at + v_0,$$

$$v_f = (-11.9)(4) + 51.33,$$

$$v_f = 3.73\ ft/s \quad \underline{Ans}.$$

3-28

Let t = driver reaction time, t' = breaking time at 50 mi/h and t" = breaking time at 30 mi/h. Then, if a is the deceleration, and since 30 mi/h = 44 ft/s and 50 mi/h = 73.33 ft/s,

$$186 = \tfrac{1}{2}at'^2 + 73.33t,$$

$$80 = \tfrac{1}{2}at''^2 + 44t,$$

$$73.33 = at',$$

$$44 = at''.$$

These are four equations for the four unknowns a, t, t', t". Use the third equation to eliminate a from the first equation, and the fourth to eliminate a from the second. Then use the third and the fourth in the form

$$\frac{73.33}{44} = \frac{t'}{t''}$$

to eliminate t', say, from the two equations resulting from the first sequence of operations. The result is a pair of equations in t, t" which can be solved easily. The results are

(a)

$$t = 0.74 \text{ s} \quad \underline{\text{Ans}};$$

(b)

$$-a = \text{acceleration} = -20 \text{ ft/s}^2 \quad \underline{\text{Ans}}.$$

<u>3-29</u>

(a) Since at^2 must have the dimensions of length, a must have dimensions of (length)/(time)^2; similarly, b must have dimensions of (length)/(time)^3; in the SI system, these are m/s^2 and m/s^3.

(b) With a = 3 and b = 1 the given equation for position x is

$$x = 3t^2 - t^3,$$

which is shown sketched, with x in meters and t in seconds.

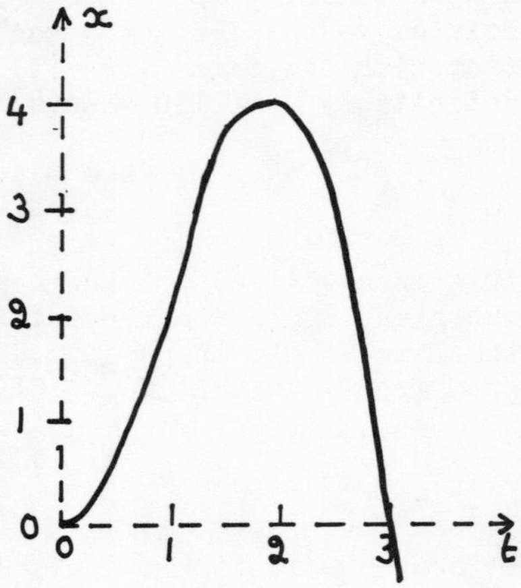

Setting dx/dt = 0 gives

$$0 = 6t - 3t^2;$$

this is satisfied by t = 0 and t = 2. The value t = 2 gives the maximum.

(c) From t = 0 to t = 2 the particle travels x(2) - x(0) = 4 m. From t = 2 to t = 3 the particle travels x(3) - x(2) = 4 m also, and from t = 3 to t = 4 it travels x(4) - x(3) = 16 m. Thus, the total distance travelled is 4 + 4 + 16 = 24 m <u>Ans</u>.

(d) Since x(4) = -16, the displacement is x(4) - x(0) = -16 m <u>Ans</u>.

(e) $v = dx/dt = 6t - 3t^2$ and (f) $a = dv/dt = 6(1 - t)$; from these the following table is derived:

t	v(m/s)	a(m/s^2)
1	3	0
2	0	-6
3	-9	-12
4	-24	-18

(g) By definition,

$$\bar{v}_{4,2} = \frac{x(4) - x(2)}{4 - 2} = -10 \text{ m/s} \quad \underline{\text{Ans}}.$$

<u>3-31</u>

Let the upward direction be - and down +; call v the speed of the ball upon striking the floor and u the speed as it leaves; the associated velocities are, then, +v and -u. If the ball is in contact with the floor for a time T, the average acceleration is, by definition,

$$\bar{a} = \frac{-u - (+v)}{T} = -\frac{u + v}{T};$$

as this is negative, the average acceleration is directed upward, as expected. Since the speed with which the ball leaves the floor is the same as the speed with which it would strike the floor if dropped from a height of 3.0 ft, both u and v can be found from

$$v^2 = 2gh,$$

using h = 4 ft for v and 3 ft for u; g = 32 ft/s^2 in each case. Therefore, v = 16 ft/s and u = 13.86 ft/s so that

$$\bar{a} = \frac{13.86 + 16}{0.01} = 3000 \text{ ft/s}^2, \text{ up} \quad \underline{\text{Ans}}.$$

(This is about 93g.)

<u>3-35</u>

Choose the positive x-axis vertically upward with origin at the ground. At t = 0 the package is released with velocity $\vec{v}_0$ directed upwards (i.e. v_0 = +12 m/s) from height x_0 = +80 m. At t = T it strikes the ground at $x = 0$. Hence,

$$x = \tfrac{1}{2}(-9.8)t^2 + 12t + 80$$

gives the position x of the package as a function of time t; a is negative since the acceleration is down, in the negative direction. At $t = T$, $x = 0$ (package strikes ground):

$$0 = \tfrac{1}{2}(-9.8)T^2 + 12T + 80,$$

$$T = 5.4 \text{ s} \quad \underline{\text{Ans}}.$$

<u>3-37</u>

(a) The initial speed v of the ball relative to the ground is 32 + 64 = 96 ft/s. The height x reached by the ball above the elevator (when thrown) is found from

$$v^2 = 2gx,$$
$$(96)^2 = 2(32)x,$$
$$x = 144 \text{ ft}.$$

Since the ball was thrown from a point 100 ft above the ground, the highest point reached by the ball is 100 + 144 = 244 ft above the ground.

(b) Let t = time for the ball to reach maximum height. Then,

$$t = \frac{v}{g} = \frac{96 \text{ ft/s}}{32 \text{ ft/s}^2} = 3.00 \text{ s}.$$

During this time the elevator has moved a distance (32 ft/s)(3 s) = 96 ft up the shaft, so that ball and elevator are separated by 144 - 96 = 48 ft at the moment the ball is at its highest point. Call T the time needed for the ball to fall back to the elevator. Relative to the elevator, the ball is projected downward, with a speed of 32 ft/s, from a height of 48 ft. Choosing down as the positive direction, origin at the maximum height,

$$48 = \tfrac{1}{2}(32)T^2 + 32T,$$

$$T = 1.00 \text{ s},$$

and therefore the total elapsed time is 3.00 + 1.00 = 4.00 s <u>Ans</u>.

3-41

Let h be the height of fall and t the time to fall this distance. Since the object falls from rest,

$$h = \tfrac{1}{2}gt^2,$$
$$\frac{h}{2} = \tfrac{1}{2}g(t - 1)^2.$$

(a) Eliminating h between these equations gives

$$\frac{t}{t - 1} = \pm\sqrt{2},$$
$$t = \frac{\sqrt{2}}{\sqrt{2} - 1} = 3.41 \text{ s} \quad \underline{\text{Ans}},$$

using $+\sqrt{2}$ in the preceding equation.

(b) The height of fall is

$$h = \tfrac{1}{2}(9.8)(3.41)^2 = 57 \text{ m} \quad \underline{\text{Ans}}.$$

(c) Use of $-\sqrt{2}$ in (a) gives $t = \sqrt{2}/(\sqrt{2} + 1) = 0.586$ s. But then $t - 1 < 0$, physically meaningless.

3-42

Let y, Y be the distances of the first and second bodies, measured from the common point of release, and let the second body be released at t = 0. Then,

$$y = \tfrac{1}{2}g(t + 1)^2; \; Y = \tfrac{1}{2}gt^2.$$

For $y - Y = 10$,

$$\tfrac{1}{2}g(t + 1)^2 - \tfrac{1}{2}gt^2 = 10,$$
$$g(t + \tfrac{1}{2}) = 10.$$

Using $g = 9.8$ m/s^2, this gives t = 0.52 s, so that the elapsed time since the first body was released is 0.52 + 1 = 1.52 s <u>Ans</u>.

3-43

Let H be the height of the building and h the distance above the ground of the bottom of the window; also, let the ball bearing be released at time t = 0 from x = 0, the downward direction being taken as positive. If T is the time taken for the ball bearing to fall to the top of the window, then at the instants when the ball bearing passes the top of the window, the bottom of the window and strikes the ground, the following equations are successively satisfied:

$$H - h - 4.0 = \tfrac{1}{2}gT^2,$$
$$H - h = \tfrac{1}{2}g(T + 0.125)^2,$$
$$H = \tfrac{1}{2}g(T + 0.125 + 1)^2.$$

From the first two equations, eliminating (H - h),

$$\tfrac{1}{2}g(T + 0.125)^2 - 4.0 = \tfrac{1}{2}gT^2,$$
$$0.125gT + 0.0078125g - 4.0 = 0.$$

Putting $g = 32\ \text{ft/s}^2$ into this last gives T = 0.9375 s. Finally, substituting this value of T into the third equation above yields

$$H = 68.1\ \text{ft} \quad \underline{\text{Ans}}.$$

3-45

Let the positive x-axis fixed on the elevator shaft be vertical, pointing upward, with origin at that point of the shaft level with the elevator floor at the moment, t = 0, the bolt left the ceiling of the elevator (see figure). Also, let x = position of bolt and X = position of the elevator floor.

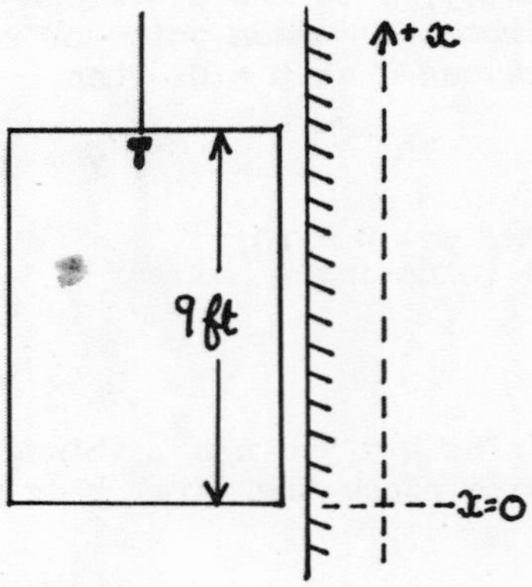

(a) At any time while the bolt is falling,

$$x = \tfrac{1}{2}(-32)t^2 + 8t + 9,$$
$$X = \tfrac{1}{2}(4)t^2 + 8t.$$

If T is the time of flight of the bolt, then at t = T, x = X:

$$-16T^2 + 8T + 9 = 2T^2 + 8T,$$

$$T = 1/\sqrt{2} = 0.71 \text{ s} \quad \underline{\text{Ans}}.$$

(b) Since the bolt started from the ceiling 9 ft above the floor, the desired distance is

$$D = 9 - X = 9 - 2T^2 - 8T = 9 - 1 - 5.68,$$

$$D = 2.3 \text{ ft} \quad \underline{\text{Ans}}.$$

<u>3-46</u>

As indicated in the sketch, let distances be measured downwards from the highest point reached. by the pot in its flight. Also, let the pot be at the highest point at t = 0 and, in its descent, pass the top of the window and bottom of the window at times t_1 and t_2. If the total time that the pot is visible is 1.0 s, then the time that it is visible in its descent is 0.5 s. Therefore,

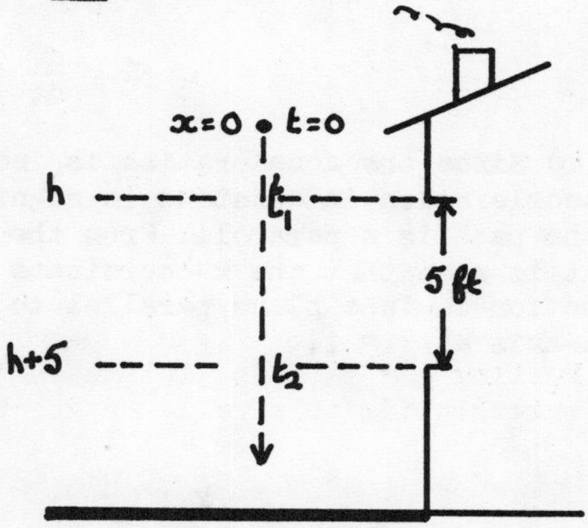

$$h = \tfrac{1}{2}gt_1^2,$$

$$h + 5 = \tfrac{1}{2}gt_2^2 = \tfrac{1}{2}g(t_1 + \tfrac{1}{2})^2.$$

Eliminating h between these equations gives

$$\tfrac{1}{2}gt_1^2 + 5 = \tfrac{1}{2}g(t_1^2 + t_1 + \tfrac{1}{4});$$

since $g = 32 \text{ ft/s}^2$, $t_1 = 1/16$ s and

$$h = \tfrac{1}{2}(32)(\tfrac{1}{16})^2 = \frac{1}{16} \text{ ft} \quad \underline{\text{Ans}}.$$

CHAPTER 4

4-2

(a) By the definitions of velocity and acceleration,

$$\vec{v} = \frac{d\vec{r}}{dt} = 8t\vec{j} + \vec{k} \quad \underline{\text{Ans}};$$

$$\vec{a} = \frac{d\vec{v}}{dt} = 8\vec{j} \quad \underline{\text{Ans}}.$$

(b) Since the acceleration is independent of the time, the acceleration is constant in magnitude and direction and, therefore, the path is a parabola. From the equation given for the position, it is seen that the x-coordinate does not vary with time, so the motion is in a plane parallel to the y-z plane and cutting the x-axis at x = 1.

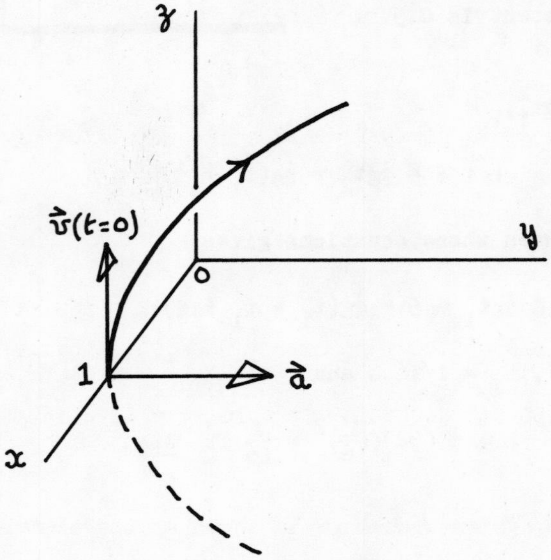

4-6

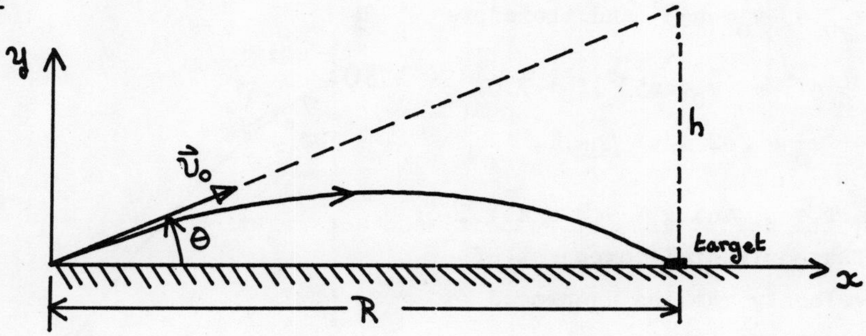

The motion of the bullet in the x and y-directions is given by

$$x = (v_0\cos\theta)t,$$
$$y = -\tfrac{1}{2}gt^2 + (v_0\sin\theta)t,$$

when the x and y axes are oriented as in the sketch above. If T is the time of flight, then $x(T) = R$ and $y(T) = 0$. Hence

$$R = v_0T\cos\theta,$$
$$0 = -\tfrac{1}{2}gT^2 + v_0T\sin\theta.$$

The first equation above gives $T = R/v_0\cos\theta$; substituting this into the second equation yields

$$\sin 2\theta = gR/v_0^2,$$

since $\sin 2\theta = 2\sin\theta\cos\theta$. Setting $g = 32\ \text{ft/s}^2$, $R = 150$ ft and $v_0 = 1500$ ft/s gives $\sin 2\theta = 0.002133$, or $\theta = 0.0611°$. Finally,

$$h = R\tan\theta = (150\ \text{ft})(0.0010667) = 0.16\ \text{ft} = 1.92\ \text{in.}\quad \underline{\text{Ans}}.$$

4-11

(a) The coordinate system used here is shown on the sketch. The bomb is released at $t = 0$ and its time of flight is T, so that when $t = T$, $y = 0$ (bomb strikes the ground). Hence, $y_0 = +730$ m

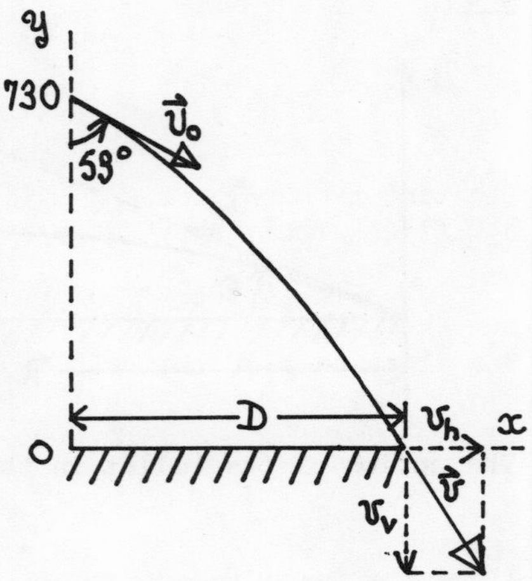

and $v_{y0} = -v_0\cos 53^\circ$ and therefore

$$0 = -\tfrac{1}{2}gT^2 - (v_0\cos 53^\circ)T + 730,$$

$$v_0 = 202 \text{ m/s} \quad \underline{\text{Ans}},$$

since $T = 5\text{s}$ and $g = 9.8 \text{ m/s}^2$.

(b) The horizontal component of the velocity remains unchanged during flight; thus,

$$D = (v_0\sin 53^\circ)T = 806 \text{ m} \quad \underline{\text{Ans}}.$$

(c) As the bomb strikes the ground,

$$v_h = v_x = v_{x0} = (202 \text{ m/s})(\sin 53^\circ) = 161 \text{ m/s} \quad \underline{\text{Ans}};$$

$$v_v = v_y = (v_0\cos 53^\circ) + gT,$$

$$v_v = (202)(\cos 53^\circ) + (9.8)(5) = 171 \text{ m/s} \quad \underline{\text{Ans}}.$$

<u>4-14</u>

The situation is shown in the sketch. The ball is "launched" from a point 4 ft above the ground so that, if the axes are arranged as shown in the figure, the position of the ball as a function of time is given by

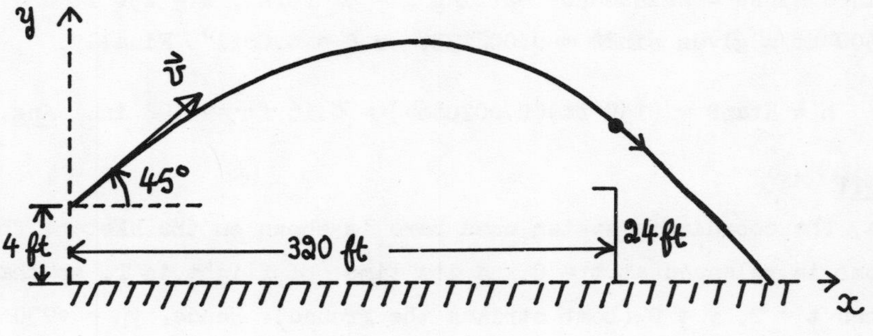

$$x = (110\cos 45^\circ)t,$$

$$y = -\tfrac{1}{2}(32)t^2 + (110\sin 45^\circ)t + 4.$$

The time required for the ball to travel a horizontal distance x = 320 ft is, from the first equation,

$$t = 320/(110\cos 45^\circ) = 4.1141 \text{ s}.$$

The distance y of the ball above the ground at this instant is

$$y = -16(4.1141)^2 + (110\sin 45^\circ)(4.1141) + 4,$$

$$y = 53.2 \text{ ft}.$$

Since the fence is only 24 ft high, the ball clears the fence by 29.2 ft Ans.

4-19

With the x,y-axes arranged as shown, the position of the shell is given by

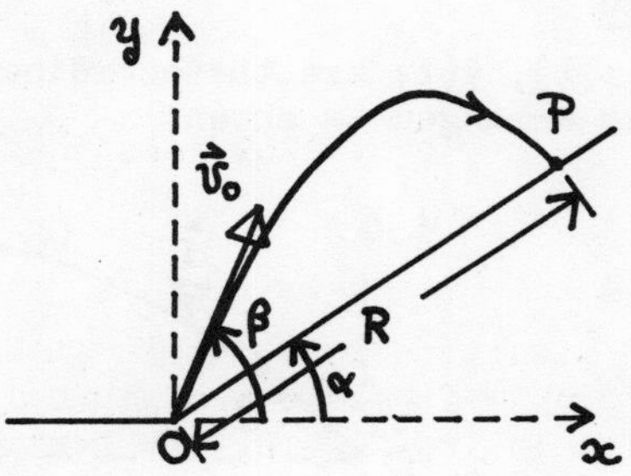

$$y = -\tfrac{1}{2}gt^2 + (v_0\sin\beta)t,$$

$$x = (v_0\cos\beta)t.$$

The coordinates of the impact point are

$$x_P = R\cos\alpha,\ y_P = R\sin\alpha.$$

Hence, if T is the time of flight to impact,

$$R\cos\alpha = (v_0\cos\beta)T;\ T = \frac{R\cos\alpha}{v_0\cos\beta}.$$

Therefore,

$$y_P = R\sin\alpha = -\tfrac{1}{2}g\left(\frac{R\cos\alpha}{v_0\cos\beta}\right)^2 + (v_0\sin\beta)\left(\frac{R\cos\alpha}{v_0\cos\beta}\right).$$

Solving for R gives

$$R = (2v_0^2/g\cos^2\alpha)\cos\beta\cdot\sin(\beta - \alpha).$$

To find the maximum range R with variations in elevation angle β set $dR/d\beta = 0$:

$$\frac{dR}{d\beta} = (2v_0^2/g\cos^2\alpha)[-\sin\beta\sin(\beta - \alpha) + \cos\beta\cos(\beta - \alpha)] = 0;$$

$$-\sin\beta\sin(\beta - \alpha) + \cos\beta\cos(\beta - \alpha) = 0;$$

$$\cos[\beta + (\beta - \alpha)] = 0;$$

$$2\beta - \alpha = \pi/2;$$

$$\beta = \frac{\alpha}{2} + \frac{\pi}{4} \quad \underline{\text{Ans}};$$

that is, fire the cannon so that the shell's path bisects the angle between the vertical and the slope of the hill.

4-20

If x(t), y(t) are the coordinates of the ball in flight then, with axes arranged as shown,

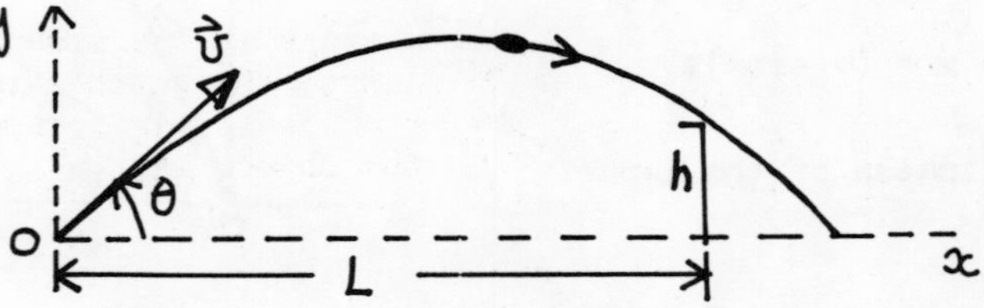

$$y = -\tfrac{1}{2}gt^2 + (v\sin\theta)t,$$

$$x = (v\cos\theta)t.$$

Let T = time interval from the kick until the ball reaches a horizontal distance L, where L = distance from kick to goal posts. Then, $T = L/v\cos\theta$ from the x-equation. When t = T, y = h (ball just clears bar):

$$h = -\tfrac{1}{2}g(L/v\cos\theta)^2 + (v\sin\theta)(L/v\cos\theta),$$

$$h = -\tfrac{1}{2}gL^2/(v^2\cos^2\theta) + L\tan\theta.$$

This can be written as

$$ax^2 - Lx + (h + a) = 0$$

if

$$a = \tfrac{1}{2}gL^2/v^2, \; x = \tan\theta.$$

Inserting the numbers gives a = 19.60 m and, since h = 3.44 m,

$$19.6x^2 - 50x + 23.04 = 0;$$

$$x = 1.9474, \; 0.6036$$

as the two solutions to the quadratic equation. With $x = \tan\theta$, these yield angles of 63° and 31° as the required limits.

4-27

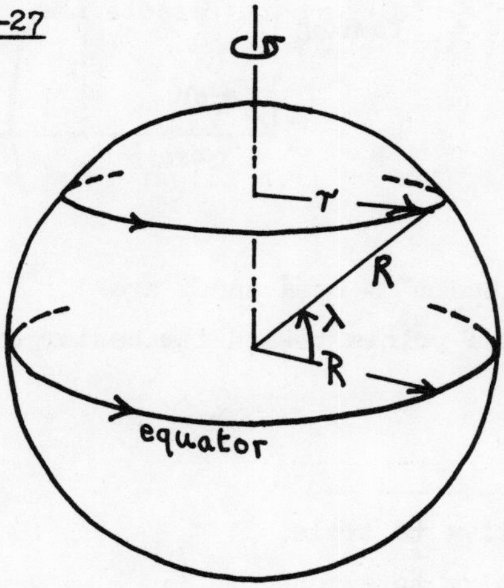

Let R = radius of the earth and λ = latitude of the object; also T = period of the earth's rotation. Then,

$$vT = 2\pi r,$$

where v is the speed of the object due to the earth's rotation. As $r = R\cos\lambda$, the acceleration of the object is

$$a = \frac{v^2}{r} = \frac{(2\pi r/T)^2}{r} = 4\pi^2 r/T^2,$$

$$a = 4\pi^2 R\cos\lambda/T^2.$$

The numbers required are $R = 6.378 \times 10^6$ m, T = 86,400 s. Thus, at

(a) the equator: λ = 0, a = 0.0337 m/s^2 Ans;

(b) λ = 60°, a = 0.0169 m/s^2 Ans.

(c) If T* is the rotation period required for the acceleration due to the rotation, at the equator, to equal g, then

$$g = 4\pi^2 R/T^{*2}$$

while in actuality

$$a = 4\pi^2 R/T^2.$$

Therefore,

$$a/g = 0.0337/9.8 = T^{*2}/T^2,$$

$$T^* = 0.0586T = T/17$$

which means that the earth would have to spin 17 times faster than its actual rate.

4-30

(a) From the figure, with $\omega = \frac{v}{r}$,

$$\vec{r} = (r\cos\omega t)\vec{i} + (r\sin\omega t)\vec{j}.$$

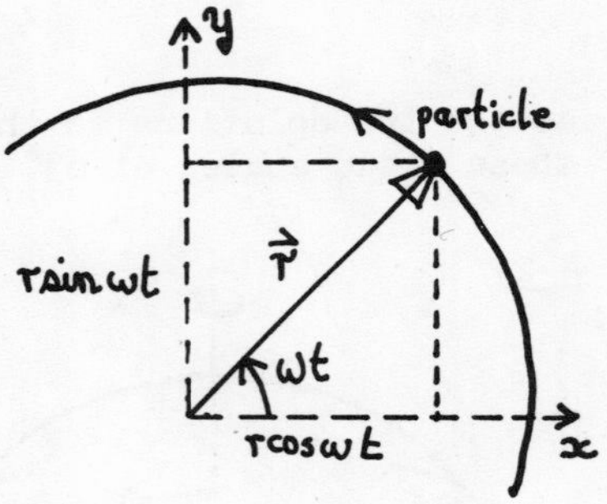

(b) Since $\vec{v} = d\vec{r}/dt$,

$$\vec{v} = -(r\omega\sin\omega t)\vec{i} + (r\omega\cos\omega t)\vec{j},$$

and $\vec{a} = d\vec{v}/dt$, so that

$$\vec{a} = -(r\omega^2\cos\omega t)\vec{i} - (r\omega^2\sin\omega t)\vec{j}.$$

(c) From (a) and (b), $\vec{a} = -\omega^2\vec{r}$; since $\omega^2 > 0$, $\vec{a}$ and $\vec{r}$ are oppositely directed, and therefore $\vec{a}$ points toward the center of the circle.

4-35

Let: $\vec{v}_{rt}$ = velocity of rain relative to train,

$\vec{v}_{rg}$ = velocity of rain relative to ground,

$\vec{v}_{tg}$ = velocity of train relative to ground.

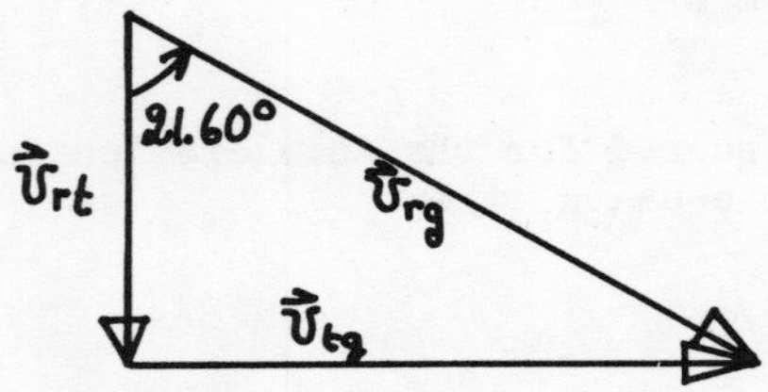

Then, as shown,

$$\vec{v}_{rt} + \vec{v}_{tg} = \vec{v}_{rg},$$

$$v_{rg} = \frac{v_{tg}}{\sin 21.60°} = \frac{88.2\text{ ft/s}}{\sin 21.60°},$$

$$v_{rg} = 240\text{ ft/s} \quad \underline{\text{Ans}}.$$

4-39

Let: $\vec{v}_{pg}$ = velocity of plane relative to the ground,
$\vec{v}_{pa}$ = velocity of plane relative to air,
$\vec{v}_{ag}$ = velocity of air relative to the ground.

By the law of relative velocities,

$$\vec{v}_{pa} + \vec{v}_{ag} = \vec{v}_{pg}.$$

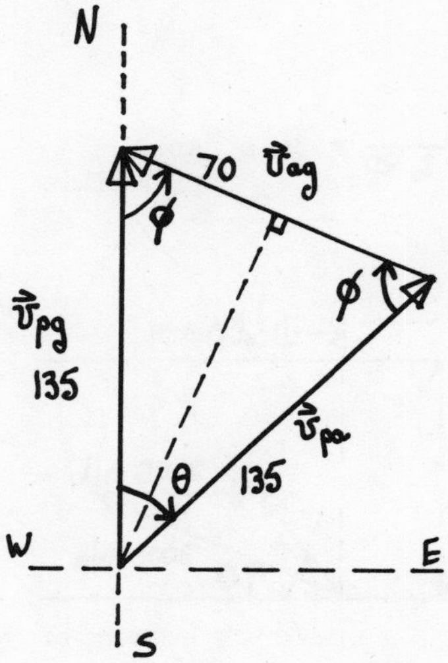

Now, the speed v_{pa} = 135 mi/h; also v_{pg} = 135 mi/h and the wind speed v_{ag} = 70 mi/h, so that the lengths of all the vectors are known. As far as directions are concerned, it is given that $\vec{v}_{pg}$ points north. This is drawn first and then the other two are arranged to obey the law for adding vectors by the geometric method. The triangle obtained contains no right angle. (A solution with θ counterclockwise from north is possible also.)

(b) Since the triangle is isosceles,

$$\sin\tfrac{1}{2}\theta = \frac{35}{135},$$

$$\theta = 30^\circ \text{ E of N} \quad \underline{\text{Ans}}.$$

(a) The angles of the triangle must add up to 180° and as the base angles are equal, each = ϕ,

$$\phi = \tfrac{1}{2}(180^\circ - \theta) = \tfrac{1}{2}(180^\circ - 30^\circ) = 75^\circ\text{E of S} \quad \underline{\text{Ans}}.$$

4-41

Let t = time required to get up the escalator and w = person's walking speed, s = escalator's speed, m = person's speed relative to the ground when walking on the escalator and L = length of the escalator. Then,

$$L/w = t_w = 90 \text{ s}; \; L/s = t_s = 60 \text{ s}; \; L/m = t_m,$$

where t_m is the time needed to walk up the moving escalator. But

$$m = s + w,$$

so that

$$t_m = \frac{L}{s + w} = \frac{L}{L/t_s + L/t_w} = \frac{1}{1/60 + 1/90} = 36 \text{ s} \quad \underline{\text{Ans}}.$$

4-42

If the boat is headed at an angle θ from the upstream shore, the boat, due to the current, will actually move relative to the shore at an angle ϕ and speed V given by

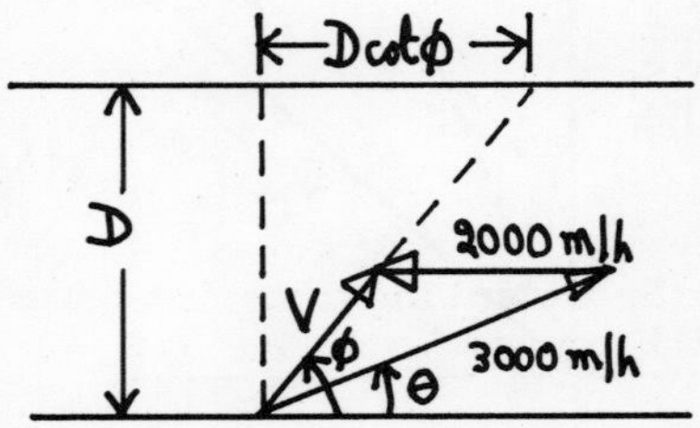

$$3000\cos\theta - 2000 = V\cos\phi,$$
$$3000\sin\theta = V\sin\phi,$$

by the law of relative velocities. Dividing these equations gives

$$\cot\phi = \frac{\cos\theta - 2/3}{\sin\theta}.$$

The distance to be walked is $D\cot\phi$. Since $\cos\theta - 2/3 < 0$ for $\theta > \cos^{-1}(2/3)$, the distance must be written

$$D\cot\phi = (500)\frac{|\cos\theta - 2/3|}{\sin\theta},$$

requiring a walking time given by

$$t_w = \frac{|\cos\theta - 2/3|}{10\sin\theta},$$

since the walking speed is 5000 m/h. The time required to row across the river is

$$t_r = \frac{(500)}{V\sin\phi} = \frac{500}{3000\ \sin\theta} = \frac{5}{30\sin\theta}.$$

Hence, the total travel time is

$$T = t_r + t_w = \frac{1}{30}\left[\frac{3|\cos\theta - 2/3| + 5}{\sin\theta}\right].$$

Substituting various values of θ, it is found that the minimum T occurs for some $\theta > \cos^{-1}(2/3)$; hence, use

$$T = \frac{1}{30}\,\frac{7 - 3\cos\theta}{\sin\theta}.$$

Then,

$$30\,\frac{dT}{d\theta} = \frac{-7\cos\theta + 3}{\sin^2\theta}.$$

(a) Setting $dT/d\theta = 0$ gives

$$\theta = \cos^{-1}\left(\tfrac{3}{7}\right) = 64.6^\circ \quad \underline{\text{Ans.}}$$

(b) Substituting this into the equation for T results in T_{min} = 0.211 hours.

CHAPTER 5

5-3

(a) Draw a free-body diagram of each block and apply Newton's second law to the horizontal forces, noting that since the blocks move together, their accelerations are the same. The blocks accelerate to the right, so that if this direction is chosen as positive, the resulting equations are

$$F - F_c = m_1 a,$$

$$F_c = m_2 a,$$

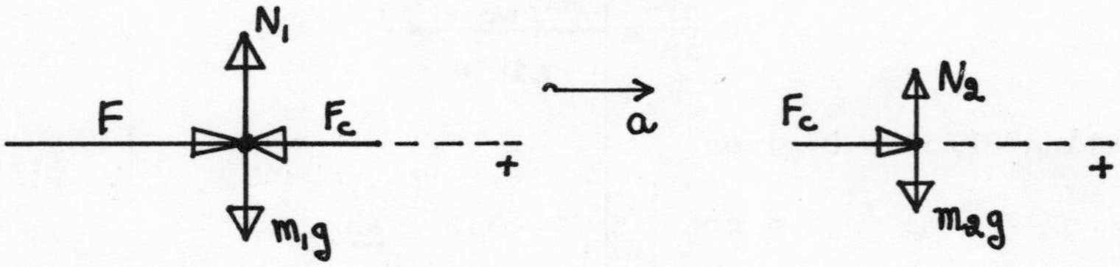

where F_c is the force of contact. Eliminating the acceleration (a) gives

$$F_c = \frac{m_2}{m_2 + m_1} F = \frac{1}{1 + 2}(3) = 1.0 \text{ N} \quad \underline{\text{Ans.}}$$

(b) Interchanging the blocks,

$$F_c = \frac{m_1}{m_1 + m_2} F = \frac{2}{2 + 1}(3) = 2.0 \text{ N} \quad \underline{\text{Ans.}}$$

The acceleration is the same in parts (a) and (b) since the net external force and total mass are the same in each case. On one of the blocks the internal force of contact is the sole force providing the acceleration and, by Newton's second law, this force is proportional to the mass it acts upon.

5-10

(a) By definition, the average acceleration upon landing is

$$\bar{a} = \frac{v_f - v_i}{t_f - t_i},$$

where v_i, v_f are the initial and final velocities with which the man strikes the patio, at times t_i, t_f. Now $v_f = 0$ and, if he jumps from height h,

$$v_i = (2gh)^{\frac{1}{2}} = [2(9.8\ m/s^2)(0.5\ m)]^{\frac{1}{2}} = 3.13\ m/s.$$

The motion is stopped in a distance d = 2 cm; if $\bar{v}$ is the average speed over this distance,

$$\bar{v}(t_f - t_i) = d.$$

A reasonable approximation is to let $\bar{v} = \frac{1}{2}v_i$, since $v_f = 0$. Then,

$$t_f - t_i = \frac{0.020\ m}{\frac{1}{2}(3.13\ m/s)} = 0.0128\ s,$$

and therefore

$$\bar{a} = \frac{3.13\ m/s}{0.0128\ s} = 245\ m/s^2,\ \text{upward}\ \ \underline{\text{Ans}}.$$

(b) The average force is

$$\bar{F} = (80\ kg)(245\ m/s^2) = 1.96 \times 10^4\ N\ \ \underline{\text{Ans}}.$$

5-12

Let the frictional force be $\vec{f}$; this acts opposite to the velocity of the body, that is upwards. The weight W of the body is W = mg and acts, as always, vertically down. Choosing down as positive, Newton's second law becomes

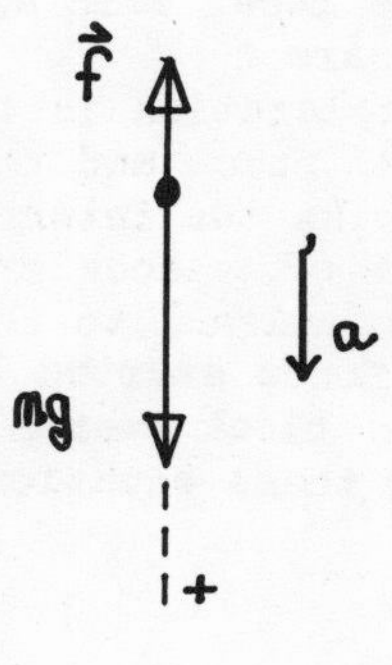

$$mg - f = m(+a),$$

since the downward acceleration $\vec{a}$ points in the chosen positive direction. Hence,

$$f = m(g - a) = (0.25)(9.8 - 9.2),$$

$$f = 0.15 \text{ N} \quad \underline{\text{Ans}}.$$

5-14

(a) Examine the forces acting on a small portion of the string: these forces are the weight W of the portion, acting vertically down, and the tensions T and T' exerted by particles of the string immediately to the left and right of the portion considered. If the string is in equilibrium, the downward acting W must be balanced by upward components of T and T'. If the string is horizontal, however, the tensions have no vertical components. Thus, the string must sag to some degree, although this sag may not be noticeable.

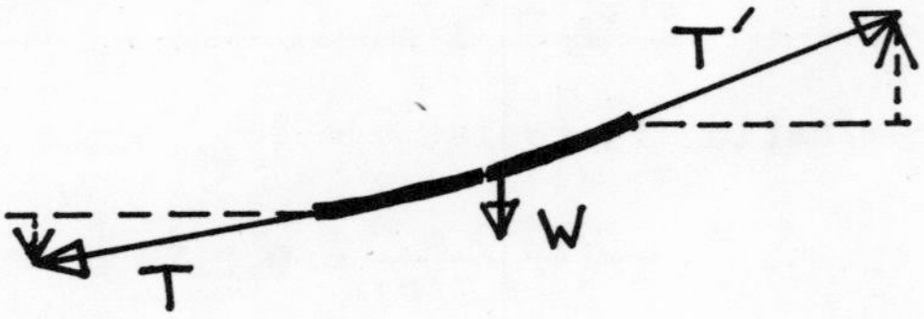

(b)

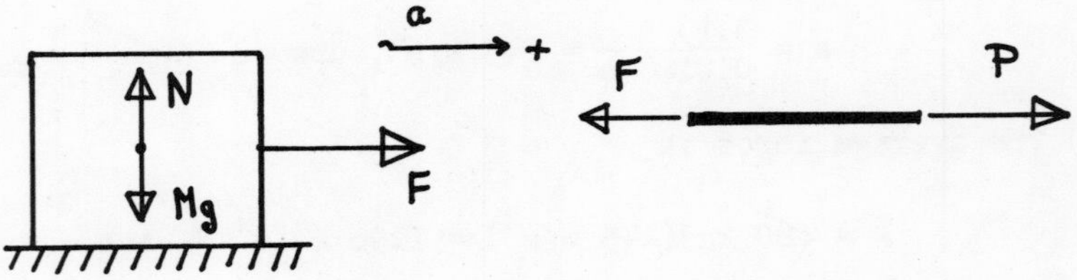

For the rope, mass m, and the block, mass M, the equations of motion are

$$P - F = ma,$$
$$F = Ma,$$

F the force exerted by the rope on the block; by Newton's third law the block exerts a force of equal magnitude on the rope. Adding these equations gives

$$P = (m + M)a,$$

$$a = \frac{P}{m + M} \quad \underline{\text{Ans.}}$$

(c) Directly from (b),

$$F = Ma = \frac{M}{m + M} P \quad \underline{\text{Ans.}}$$

(d) Let T be the tension at the midpoint of the rope. Draw a free-body diagram of the leading half of the rope. From this, it is clear that

$$P - T = \frac{m}{2} a = \frac{m}{2} \frac{P}{m + M},$$

$$T = \frac{P}{2} \frac{m + 2M}{m + M} \quad \text{Ans.}$$

5-19

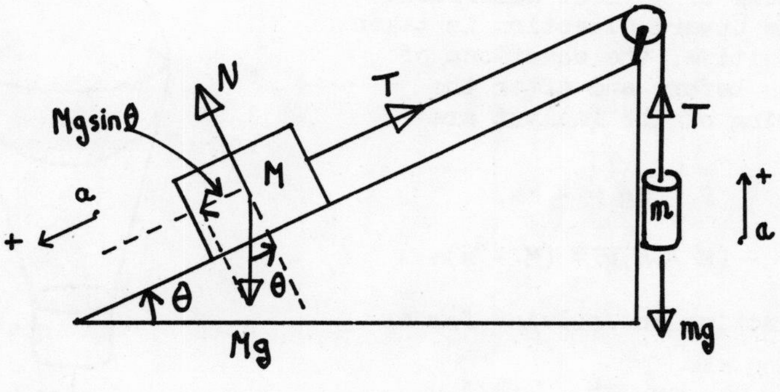

Let T be the tension in the cord. A choice must be made for the direction of the acceleration, as it is not given; the assumed direction is shown on the sketch. (The direction of acceleration must be consistent with the constraints of the problem: for example it cannot be assumed that M accelerates down the plane and that m falls, for this would require the string to stretch.) Since the pulley, by implication, is frictionless and massless, the tensions on the two sides of the pulley are equal. The equations of motion for M, m are

$$Mg\sin\theta - T = Ma,$$

$$T - mg = ma.$$

(a) Adding the equations gives

$$a = \frac{M\sin\theta - m}{M + m}\, g = \frac{(3)(\frac{1}{2}) - 2}{3 + 2}(32) = -3.2 \text{ ft/s}^2.$$

As this is negative, the acceleration is actually in the direction opposite to that which was assumed: to wit, M accelerates up the incline.

(b) From the second equation above, and using the result for a:

$$T = m(g + a) = (2)(32 - 3.2) = 57.6 \text{ lb} \quad \underline{\text{Ans.}}$$

<u>5-24</u>

Let F be the upward lift of the balloon (assumed constant) and m the mass of ballast discarded. If the upward direction is taken as positive, the equations of motion before and after the dropping of the ballast are

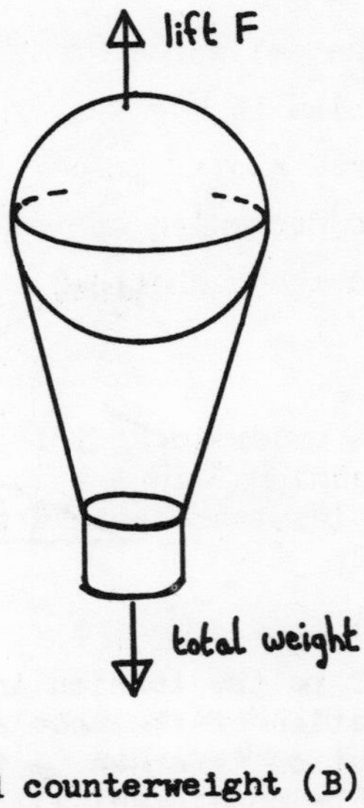

$$F - Mg = -Ma,$$

$$F - (M - m)g = (M - m)a.$$

Subtracting and solving for m:

$$m = 2M\frac{a}{a + g} \quad \underline{\text{Ans.}}$$

<u>5-25</u>

Draw free-body diagrams of the cage (A) and counterweight (B); also shown is the mechanism (C) and the forces exerted on it by the cables.

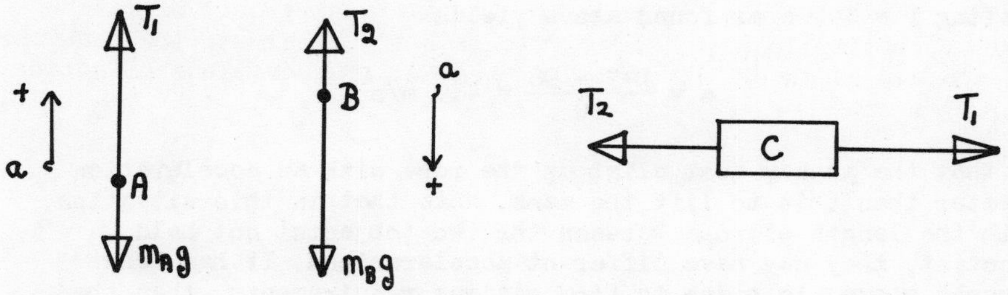

(a) For the cage,

$$T_1 - m_A g = m_A a,$$

$$T_1 = m_A(g + a) = (1100)(9.8 + 2) = 12980 \text{ N} \quad \underline{\text{Ans}}.$$

(b) For the counterweight,

$$m_B g - T_2 = m_B a,$$

$$T_2 = m_B(g - a) = (1000)(9.8 - 2) = 7800 \text{ N} \quad \underline{\text{Ans}}.$$

(c) The net force $\vec{F} = \vec{T}_1 + \vec{T}_2$ exerted by the cable on the mechanism is $F = T_1 - T_2 = 12980 - 7800 = 5180$ N, to the right as pictured above. Hence, by Newton's third law, the force exerted by the mechanism on the cable is 5180 N to the left, that is, toward the mechanism.

5-28

(a) In order to lift the mass, the tension T in the rope must be at least the weight of the object:

$$T = W = Mg = (15)(9.8) = 147 \text{ N}$$

as the minimum requirement. As for the monkey, his equation of motion is

$$T - w = ma,$$

w the monkey's weight, 98 N.

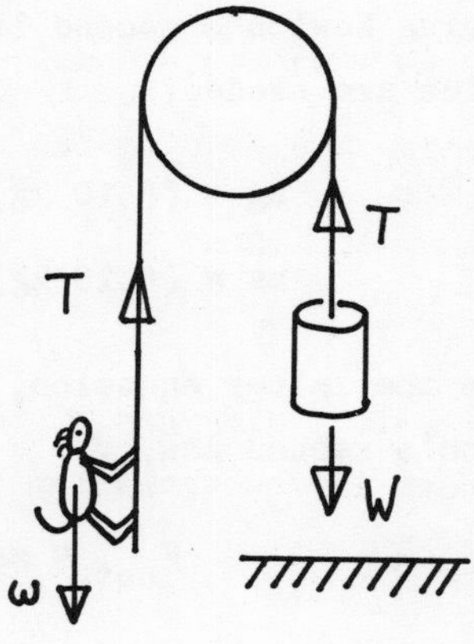

Putting T = 147 N as found above yields

$$a = \frac{147 - 98}{10} = 4.9 \text{ m/s}^2,$$

so that the monkey must climb up the rope with an acceleration greater than this to lift the mass. Note that in this situation, with the length of rope between the two 'objects' not held constant, they may have different accelerations. It has been assumed above, in order to find minimum requirements, that the acceleration of the mass to be lifted is virtually zero.

(b) When the monkey stops climbing, the system forms an Atwood's machine. By Example 8,

$$a = \frac{15 - 10}{15 + 10}(9.8) = 1.97 \text{ m/s}^2 \quad \underline{\text{Ans}};$$

(c)

$$T = \frac{(2)(10)(15)}{15 + 10}(9.8) = 117.6 \text{ N} \quad \underline{\text{Ans}}.$$

5-32

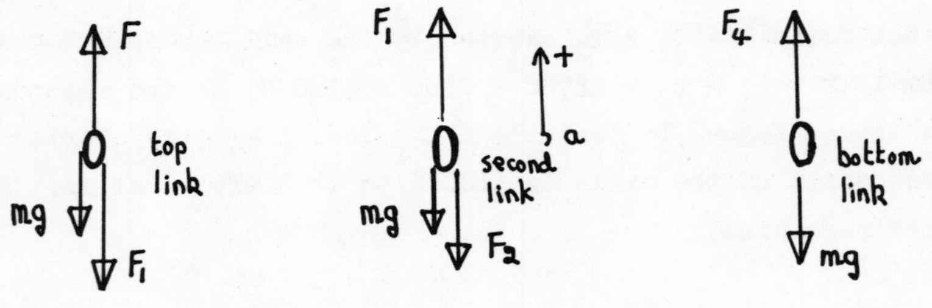

In applying Newton's second law to each link, the following quantities are needed:

$$mg = (0.10 \text{ kg})(9.8 \text{ m/s}^2) = 0.98 \text{ N},$$

$$ma = (0.10 \text{ kg})(2.5 \text{ m/s}^2) = 0.25 \text{ N}.$$

(c) From the latter equation, the net force F_{net} on each link is, by Newton's second law,

$$F_{net} = ma = 0.25 \text{ N} \quad \underline{\text{Ans}}.$$

(a) Now let F_i = forces between adjacent links, counting from the top pair. Then, using Newton's third law, and realizing that the external force F lifting the chain acts only on the top link, the equations for each link, starting at the top, are

$$F - F_1 - mg = ma\ ,$$

$$F_1 - F_2 - mg = ma,$$

$$F_2 - F_3 - mg = ma,$$

$$F_3 - F_4 - mg = ma,$$

$$F_4 - mg = ma\ .$$

The last equation gives

$$F_4 = mg + ma = 0.98 + 0.25 = 1.23\ \text{N} \quad \text{Ans.}$$

Substituting this into the fourth equation gives F_3, which can then be put into the third equation to give F_2, and so on. The results obtained are

$$F_3 = 2.46\ \text{N};\ F_2 = 3.69\ \text{N};\ F_1 = 4.92\ \text{N} \quad \text{Ans};$$

and (b)

$$F = 6.15\ \text{N} \quad \underline{\text{Ans}}.$$

This result for F can be obtained also by considering the chain as a single object of mass 5m; then,

$$F - W = (5m)a,$$

$$F - (5m)g = (5m)a,$$

$$F = (5m)(g + a) = 6.15\ \text{N}.$$

<u>5-35</u>

The x,y-frame is attached to the earth and is an inertial frame. Let

$\vec{a}'$ = acceleration of block relative to elevator;

$\vec{a}''$ = acceleration of block relative to the earth;

$\vec{a}$ = acceleration of elevator relative to the earth.

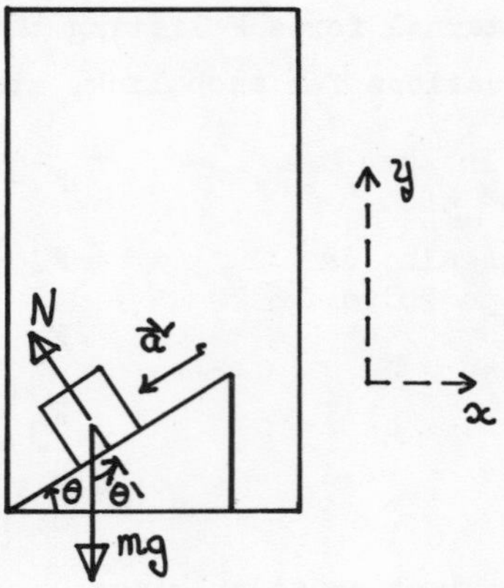

(a),(b) In these cases, $a = 0$, and the elevator is also an inertial frame; hence,

$\vec{a}' = g\sin\theta$, down the incline <u>Ans</u>.

(c) The elevator is no longer an inertial frame, so that Newton's second law must be applied in the x,y-frame:

$$N\cos\theta - mg = ma''_y,$$

$$-N\sin\theta = ma''_x.$$

But

$$\vec{a}' + \vec{a} = \vec{a}''.$$

Now $\vec{a}'$ is parallel to the incline, so that

$$\vec{a}' = -a'\cos\theta\vec{i} - a'\sin\theta\vec{j}.$$

Then, since $\vec{a} = -a\vec{j}$,

$$\vec{a}'' = -a'\cos\theta\vec{i} - (a + a'\sin\theta)\vec{j}.$$

Substitute this into Newton's second law, above, and obtain

$$N\cos\theta - mg = -m(a + a'\sin\theta),$$

$$-N\sin\theta = -ma'\cos\theta.$$

These equations may be solved for a' by eliminating N:

$a' = (g - a)\sin\theta$, down the incline <u>Ans</u>.

(d) Here the analysis is the same as in (c), except that a is replaced with -a, to give

$$a' = (g + a)\sin\theta, \text{ down the incline} \quad \underline{\text{Ans}}.$$

(e) In this case, use the result from (c) but put a = g. The result is a' = 0.

(f) Again, use the equations in part (c), but this time eliminate a' and solve for N to get

$$N = m(g - a)\cos\theta \quad \underline{\text{Ans}}.$$

CHAPTER 6

6-1

(a) Let v be the initial speed and x the distance required to come to a stop. The only force acting in the horizontal direction is the force f_k of kinetic friction. Therefore,

$$f_k = ma.$$

But $v^2 = 2ax$, giving

$$a = \frac{v^2}{2x} = \frac{(6.10)^2}{2(15)} = 1.24 \text{ m/s}^2.$$

The mass m of the puck is

$$m = \frac{W}{g} = \frac{1.1}{9.8} = 0.112 \text{ kg}.$$

Therefore,

$$f_k = (0.112 \text{ kg})(1.24 \text{ m/s}^2) = 0.139 \text{ N} \quad \underline{\text{Ans}}.$$

(b) Since $f_k = \mu_k N$ and $N = mg$ in this situation,

$$f_k = \mu_k N = \mu_k mg = ma,$$

$$\mu_k = \frac{a}{g} = \frac{1.24}{9.8} = 0.127 \quad \underline{\text{Ans}}.$$

6-5

(a) The tension T in the rope is just the force exerted by the man in pulling the crate; f is the force of friction. With the crate on the verge of moving, $f = \mu_s N$. Note that $N \neq W$ in this case

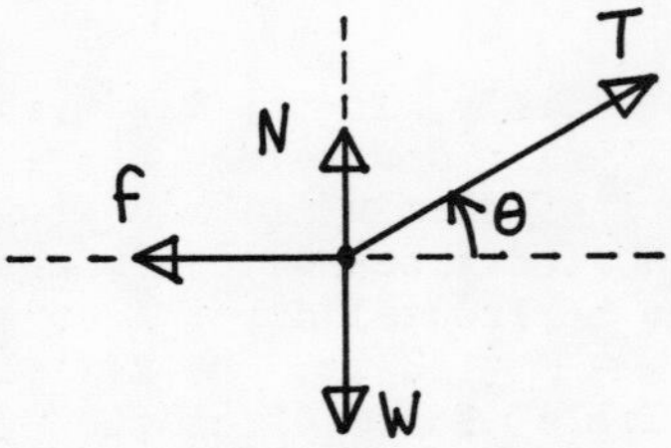

since part of the weight W is balanced by the vertical component of T. Applying the second law:

$$T\cos\theta - \mu_s N = 0,$$

$$N + T\sin\theta - W = 0.$$

Eliminate the normal force N and solve for T:

$$T = \frac{\mu_s W}{\cos\theta + \mu_s \sin\theta},$$

$$T = \frac{(0.5)\cdot(150\text{ lb})}{\cos 15° + (0.5)\sin 15°} = 68.47\text{ lb}\quad \underline{\text{Ans}}.$$

(b) With the crate moving the force of friction becomes $\mu_k N$; also, the acceleration is no longer zero:

$$T\cos\theta - \mu_k N = ma,$$

$$N + T\sin\theta - W = 0.$$

Set m = W/g and solve for a by eliminating N, as before; the result can be written

$$\frac{a}{g} = \frac{T}{W}(\cos\theta + \mu_k \sin\theta) - \mu_k,$$

$$a = 4.23\text{ ft/s}^2\quad \underline{\text{Ans}},$$

using the value of T found in part (a).

<u>6-11</u>

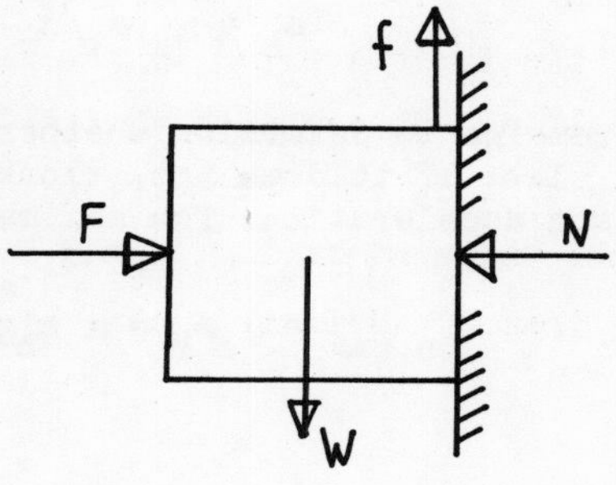

(a) Obviously, if the block slips it will slip downwards; hence, the force $\vec{f}$ of static friction points upward. The block will not move in the horizontal direction, so that N = F = 12 lb. The maximum "available" force of static friction is $\mu_s N$ = (0.6)(12 lb) =

7.2 lb. As this is greater than the weight of the block, which force it opposes, the block will not move.

(b) The force of friction can be no greater than the weight of the block, for if it was, the block would accelerate up the wall. Thus, the forces exerted by the wall on the block are a force N = 12 lb to the left and a force f of static friction equal to 5 lb and directed upward.

6-17

Assume, for the moment, that the block slips on the slab. It is essential to note that, by Newton's third law, forces of friction of equal magnitudes act on each object: indeed, it is solely a force of friction that causes the slab to accelerate. Let F be the 100 N force acting on the block. As usual, the equations of motion follow from the free-body diagrams; since $N_b = m_b g$, the equations are

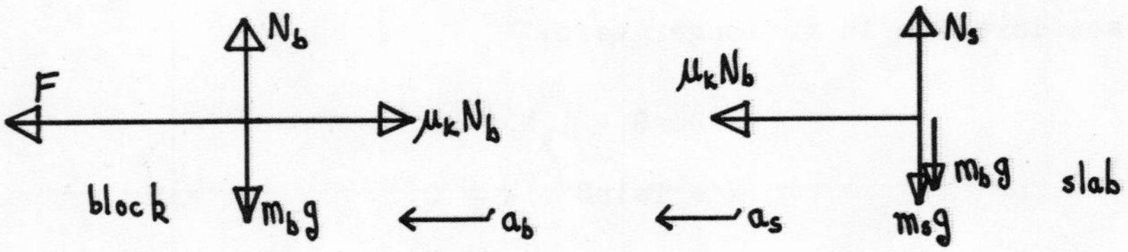

(a) for the block,

$$F - \mu_k m_b g = m_b a_b,$$

$$100 - (0.4)(10)(9.8) = (10)a_b,$$

$$a_b = 6.08 \text{ m/s}^2 \quad \underline{\text{Ans}}.$$

(b) For the slab,

$$\mu_k m_b g = m_s a_s,$$

$$a_s = \mu_k (m_b/m_s) g = 0.98 \text{ m/s}^2 \quad \underline{\text{Ans}}.$$

It remains to determine whether the block does, in fact, slip on the slab: if it does not, block and slab move together with a common acceleration. The maximum force of static friction is

$$f_{s,max} = \mu_s N_b = \mu_s m_b g = (0.6)(10 \text{ kg})(9.8 \text{ m/s}^2),$$

$$f_{s,max} = 58.8 \text{ N} < 100 \text{ N},$$

indicating that static friction cannot hold the block and the motion of the system is, indeed, as was outlined above.

6-19

(a) Draw a diagram showing the forces acting, and with the 8-lb block leading.

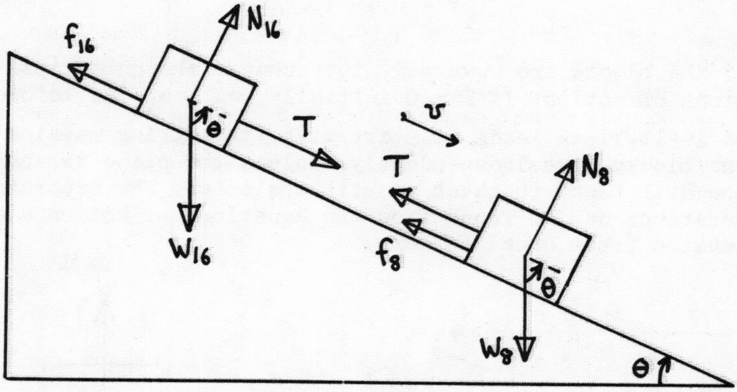

Since

$$f_{16} = \mu_{16}m_{16}g\cos\theta,$$
$$f_8 = \mu_8 m_8 g\cos\theta,$$

the equations of motion are

$$m_8 a_8 = m_8 g\sin\theta - T - \mu_8 m_8 g\cos\theta,$$
$$m_{16}a_{16} = m_{16}g\sin\theta + T - \mu_{16}m_{16}g\cos\theta.$$

Now $\mu_{16} = 2\mu_8$ so that if $T = 0$ initially, then $a_8 > a_{16}$ and the leading block eventually will pull on the following block, with the blocks subsequently moving with a common acceleration, a tension present in the string. Set $a_{16} = a_8 = a$ and add the equations:

$$(m_8 + m_{16})a = (m_8 + m_{16})g\sin\theta - (m_8\mu_8 + m_{16}\mu_{16})g\cos\theta,$$

$$a = 11.38 \text{ ft/s}^2 \quad \underline{\text{Ans}},$$

since $m_8 = \frac{1}{4}$ slug, $m_{16} = \frac{1}{2}$ slug, etc.

(b) From the equation of motion for the heavier block,

$$T = m_{16}a - m_{16}g\sin\theta + \mu_{16}m_{16}g\cos\theta,$$

$$T = 0.46 \text{ lb} \quad \underline{\text{Ans}}.$$

(c) If the blocks are reversed, interchange the subscripts in the equations of motion. If T = 0 initially, $a_8 > a_{16}$ as before. But as the 16-lb block leads to start with, the string remains slack and the blocks move independently, unless the plane is long enough for the 8-lb block to catch up with the other. The separate accelerations can be found from the equations of motion above with the tension T set equal to zero.

6-20

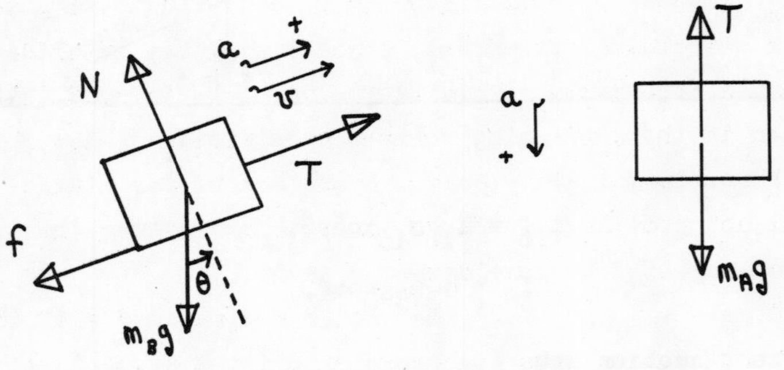

The free-body diagrams are drawn for block B moving up the plane, and also accelerating up the plane so that A accelerates downward. For the opposite direction of acceleration, substitute -a for a in the equations that follow. Note that there is no correlation between the directions of velocity and acceleration. When friction is present, the direction of the velocity must be specified, for the force of friction is opposite to the velocity and not necessarily to the acceleration. In general, reversing the direction of the velocity leads to a different acceleration, so that the two directions of velocity must be treated separately. In writing the equations of motion below, acceleration up the plane is considered as positive. The equations of motion are

$$m_A g - T = m_A a,$$

$$T \mp f - m_B g\sin\theta = m_B a,$$

$$f = \mu N = \mu m_B g\cos\theta.$$

In the middle equation the upper sign applies if B is moving up the plane, as pictured, since then f, opposing v, points downward, in the negative direction, up being chosen as positive. Substitute for f as indicated in the last equation and then add the first two equations together to obtain

$$a = \frac{m_A g - m_B g\sin\theta \mp \mu m_B g\cos\theta}{(m_A + m_B)g}\,g.$$

(a) For $v = 0$ use $\mu = \mu_s$, the static coefficient. Numerically, the terms in the numerator are $m_A g = 32$ lb, $m_B g\sin\theta = 70.71$ lb since $\theta = 45°$, and $\mu_s m_B g\cos\theta = 39.60$ lb. For $f = 0$, $a < 0$ indicating that if the system is released from rest, block B will try to slide down the plane. But if the maximum possible value of f_s is used (with the lower sign in the expression for the acceleration), i.e., $\mu_s N$, then $a > 0$. Thus, in actuality, $f < \mu_s N$ and $a = 0$, for static friction will not provide a force so large as to reverse the "natural" tendency of B to move down the plane.

(b) For $v \neq 0$, use $\mu = \mu_k = 0.25$, so that $f = \mu_k m_B g\cos\theta = 17.68$ lb. Substituting the numbers into the expression for the acceleration gives

$$a = \frac{32 - 70.71 \mp 17.68}{132}(32),$$

$$a = -13.7 \text{ ft/s}^2 \quad \text{(B moving up the plane)},$$

$$a = -5.10 \text{ ft/s}^2 \quad \text{(B moving down the plane)}.$$

The negative signs attached to the final answers indicate that the acceleration of block B is directed down the plane, that of block A directed vertically upward, in both cases.

6-30

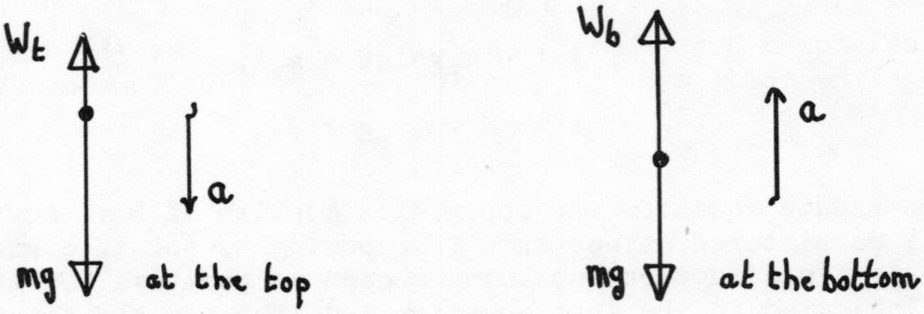

(a) The student's apparent weight W is the force exerted by him on the Ferris wheel. By Newton's third law this, in magnitude, equals the force exerted on him by the wheel. Note that $W \neq mg$ in this situation, for W is used to represent his apparent weight. Let b refer to the bottom of the wheel and t to the top. The student's acceleration is always directed to the center of the wheel. Thus, at the top,

$$mg - W_t = mv^2/R,$$

$$150 - 125 = mv^2/R,$$

$$mv^2/R = 25 \text{ lb}.$$

At the bottom,

$$W_b - mg = mv^2/R,$$

$$W_b - 150 = 25,$$

$$W_b = 175 \text{ lb} \quad \underline{\text{Ans.}}$$

(b) If the new speed $u = 2v$, then $mu^2/R = m(2v)^2/R = 4mv^2/R = 100$ lb. Under these conditions, $W_t = mg - mu^2/R = 150 - 100 = 50$ lb.

6-34

(a)

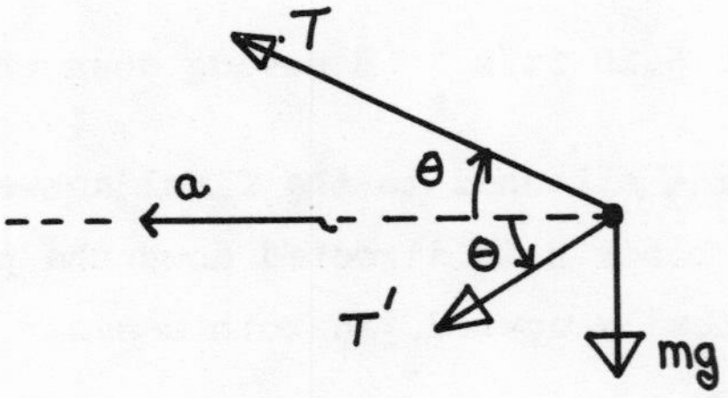

(b) The acceleration of the ball is directed horizontally toward the rod, that is, to the center of its circular path. Hence, the vertical force components must sum to zero; since the angle between the cords and the horizontal is $\theta = 30^\circ$, the sum gives

$$T\sin 30^\circ - T'\sin 30^\circ - mg = 0,$$

$$(25)(\tfrac{1}{2}) - T'(\tfrac{1}{2}) - 9.8 = 0,$$

$$T' = 5.4\ \text{N}\quad \underline{\text{Ans.}}$$

(c) The net force F acts in the direction of the acceleration, therefore toward the rod. Thus,

$$F = T\cos 30^\circ + T'\cos 30^\circ,$$

$$F = (30.4)\sqrt{3}/2 = 26.3\ \text{N}\quad \underline{\text{Ans.}}$$

(d) $F = ma = mv^2/r$, where r is the perpendicular distance of the ball to the rod (i.e., the radius of the circular path). This distance is $\sqrt{3}/2$ m and therefore

$$F = \frac{mv^2}{r} = (1)\,\frac{v^2}{\sqrt{3}/2},$$

$$v = 4.77\ \text{m/s}\quad \underline{\text{Ans.}}$$

6-36

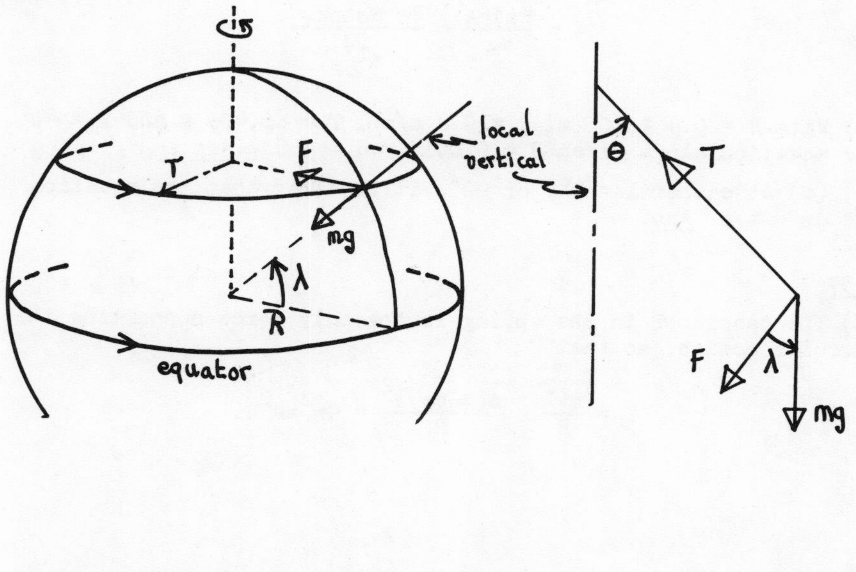

Due to the rotation of the earth, the plumb bob is moving at constant speed in a circle of radius $r = R\cos\lambda$, where R is the radius of the earth and λ is the latitude. Consequently, there must be a resultant force F on the bob directed at the axis of rotation. F makes an angle λ with the local vertical and has a magnitude

$$F = m\cdot\frac{v^2}{r} = m\cdot\frac{(2\pi R\cos\lambda/T)^2}{R\cos\lambda},$$

where m is the mass of the bob and T = 86,400 seconds is the period of rotation of the earth. The local situation is shown in the figure to the right, above. Resolving the forces into vertical and horizontal components,

$$mg - T\cos\theta = F\cos\lambda,$$

$$T\sin\theta = F\sin\lambda.$$

Transpose mg in the first equation and then divide the second equation by the first. The result is

$$\tan\theta = \frac{F\sin\lambda}{mg - F\cos\lambda}.$$

Now θ is a small angle, so that $\tan\theta \approx \theta$. Also, mg is much larger than $F\cos\lambda$. To a good degree of accuracy, then,

$$\theta = \frac{F\sin\lambda}{mg} = \frac{2\pi^2 R\sin 2\lambda}{gT^2}.$$

(a) With $R = 6.4 \times 10^6$ m, $g = 9.8$ m/s^2, T = 86,400 s and $\lambda = 40°$, the equation above gives $\theta = 0.0017$ rad = 5.8' Ans.

(b),(c) At either $\lambda = 0°$, or 90°, it is clear that the equation yields $\theta = 0$ Ans.

6-37

(a) The tension T in the spring is the only force supporting the circular motion, so that

$$T = \frac{mv^2}{R} = \frac{m(2\pi R\nu)^2}{R} = 4\pi^2 m\nu^2 R.$$

But

$$T = k(\ell - \ell_0) = k(R - \ell_0),$$

the tension T being proportional to the elongation of the spring, and the radius R equalling the length ℓ of the spring. Hence,

$$k(R - \ell_0) = 4\pi^2 m\nu^2 R,$$

$$R = k\ell_0/(k - 4\pi^2 m\nu^2).$$

(b) Substitute this result for R into the expression for T above to obtain

$$T = k(R - \ell_0) = 4\pi^2 mk\ell_0\nu^2/(k - 4\pi^2 m\nu^2).$$

6-38

If the funnel rotates too slowly, the cube will slide down the funnel surface; if it rotates too rapidly, the cube will move up on the funnel surface. In the first case, the force of static friction f will be directed upward parallel to the surface, in the second case downward. Consider the former situation. Newton's second law, applied in the horizontal (x) and vertical (y) directions gives

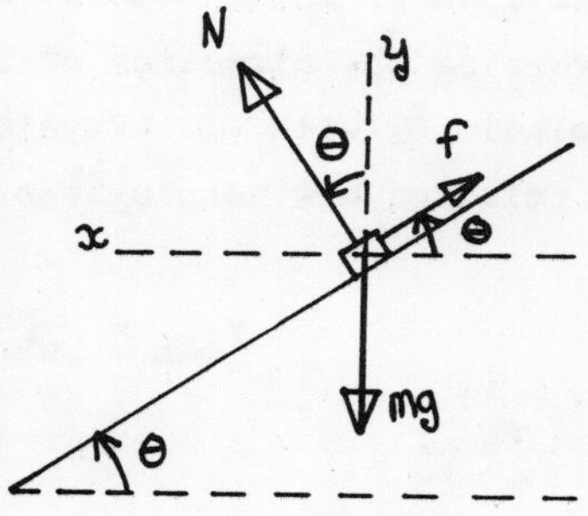

$$N\sin\theta - f\cos\theta = \frac{mv^2}{r} = \frac{m(2\pi r\nu)^2}{r} = 4\pi^2 mr\nu^2;$$

$$N\cos\theta + f\sin\theta - mg = 0.$$

To obtain the limiting value of ν, set $f = \mu N$ and obtain

$$N(\sin\theta - \mu\cos\theta) = 4\pi^2 mr\nu^2,$$

$$N(\cos\theta + \mu\sin\theta) = mg.$$

Divide the equations to eliminate N:

$$\frac{\sin\theta - \mu\cos\theta}{\cos\theta + \mu\sin\theta} = 4\pi^2 r\nu^2/g,$$

(b) $$\nu_{min} = \frac{1}{2\pi}\left[\frac{g}{r}\cdot\frac{\sin\theta - \mu\cos\theta}{\cos\theta + \mu\sin\theta}\right]^{\frac{1}{2}} \quad \underline{\text{Ans}}.$$

The solution above assumes that static friction is not so large as to hold the cube in place if the funnel is not rotating: that is, $\nu_{min} = 0$ if $\theta < \tan^{-1}(1/\mu)$.

(a) If the funnel is rotating rapidly, the friction force f reverses direction, changing the signs of its components in the equations of motion above. But, with $\vec{f} = \mu\vec{N}$ in the limiting case, reversing the direction of f is mathematically equivalent to replacing μ with $-\mu$. (Physically, of course, negative coefficients of friction are meaningless.) Hence

$$\nu_{max} = \frac{1}{2\pi}\left[\frac{g}{r}\cdot\frac{\sin\theta + \mu\cos\theta}{\cos\theta - \mu\sin\theta}\right]^{\frac{1}{2}} \quad \underline{\text{Ans}}.$$

7-4

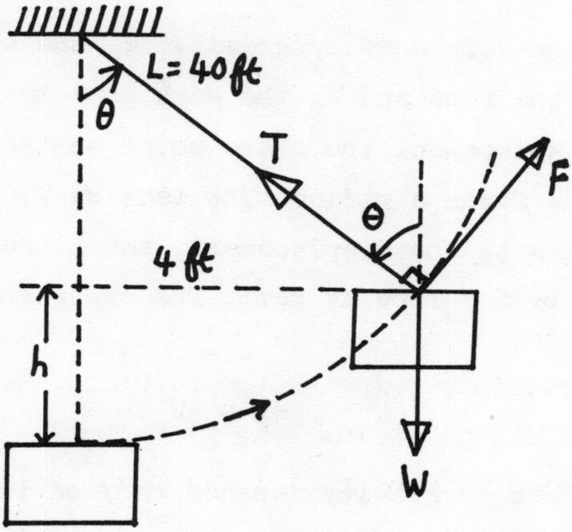

(a) In the final position, the sums of the vertical and horizontal components of the forces acting on the crate each must be zero:

$$T\sin\theta - F\cos\theta = 0,$$

$$T\cos\theta + F\sin\theta - W = 0.$$

Multiply the first equation by cosθ, the second by sinθ and subtract the equations to obtain

$$-F\cos^2\theta - F\sin^2\theta + W\sin\theta = 0,$$

$$F = W\sin\theta = (500 \text{ lb})(\frac{4}{40}) = 50 \text{ lb} \quad \underline{\text{Ans}}.$$

(b) Once the block has reached its final position, the work done in holding it there is zero, since there is no further movement.

(c) By the work-energy theorem, the total work W_t done on the crate in moving it aside is

$$W_t = W_g + W_T + W_F = \tfrac{1}{2}mv_f^2 - \tfrac{1}{2}mv_i^2 = 0 - 0 = 0,$$

where W_g is the work done by gravity, W_T the work done by the tension T in the rope and W_F the work done by the force F. Now, during the displacement the crate moves on the arc of a circle of which the rope forms a radius. The tension T, therefore, is always at right angles to the displacement, and since $\cos 90° = 0$, the work W_T done by the rope is zero. The preceding equation then becomes

$$W_F = -W_g.$$

The work W_g done by gravity depends only on the displacement h in the vertical direction. Gravity acts vertically downward; the block is lifted upward vertically; with $\cos 180° = -1$, then, the work done by gravity is

$$W_g = -mgh,$$

so that

$$W_F = +mgh.$$

It remains to calculate h; from the sketch,

$$h = L - L\cos\theta = (40)(1 - \cos\theta);$$

but,

$$\cos\theta = [1 - \sin^2\theta]^{\frac{1}{2}} = [1 - (4/40)^2]^{\frac{1}{2}},$$

giving

$$h = 0.2005 \text{ ft}.$$

Hence,

$$W_F = (500 \text{ lb})(0.2005 \text{ ft}) = 100.25 \text{ ft·lb} \quad \underline{\text{Ans}}.$$

(d) $W_T = 0$: see part (c).

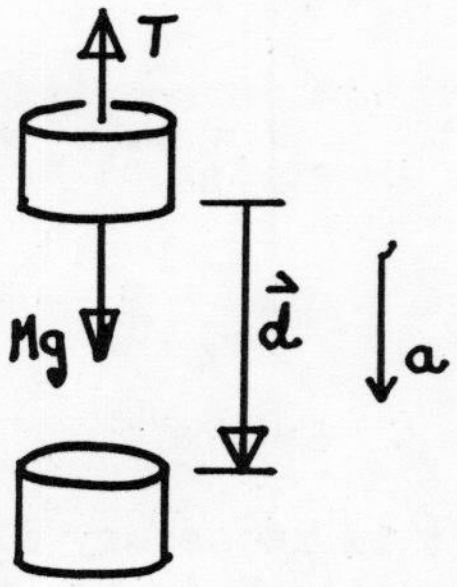

7-5

First, find the tension T. Since $a = g/4$,

$$Mg - T = M\left(\frac{g}{4}\right),$$

$$T = \frac{3}{4}\,Mg.$$

The work done by the tension is

$$W = \vec{T}\cdot\vec{d} = Td\cos 180^\circ = \left(\frac{3}{4}\,Mg\right)(d)(-1),$$

$$W = -\frac{3}{4}\,Mgd \quad \underline{\text{Ans}}.$$

7-8

The total work W is obtained by integrating dW from the initial point i to the final point f.

$$W = \int_i^f \vec{F}\cdot\vec{dr} = m\int_i^f \left(\frac{d^2x}{dt^2}\,\vec{i} + \frac{d^2y}{dt^2}\,\vec{j} + \frac{d^2z}{dt^2}\,\vec{k}\right)\cdot\left(dx\,\vec{i} + dy\,\vec{j} + dz\,\vec{k}\right),$$

$$W = m\int_i^f \left(\frac{d^2x}{dt^2}\,dx + \frac{d^2y}{dt^2}\,dy + \frac{d^2z}{dt^2}\,dz\right).$$

Consider a representative term:

$$\frac{d^2x}{dt^2}\,dx = \frac{d}{dt}\left(\frac{dx}{dt}\right)dx = \frac{d}{dx}\left(\frac{dx}{dt}\right)\frac{dx}{dt}\,dx = \frac{dx}{dt}\,d\left(\frac{dx}{dt}\right).$$

Since $dx/dt = v_x$, the last expression is $v_x dv_x$, which integrates to $\frac{1}{2}v_x^2 - \frac{1}{2}v_{x0}^2$. Therefore,

$$W = m\left(\tfrac{1}{2}v_x^2 - \tfrac{1}{2}v_{x0}^2 + \tfrac{1}{2}v_y^2 - \tfrac{1}{2}v_{y0}^2 + \tfrac{1}{2}v_z^2 - \tfrac{1}{2}v_{z0}^2\right),$$

$$W = \tfrac{1}{2}m(v_x^2 + v_y^2 + v_z^2) - \tfrac{1}{2}m(v_{x0}^2 + v_{y0}^2 + v_{z0}^2),$$

$$W = \tfrac{1}{2}mv^2 - \tfrac{1}{2}mv_0^2.$$

7-18

(a) It is necessary first to find the tension T in the cable. From the free-body diagram of the astronaut,

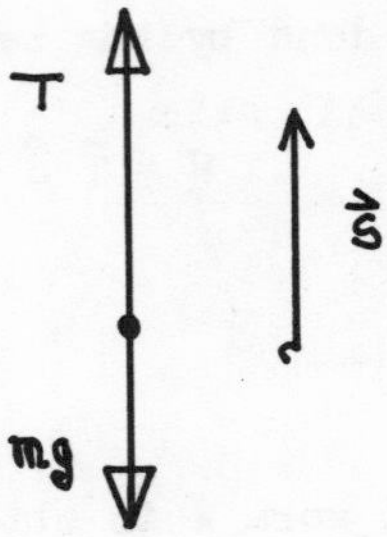

$$T - mg = ma = m(\frac{g}{10}),$$

$$T = \frac{11}{10}\, mg.$$

Let the total displacement of the astronaut be s. The work W done by the helicopter cable is

$$W = \vec{T}\cdot\vec{s} = (\frac{11}{10}\, mg)s = \frac{11}{10}(160\ \text{lb})(50\ \text{ft}) = 8800\ \text{ft}\cdot\text{lb}\quad \underline{\text{Ans}}.$$

(b) The work W_g done by gravity is

$$W_g = m\vec{g}\cdot\vec{s} = mgs\cos 180^\circ = -mgs,$$

$$W_g = -(160\ \text{lb})(50\ \text{ft}) = -8000\ \text{ft}\cdot\text{lb}\quad \underline{\text{Ans}}.$$

(c) The net work done on the astronaut is the sum of (a) and (b), to wit, 800 ft·lb. But this is equal to the change ΔK in the astronaut's kinetic energy. Presumably the initial kinetic energy (astronaut in the ocean) is zero, so that if v is the speed with which he reaches the helicopter,

$$800\ \text{ft}\cdot\text{lb} = \tfrac{1}{2}mv^2 = \tfrac{1}{2}(\frac{160}{32}\ \text{slug})v^2,$$

$$v = 17.9\ \text{ft/s}\quad \underline{\text{Ans}}.$$

7-19

(a) The free-body diagram shows the block at some moment before it has been brought to rest. The spring exerts a force F_s against the block. The force of friction f is given by $f = \mu_k Mg$ since N = Mg in this situation. Let L be the total displacement of the block before being brought to rest. The work W_f done by the force of friction is

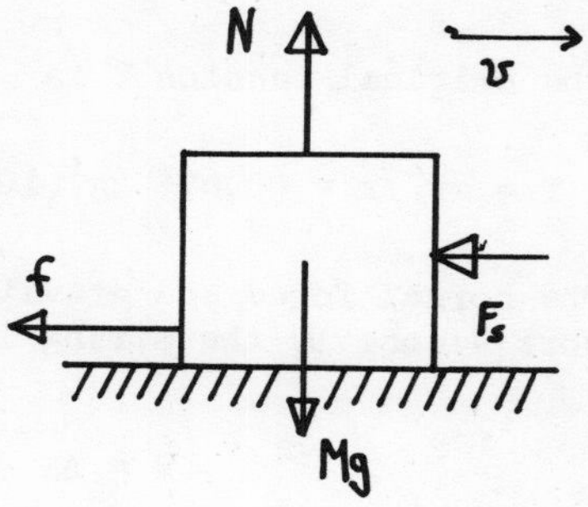

$$W_f = \vec{f}\cdot\vec{L} = fL\cos 180^\circ = -\mu_k MgL \quad \underline{\text{Ans.}}$$

(b) The work W_s done by the spring force in displacing the mass from an initial position x_i to a final position x_f is $\frac{1}{2}kx_i^2 - \frac{1}{2}kx_f^2$. Set $x_i = 0$ and $x_f = L$; then $W_s = -\frac{1}{2}kL^2$ Ans.

(c) The normal force N and the weight Mg act on the block, but each is perpendicular to the displacement of the block and hence do no work on it, since $\cos 90^\circ = 0$.

(d) The total work done is the sum of (a) and (b):

$$W_t = -\mu_k MgL - \tfrac{1}{2}kL^2 \quad \underline{\text{Ans.}}$$

(e) By the work-energy theorem, $W_t = \Delta K = K_f - K_i$:

$$-\mu_k MgL - \tfrac{1}{2}kL^2 = 0 - \tfrac{1}{2}Mv_0^2;$$

solving the equation, a quadratic, for the root with L > 0 (for L is a distance and must be positive) yields

$$L = \frac{1}{k}[(\mu_k^2M^2g^2 + v_0^2kM)^{\frac{1}{2}} - \mu_k Mg] \quad \underline{\text{Ans.}}$$

7-20

(a) The original tension T is

$$T = mv^2/r = (0.675\ \text{kg})(10.0\ \text{m/s})^2/(0.5\ \text{m}) = 135\ \text{N} \quad \underline{\text{Ans}}.$$

(b) The normal force and gravity do no work on the mass. Hence, the work W done by the string in reducing the radius of the circle is

$$W = \Delta K = \tfrac{1}{2}mv_f^2 - \tfrac{1}{2}mv^2.$$

But the final tension T_f is

$$T_f = mv_f^2/r_f.$$

Therefore,

$$W = \tfrac{1}{2}(T_f r_f - Tr).$$

Since $T_f = 4.63T$, $r_f = 0.3$ m, $r = 0.5$ m,

$$W = \tfrac{1}{2}(135\ \text{N})[(4.63)(0.3\ \text{m}) - (0.5\ \text{m})],$$

$$W = 60.0\ \text{J} \quad \underline{\text{Ans}}.$$

7-26

The force F applied to the tool must be directed at an angle to the radial direction in order to supply a tangential component to counteract friction: $F_t = f_k$. The radial component F_r of F is the force actually exerted against the wheel, and this equals 40 lb. The wheel must do work against the friction force exerted by the tool against it. By Newton's third law, this in magnitude equals the force of friction exerted by the wheel against the tool. Therefore,

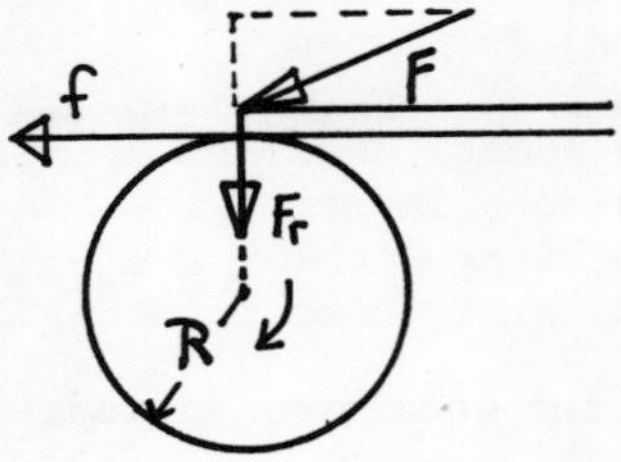

$$P = fv = \mu_k F_r v.$$

In making N revolutions in time t, a point on the rim of the wheel moves with a speed v found from

$$vt = N(2\pi R),$$

$$v = \left(\frac{N}{t}\right)2\pi R = (2.5\ s^{-1})2\pi\left(\frac{8}{12}\ ft\right) = 10.47\ ft/s.$$

This gives

$$P = (0.32)(40\ lb)(10.47\ ft/s) = 134\ ft\cdot lb/s,$$

$$P = \frac{134}{550} = 0.24\ hp\ \ \underline{Ans}.$$

7-32

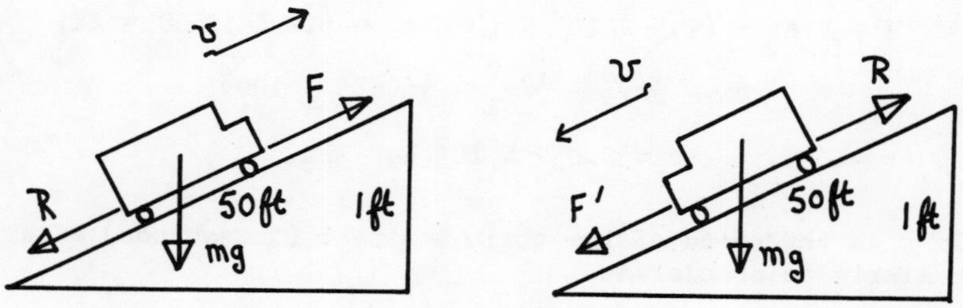

Let F,F' be the forces pulling the truck up the hill with speed v and then down with speed v' (these are friction forces exerted by the road). The resisting force in each direction (wind, rolling friction) is R = mg/25. In moving up and down the hill, the total work done on the truck is zero, for with the speed constant the kinetic energy does not change and W = ΔK. Thus, the rate at which work is done is zero also. The forces that do work are F (or F'), the weight mg and the resisting force R. Taking note of the directions of these forces, for motion up the hill

$$P - mg(v/50) - Rv = 0,$$

where P = Fv, and v/50 is the vertical component of the truck's velocity. Since R = mg/25,

$$P = \frac{3}{50}\,mgv.$$

Moving down the hill, with $P' = F'v'$,

$$P' + mg(v'/50) - Rv' = 0,$$
$$P' = \frac{1}{50}\,mgv'.$$

But $P = P'$ by assumption, so that

$$\frac{3}{50}\,mgv = \frac{1}{50}\,mgv',$$
$$v' = 3v = 3(15) = 45 \text{ mph} \quad \underline{\text{Ans}}.$$

7-33

(a) By the work-energy theorem,

$$W = P\cdot\Delta t = (1.5 \times 10^6 \text{ W})(360 \text{ s}) = 5.4 \times 10^8 \text{ J} = \Delta K,$$
$$\Delta K = \tfrac{1}{2}mv_f^2 - \tfrac{1}{2}mv_i^2 = \tfrac{1}{2}m(625 - 100),$$
$$m = 2.057 \times 10^6 \text{ kg} \quad \underline{\text{Ans}}.$$

(b) If v is the speed of the train a time t (in seconds) after the train starts to accelerate,

$$Pt = \tfrac{1}{2}mv^2 - \tfrac{1}{2}mv_i^2,$$
$$v^2 = v_i^2 + \frac{2P}{m}\,t,$$
$$v = (100 + 1.458t)^{\frac{1}{2}} \quad \underline{\text{Ans}},$$

v in m/s and t in s.

(c) From (b), and differentiating,

$$\tfrac{1}{2}mv^2 - \tfrac{1}{2}mv_i^2 = Pt,$$
$$mv\,\frac{dv}{dt} = mva = P,$$
$$a = \frac{P}{mv},$$

since v_i and P are constants. The force is $F = ma$, giving for the force F,

$$F = \frac{P}{v} = \frac{1.5 \times 10^6}{(100 + 1.458t)^{\frac{1}{2}}} \quad \underline{\text{Ans}},$$

F in newtons.

(d) The total distance x moved is

$$x = \int v(t)dt = \int_0^{360} (100 + 1.458t)^{\frac{1}{2}} dt,$$

$$x = 6.7 \text{ km} \quad \underline{\text{Ans}}.$$

CHAPTER 8

8-3

The minimum work W required is that work performed by exerting a force F on the chain just equal to the weight of chain still hanging over the table's edge. If at some instant a length x is hanging over, then $F(x) = (m/\ell)gx$. Since the chain is being moved in a way to decrease x,

$$W = -\int F\,dx = -\int_{\ell/5}^{0}(m/\ell)gx\,dx = mg\ell/50 \quad \underline{\text{Ans}}.$$

8-6

The initial and final positions of the block are shown in the diagram, with the zero-level of gravitational potential energy marked. In both these positions the speed and hence kinetic energy of the block are zero. By conservation of energy, with U_s the spring's potential energy,

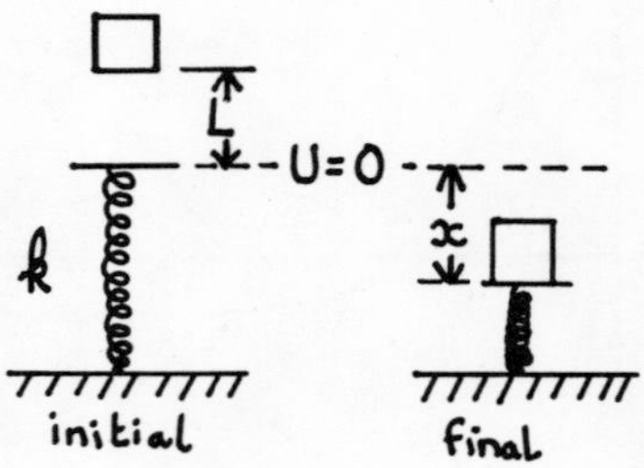

$$U_i + U_{si} + K_i = U_f + U_{sf} + K_f,$$

$$mg(+L) + \tfrac{1}{2}k(0)^2 + 0 = mg(-x) + \tfrac{1}{2}kx^2 + 0.$$

This can be rearranged as

$$x^2 - (2mg/k)x - (2mgL/k) = 0.$$

But $mg/k = (2\text{ kg})(9.8\text{ m/s}^2)/(1960\text{ N/m}) = 0.01\text{ m}$; $mgL/k = 0.004\text{ m}^2$. With these substitutions the equation becomes

$$x^2 - (0.02)x - 0.008 = 0,$$

$$x = 0.10\text{ m} \quad \underline{\text{Ans}},$$

taking the positive root.

8-12

(a) The force F is given by

$$F = -\frac{dU}{dx} = -(\text{slope of U vs. x graph}).$$

Near x = 0 the slope is about -4, so $F(0) \approx 4$ N. U has zero slope at x = 2 m and therefore F(2) = 0. Between x = 2 and x = 6 m, the graph of U is almost a straight line with slope 2/4, giving F roughly constant at $F = -\frac{1}{2}$ N in this region. The negative sign indicates that the force points in the -x-direction.

(b) The kinetic energy is K = 4 - U, so the graph of K vs. x is just a reflection of U vs. x about the line 4 J.

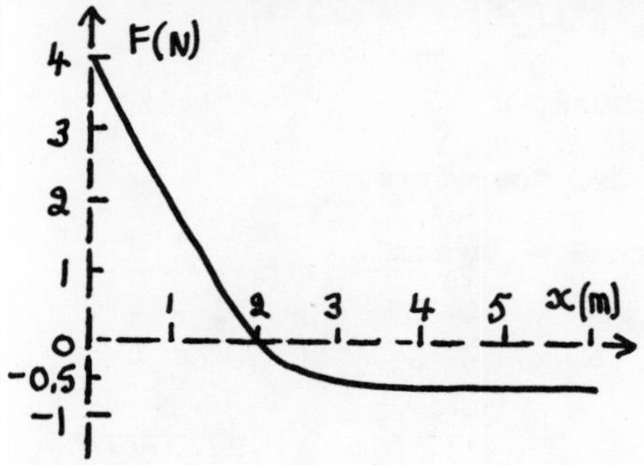

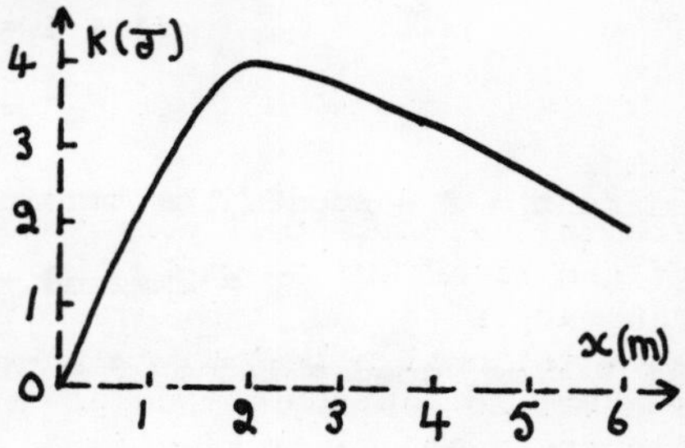

8-21

(a) By conservation of energy,

$$U_R + K_R = U_Q + K_Q,$$

$$mg(2\ell) + 0 = 0 + \tfrac{1}{2}mv^2,$$

$$v = 2(g\ell)^{\frac{1}{2}} \quad \underline{\text{Ans}}.$$

From the free-body diagram of the mass when at Q,

$$T - mg = mv^2/\ell,$$

since ℓ is the radius of the circular path. But the second preceding equation gives

$$mv^2/\ell = 4mg,$$

so that

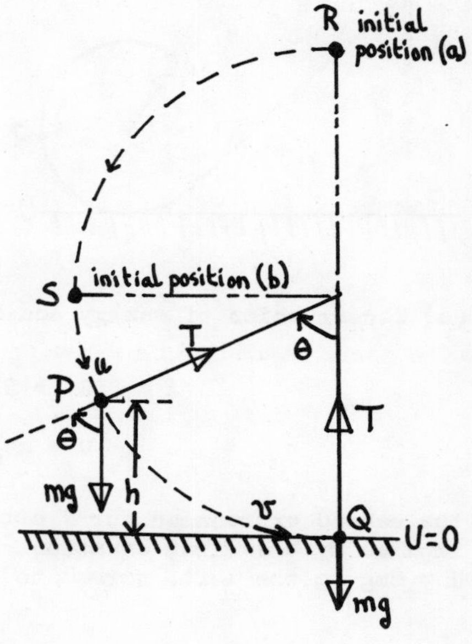

$$T = mg + 4mg = 5\ mg \quad \underline{\text{Ans}}.$$

(b) Applying Newton's second law along the direction of the suspension (i.e., along the direction to the center of the circle) to the mass at P, where the angle with the vertical takes on the desired value,

$$T - mg\cos\theta = mu^2/\ell,$$

u the speed at P. Energy conservation gives

$$U_S + K_S = U_P + K_P,$$

$$mg\ell + 0 = mgh + \tfrac{1}{2}mu^2,$$

$$u^2 = 2g\ell\cos\theta,$$

since $h = \ell - \ell\cos\theta$. The tension T is, therefore,

$$T = 2mg\cos\theta + mg\cos\theta = 3mg\cos\theta.$$

If $T = mg$, $\cos\theta = 1/3$ or $\theta = 71^\circ$.

<u>8-24</u>

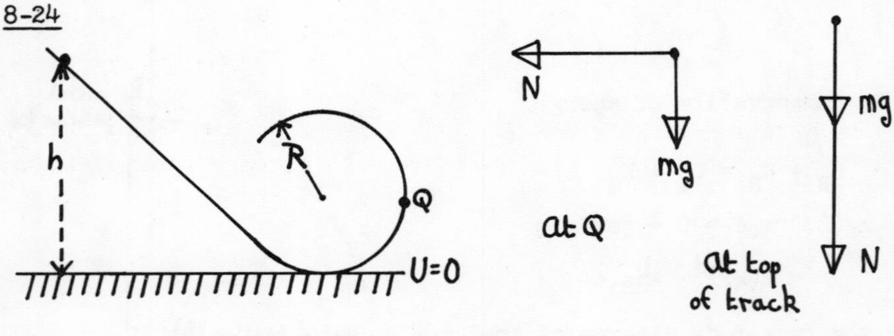

(a) Conservation of energy and Newton's second law give

$$mg(5R) + 0 = mgR + \tfrac{1}{2}mv_Q^2,$$

$$N = mv_Q^2/R = 8mg,$$

the second expression for N obtained by using the value of v_Q implied by the first equation. Hence, the forces acting at Q are $N = 8mg$ to the left, normal to the track, and the weight mg down.

(b) At the top of the track,

$$N + mg = mv_t^2/R,$$

since N, normal to the track surface, here is in the same direction as the weight. By energy conservation,

$$mgh + 0 = mg(2R) + \tfrac{1}{2}mv_t^2.$$

Set N = mg as requested; the former equation gives $\frac{1}{2}mv_t^2 = mgR$; then the latter equation yields h = 3R.

8-26

(a) At the top, Newton's second law and energy conservation require that

$$N + mg = mv^2/R,$$

$$\tfrac{1}{2}mv_0^2 + 0 = \tfrac{1}{2}mv^2 + mg(2R).$$

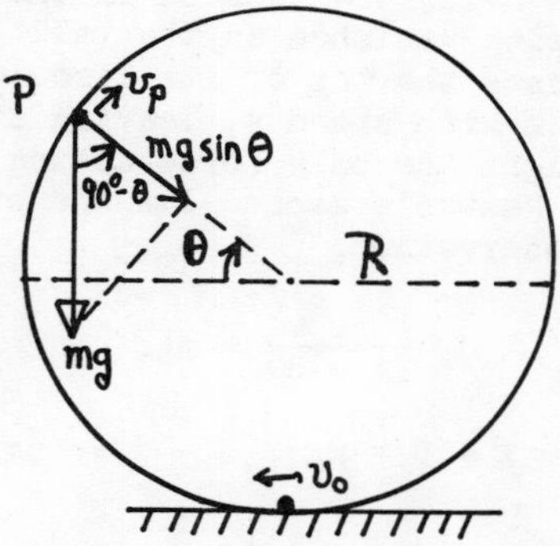

Combining these yields

$$mv_0^2/R = N + 5mg.$$

If $v_0 = v_m$, the normal force N = 0 at the top, for if the particle loses contact with the track, the track cannot exert a normal force on the particle. Therefore,

$$mv_m^2/R = 5mg,$$

$$v_m = (5gR)^{\frac{1}{2}} \quad \underline{\text{Ans}}.$$

(b) This time, with $v_0 < v_m$, the particle loses contact and N goes to zero at some point P before the top is reached. Again, by Newton's second law and energy conservation,

$$mg\sin\theta = mv_p^2/R,$$

$$\tfrac{1}{2}mv_0^2 + 0 = \tfrac{1}{2}mv_p^2 + mg(R + R\sin\theta).$$

Eliminating v_p, the speed at P, gives

$$mg\sin\theta = mv_0^2/R - 2mg(1 + \sin\theta),$$

$$3g\sin\theta = \frac{1}{R}\cdot v_0^2 - 2g,$$

$$3g\sin\theta = [(0.775)(5gR)^{\frac{1}{2}}]^2/R - 2g = g,$$

$$\theta = \sin^{-1}(1/3) \quad \underline{\text{Ans}}.$$

This is about 19.5°.

<u>8-27</u>

If the ball just swings around the nail, the tension in the string vanishes as the ball passes the top of its circular path with speed v, leaving its weight the only force acting. By Newton's second law and energy conservation,

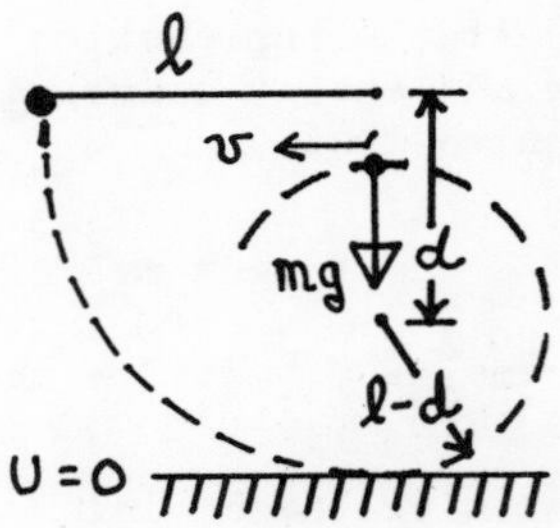

$$\frac{mv^2}{(\ell - d)} = mg,$$

$$mg\ell + 0 = mg\cdot 2(\ell - d) + \tfrac{1}{2}mv^2.$$

Eliminating v between these and solving for d gives

$$d = \frac{3}{5}\ell \quad \underline{\text{Ans}}.$$

<u>8-28</u>

Call h the height of the table, m the mass of each marble and L the horizontal distance to impact. The horizontal speed v needed to project the marble a distance L can be found from

$$L = vt,$$

$$h = \tfrac{1}{2}gt^2,$$

giving

$$v^2 = L^2g/2h.$$

If the spring is compressed a distance x, the speed v with which the marble leaves the spring gun is given from

$$\tfrac{1}{2}kx^2 = \tfrac{1}{2}mv^2.$$

Substituting v^2 from the previous equation leads to

$$kx^2 = \tfrac{1}{2}mL^2 g/h.$$

Call x_1, L_1 and x_2, L_2 the values of x, L for the two attempts. The equation above implies that

$$x_1/x_2 = L_1/L_2,$$

$$\frac{1 \text{ cm}}{x_2} = \frac{1.8 \text{ m}}{2.0 \text{ m}},$$

$$x_2 = 1.11 \text{ cm} \quad \underline{\text{Ans}}.$$

8-29

(a) The work done in one minute = mgh, where m is the mass carried in one minute and h is the vertical displacement; m = 7300 kg and h = 7.6 m. Therefore, the power "consumed" is

$$P = \frac{mgh}{t} = \frac{(7300)(9.8)(7.6)}{60} = 9062 \text{ W} \quad \underline{\text{Ans}}.$$

(b) The work that must be done against gravity is (mg)h = (710 N)·(7.6 m) = 5396 J. The escalator would require (12 m)/(0.61 m/s) = 19.67 s to carry him up if he stood still. Since he reaches the top in 10 s, apparently he is supplying about half the required work and the escalator is providing the other half, or 2698 J.

(c),(d) In walking to stay at the same level in space, at least part of the man (his feet) are alteratively being carried upward by the escalator and then allowed to fall back. Thus, there is some upward displacement and work is done by the motor. If at any instant, at least one foot is in contact with the step (normal walking), then work is being done continually. On the other hand, if the man jumps down after each short upward movement, then he is in contact with the escalator for only part of the time, affecting the power "consumed" from the motor. Evidently, the power "consumed" from an escalator motor depends on the details of "how" one walks.

8-30

(a) The string is expected to stretch under the centrifugal forces that develop as a result of the circular motion.

(b) Let U = 0 at the bottom of the swing. Energy conservation and Newton's second law require

$$mg(\ell + \Delta\ell) = \tfrac{1}{2}mv^2 + \tfrac{1}{2}k(\Delta\ell)^2,$$

$$k(\Delta\ell) - mg = mv^2/(\ell + \Delta\ell),$$

where v is the speed at the bottom. Eliminating this quantity gives

$$\frac{3}{2}\cdot mg\ell(1 + \frac{\Delta\ell}{\ell}) = \tfrac{1}{2}k\ell^2(\frac{\Delta\ell}{\ell})[1 + 2(\frac{\Delta\ell}{\ell})].$$

If $\Delta\ell/\ell \ll 1$, this equation becomes

$$\frac{3}{2}\cdot mg\ell = \tfrac{1}{2}k\ell^2(\frac{\Delta\ell}{\ell}),$$

$$\Delta\ell = \frac{3mg}{k} \quad \underline{\text{Ans}}.$$

(c) From the equation of motion (second equation),

$$v^2 = \frac{k\ell}{m}(1 + \frac{\Delta\ell}{\ell})\Delta\ell - g\ell(1 + \frac{\Delta\ell}{\ell}) = \frac{k\ell}{m}\cdot\Delta\ell - g\ell - g\cdot\Delta\ell,$$

the last step only if $\Delta\ell/\ell \ll 1$, as is being assumed. Putting in the results from (a) for $\Delta\ell$ leads to

$$v^2 = 2g(\ell - \frac{3mg}{2k}).$$

Evidently, some of the original gravitational potential energy of the system ends up as potential energy of the stretched string, leaving less for kinetic energy.

8-34

(a) Let v be the speed with which the boy leaves the ice; as he is given a very small push, let the initial speed (at the top) be zero. Energy conservation gives

$$0 = \tfrac{1}{2}mv^2 - mg(R - h),$$

From Newton's second law,

$$mg\cos\theta - N = mv^2/R.$$

Combining these equations yields

$$mg\cos\theta - N = 2mg.\frac{R - h}{R}.$$

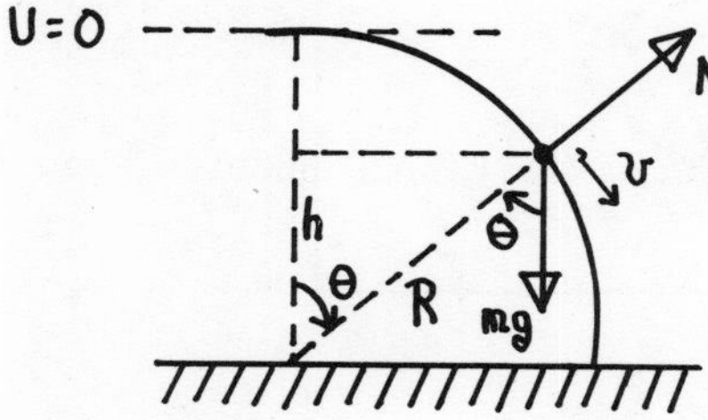

As the boy leaves the ice mound, the normal force N vanishes; setting N = 0 in the previous equation gives

$$\cos\theta = 2 - 2\,\frac{h}{R},$$

$$h = 2R/3,$$

since $\cos\theta = h/R$.

(b) With friction present the energy conservation equation is no longer valid, but Newton's second law for the radial direction,

$$mg\cos\theta - N = mv^2/R,$$

still holds. Now N = mg at the top of the mound where v = 0. As θ increases, v increases and N diminishes. But with friction present v will increase more slowly with θ than for the frictionless case. Hence, when $\theta = \cos^{-1}(2/3)$ (the "departure" angle for frictionless ice) is reached, the value of the speed v is smaller than for the smooth mound. For the latter N = 0 here. But the term $mg\cos\theta$ in the above equation is the same at this angle for both cases. Thus, to get a smaller v requires $N > 0$ in the equation. This means that N has not yet reached zero, so that the boy leaves at a larger angle than for the frictionless mound.

8-38

(a) The mass of the body is m = W/g = 40/32 = 1.25 slug. By the work-energy theorem, with only gravity and the force F doing work on the body (the track being frictionless in this part),

$$W_g + W_F = \Delta K = \tfrac{1}{2}mv_f^2 - \tfrac{1}{2}mv_i^2,$$

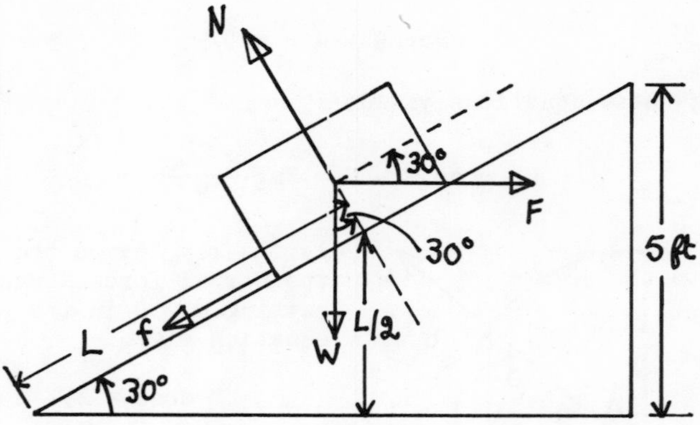

$$- mgh + W_F = \tfrac{1}{2}m(v_f^2 - v_i^2),$$

$$-(1.25)(32)(5) + W_F = \tfrac{1}{2}(1.25)(10^2 - 2^2),$$

$$W_F = 260 \text{ ft·lb} \quad \underline{\text{Ans.}}$$

(b) By definition, the work done by F is

$$W_F = \vec{F}\cdot\vec{d} = Fd\cos\theta = F(10)\cos 30^\circ = 260,$$

$$F = 30.02 \text{ lb} \quad \underline{\text{Ans.}}$$

(c) Apply the work-energy theorem. Friction is present now and does work on the body. Let L be the distance the body moves on the plane before coming to rest. Then,

$$W_g + W_F + W_f = \Delta K.$$

The problem implies that the block does not travel all the way up the plane. Hence $v_f = 0$ so that

$$\Delta K = 0 - \tfrac{1}{2}(1.25)(2)^2 = -2.50 \text{ ft·lb}.$$

The work done by gravity is $W_g = -(40 \text{ lb})(\tfrac{1}{2}L) = -20.00L$ ft·lb; the work done by F is $W_F = +(30.02 \text{ lb})(L\cdot\tfrac{1}{2}\sqrt{3}) = +25.98L$ ft·lb. Finally, the work done by friction is $W_f = -\mu_k NL$. But, the normal force N is given by

$$N = W\cos 30° + N\sin 30° = (40)\cos 30° + (30.02)\sin 30°,$$

$$N = 49.65 \text{ lb}.$$

Therefore $W_f = -(0.15)(49.65 \text{ lb})L = -7.45L$ ft·lb. Putting W_g, W_F and W_f into the work-energy theorem gives

$$-2.50 = -20.00L - 7.45L + 25.98L,$$

$$L = 1.70 \text{ ft} \quad \underline{\text{Ans}}.$$

8-41

(a) Let f be the friction force. The mass of the elevator is m = (4000)/(32) = 125 slug. If L is the distance traveled before hitting the spring, the work-energy theorem says that

$$\tfrac{1}{2}mv^2 - 0 = mgL - fL,$$

$$\tfrac{1}{2}(125)v^2 = (4000 - 1000)(12),$$

$$v = 24 \text{ ft/s} \quad \underline{\text{Ans}}.$$

(b) At the moment of maximum compression of the spring, the elevator is at rest momentarily. Thus,

$$0 - \tfrac{1}{2}mv^2 = mgx - \tfrac{1}{2}kx^2 - fx,$$

since the spring does negative work in trying to push the elevator back up the shaft; the friction force f always does negative work. Numerically,

$$-36{,}000 = (4000)x - \tfrac{1}{2}(10{,}000)x^2 - (1000)x,$$

$$5x^2 - 3x - 36 = 0,$$

$$x = 3 \text{ ft} \quad \underline{\text{Ans}}.$$

(c) Let the distance be s. The elevator is at rest momentarily at the highest point reached and also at the point of maximum compression of the spring. Hence,

$$0 = -mgs + \tfrac{1}{2}kx^2 - fs,$$

the spring force now acting in the same direction the elevator moves; it is assumed that $s > x$. Using the result from (b),

$$0 = -(4000)s + \tfrac{1}{2}(10{,}000)(3)^2 - (1000)s,$$

$$s = 9 \text{ ft} \quad \underline{\text{Ans}}.$$

(d) Gravity and the spring force are conservative forces; all of the energy dissipated is dissipated by friction. If static friction can be neglected (f represents kinetic friction), the spring will be compressed a distance z when the elevator finally comes to rest, where $kz = mg$, or $z = (4000\text{ lb})/(10{,}000\text{ lb/ft}) = 0.4$ ft. The total energy of the system the instant the cable snapped is

$$E_i = (4000\text{ lb})(12.4\text{ ft}) = 49{,}600\text{ ft}\cdot\text{lb},$$

taking the final position as the zero level of gravitational potential energy. The final energy is

$$E_f = \tfrac{1}{2}kz^2 = \tfrac{1}{2}(10{,}000)(0.4)^2 = 800\text{ ft}\cdot\text{lb}.$$

The difference was removed by friction doing work on the elevator. If the elevator moved a total distance y,

$$(49{,}600 - 800) = fy = 1000y,$$

$$y = 48.8 \text{ ft} \quad \underline{\text{Ans}}.$$

The answer is exact to the extent that static friction can be neglected in determining the elevator's final resting place.

<u>8-48</u>

(a) Relativistically, $m = m_0/(1 - \beta^2)^{\frac{1}{2}}$ with $\beta = v/c$. Hence, $m_0 = 0.010$ kg and $\beta = 3 \times 10^7/3 \times 10^8 = 0.1$ in one case, and $\beta = 2.7 \times 10^8/3 \times 10^8 = 0.9$ in the other. By direct substitution,

$$m = 0.01005 \text{ kg } (\beta = 0.1); \quad m = 0.023 \text{ kg } (\beta = 0.9).$$

(b) Classically, $K = \frac{1}{2}m_0v^2 = \frac{1}{2}m_0\beta^2c^2$. Again, by direct substitution,

$$K = 4.5 \times 10^{12} \text{ J } (\beta = 0.1);$$

$$K = 3.6 \times 10^{14} \text{ J } (\beta = 0.9).$$

In the relativistic calculation, $K = (m - m_0)c^2$. Using the results from (a) this gives

$$K = 4.5 \times 10^{12} \text{ J } (\beta = 0.1);$$

$$K = 1.2 \times 10^{15} \text{ J } (\beta = 0.9).$$

(c) In this case $v = 0$ and therefore $m = m_0 = 0.010$ kg; also $K = 0$ both classically and relativistically.

8-49

(a) Solving $m = m_0(1 - v^2/c^2)^{-\frac{1}{2}}$ for v^2 gives

$$v^2 = c^2(1 - m_0^2/m^2).$$

Therefore,

$$K = (m - m_0)c^2 = (m - m_0)v^2/(1 - m_0^2/m^2) = \frac{m^2v^2}{m + m_0}.$$

(b) For the usual expression,

$$K = (m - m_0)c^2 = m_0[(1 - \beta^2)^{-\frac{1}{2}} - 1)]c^2,$$

$$K = m_0[(1 + \tfrac{1}{2}\beta^2 + \ldots) - 1]c^2$$

$$K = \tfrac{1}{2}m_0\beta^2c^2 = \tfrac{1}{2}m_0v^2.$$

For the new expression, as m approaches m_0,

$$K = \frac{m^2v^2}{m_0 + m} = m_0^2v^2/(m_0 + m_0) = \tfrac{1}{2}m_0v^2.$$

CHAPTER 9

9-6

If the plate is arranged as shown with respect to the x,y-axes, then the x-coordinate of the center of mass is zero, that is, the center of mass lies on the y-axis, the axis of symmetry of the plate. For the y-coordinate of the center of mass,

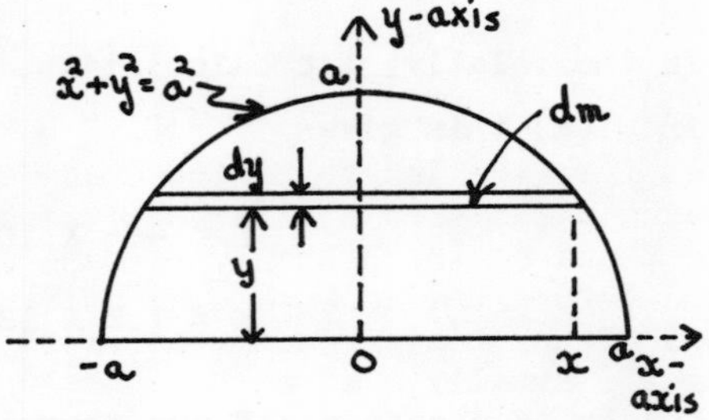

$$My_{cm} = \int y\ dm,$$

by definition; M is the mass of the plate. Consider a thin strip of the plate parallel to the x-axis; its mass dm is

$$dm = \left(\frac{M}{A}\right)(2x\ dy),$$

where A is the area of the plate and 2xdy is the area of the strip. The lower boundary of the plate is y = 0 and the upper boundary is the curve given by

$$x^2 + y^2 = a^2.$$

Hence,

$$My_{cm} = \frac{2M}{A}\int_0^a y(a^2 - y^2)^{\frac{1}{2}}dy.$$

The integral can be put into the form $\int u^{\frac{1}{2}}du$, giving

$$y_{cm} = \frac{4a}{3\pi} \quad \underline{\text{Ans}},$$

since $A = \frac{1}{2}\pi a^2$.

9-11

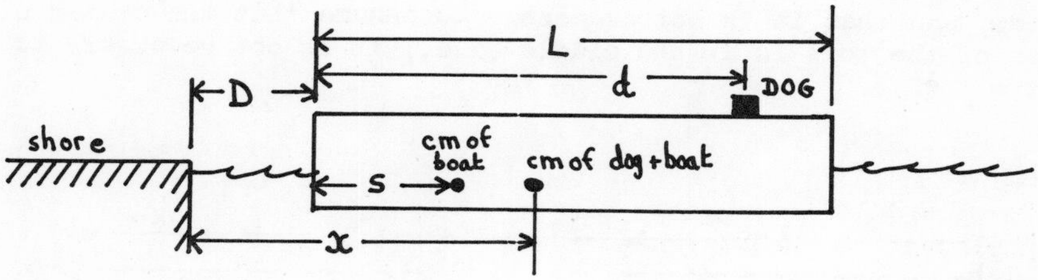

Let L be the length of the boat, the center of mass of the boat being a distance s from that end of the boat closer to shore (the left end on the sketch). Let the dog be a distance d from the same end before the walk, and d' afterwards. The distance of the left end of the boat from shore before the walk is D and after the walk is D'. Finally, let x,x' be the distances from shore of the center of mass of the dog+boat system before and after the walk. Then,

$$(W + w)x = W(D + s) + w(D + d),$$

$$(W + w)x' = W(D' + s) + w(D' + d'),$$

W the weight of the boat, w the weight of the dog. Since the distance the dog walks on the boat is 8 ft,

$$d = d' + 8.$$

Also x = x', since the center of mass of the system does not move, provided the water offers no resistance. Therefore,

$$W(D + s) + w(D + d) = W(D' + s) + w[D' + (d - 8)],$$

$$(W + w)D = (W + w)D' - 8w,$$

$$D' - D = 8\frac{w}{W + w} = 8\frac{10}{40 + 10} = 1.6 \text{ ft}.$$

That is, the boat moves 1.6 ft further from shore. The distance of the dog from shore is now D' + d'; but

$$D' + d' = (D + 1.6) + (d - 8) = (D + d) - 6.4,$$

$$D' + d' = 20 - 6.4 = 13.6 \text{ ft} \quad \underline{\text{Ans}}.$$

This last result follows because the dog was originally D + d = 20 ft from shore. Notice that the length L of the boat does not enter, and that it is not necessary to assume that the center of mass of the boat is at the middle (i.e., it is not necessary to set $s = \frac{1}{2}L$).

9-14

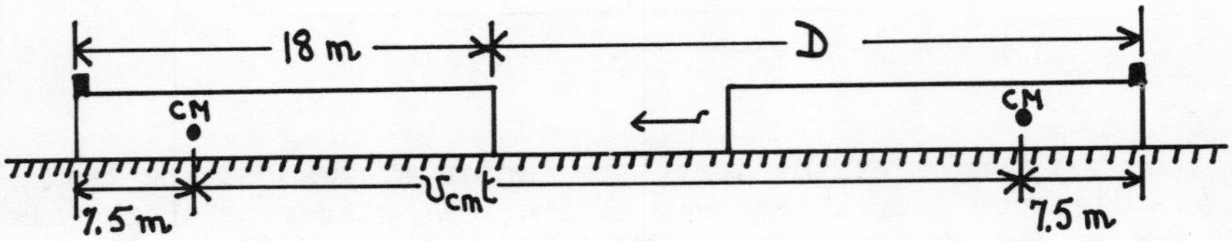

With the ice frictionless, the center of mass of iceboat+man must move with constant velocity even as the man changes his position relative to the iceboat. The center of mass of iceboat+man is at a distance

$$\frac{(80)(0) + (400)(9)}{400 + 80} = 7.5 \text{ m}$$

from the end on which the man is standing, before and after his walk. The man takes a time t = (18 m)/(2 m/s) = 9 s to walk from one end of the boat to the other. In that time, the center of mass of the system moves across the ice a distance

$$d = v_{cm}t = (4 \text{ m/s})(9 \text{ s}) = 36 \text{ m}.$$

The distance D the boat moved in this time measured, say, by the motion of the end of the boat, is

$$D = d + 2(7.5) - 18 = 36 + 15 - 18 = 33 \text{ m} \quad \underline{\text{Ans}}.$$

9-18

(a) Clearly, the center of mass lies midway between the bodies, 2.5 cm from each.

(b) Let x,y be the distances of the center of mass from the 520 g, 480 g bodies. Then,

$$x + y = 5,$$
$$520x = 480y.$$

These equations give x = 2.4 cm, or the center of mass has moved 1.0 mm closer to the heavier body.

(c) In terms of the accelerations of the two bodies, the acceleration of the center of mass is

$$a_{cm} = \frac{m_1a_1 + m_2a_2}{m_1 + m_2} = \frac{m_1a + m_2(-a)}{m_1 + m_2},$$

$$a_{cm} = \frac{m_1 - m_2}{m_1 + m_2}a,$$

where m_1 = 520 g, and a is the acceleration of m_1 (directed down, as this is the heavier body). From Example 7, Chapter 5,

$$a = \frac{m_1 - m_2}{m_1 + m_2}g,$$

and therefore,

$$a_{cm} = \left[\frac{m_1 - m_2}{m_1 + m_2}\right]^2 g = (0.04)^2 g,$$

$$a_{cm} = 0.0016g \quad \underline{\text{Ans}},$$

directed down.

9-22

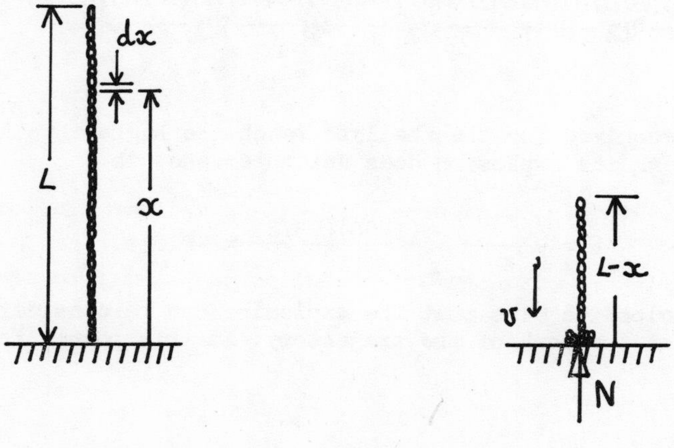

In magnitude, the force N exerted by the table on the chain is equal to the force exerted on the table by the chain. This is

$$N = (\frac{M}{L}x)g + \frac{dp}{dt}.$$

M is the mass of the chain and L is its length. The first term is the weight of that part of the chain already on the table and the second term is the force exerted by the link just landing and being brought to rest. If dM, dx are the mass and length of a link and v the speed with which it strikes the table,

$$\frac{dp}{dt} = \frac{(dM)v}{dt} = \frac{(\frac{M}{L}dx)v}{dt} = \frac{M}{L}v^2 = \frac{M}{L}(2gx),$$

x being the original height above the table of the link just now landing. With this, N becomes

$$N = \frac{M}{L}xg + 2\frac{M}{L}xg = 3(\frac{M}{L}x)g \quad \underline{\text{Ans}},$$

or three times the weight of chain already on the table.

9-28

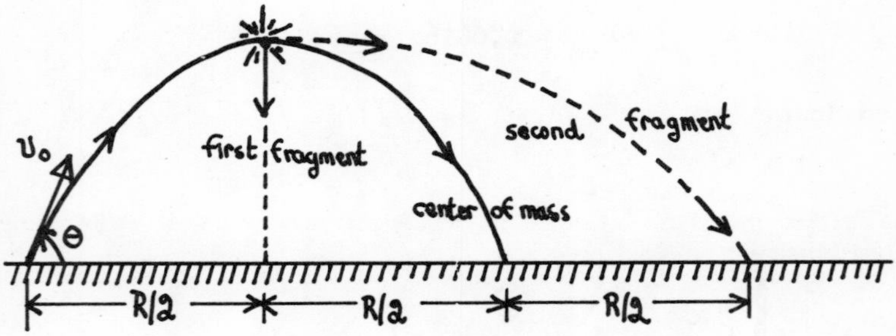

The time t required for the shell to reach the highest point of its path, assuming the explosion does not supervene, is

$$t = \frac{v_0 \sin\theta}{g} = \frac{(1500)(\sqrt{3}/2)}{32} = 40.6 \text{ s}.$$

This is so close to 41 s that the explosion can be considered to take place at the peak of the trajectory. At this point, the vertical component of the momentum of the shell just before the explosion is zero, and therefore the total vertical component of the momentum of the fragments immediately after the explosion must be zero also. Since one fragment is at rest just after the catastrophe, the other must have been projected horizontally. Thus, the fragments strike the ground simultaneously and their center of mass must strike the ground as they do. But the fragments have the same mass and therefore the center of mass impacts at a point equidistant from the impact points of the fragments. As the center of mass moves as though no explosion took place, the situation is as presented in the sketch, with the fragments landing R/2 from the center of mass, R the range of the gun. The second fragment, then, lands from the gun a distance

$$\frac{3}{2}R = \frac{3}{2}(v_0^2 \sin 2\theta/g) = 91{,}340 \text{ ft} \quad \underline{\text{Ans}}.$$

9-29

Let

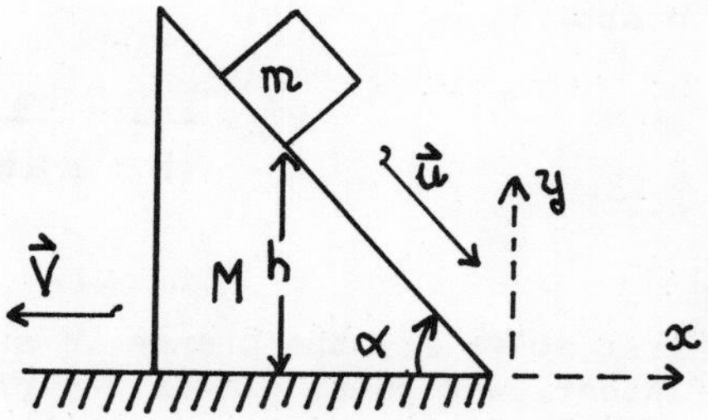

$\vec{V}$ = velocity of wedge relative to the table;

$\vec{u}$ = velocity of block relative to wedge;

$\vec{v}$ = velocity of block relative to table.

Then,

$$\vec{u} + \vec{V} = \vec{v}.$$

As the block remains in contact with the wedge, and referring to the coordinate system shown,

$$\vec{u} = (u\cos\alpha)\vec{i} - (u\sin\alpha)\vec{j};$$

since the block remains in contact with the table,

$$\vec{V} = -V\vec{i}.$$

Therefore,

$$\vec{v} = (u\cos\alpha - V)\vec{i} - (u\sin\alpha)\vec{j},$$

which implies that

$$v^2 = u^2 + V^2 - 2uV\cos\alpha.$$

Momentum is conserved in the horizontal direction:

$$MV = m(u\cos\alpha - V),$$

so that

$$u = \frac{m + M}{m\cos\alpha}V.$$

If u, v, V now are taken as applying at the instant the block makes contact with the table, conservation of energy gives

$$mgh = \tfrac{1}{2}MV^2 + \tfrac{1}{2}mv^2;$$

hence,

$$v^2 = 2gh - \frac{M}{m}V^2.$$

The last equation and the second preceding equation give v and u in terms of V; substitute them into the previous equation for v^2 to obtain

$$V^2 = \frac{2ghm^2\cos^2\alpha}{(M + m)(M + m\sin^2\alpha)} \quad \underline{\text{Ans}}.$$

9-36

First solve for the change in speed for only one man on the flatcar, who jumps off in the prescribed manner. Initially, the velocity of the car is v_0 to the right, and after the man has jumped off it is v (assuming that the man simply "runs off"). Then, since the man runs to the left, his velocity relative to the ground is $v - v_{rel}$ after jumping. The velocity of the center of mass of the system (car + man) is unaffected by these goings-on, so that

$$v_{cm} = v_0 = \frac{Wv + w(v - v_{rel})}{W + w}.$$

Therefore,

$$\Delta v = v - v_0 = \frac{wv_{rel}}{W + w}.$$

(i) If there are n men and all go at once, use the result above with w replaced by nw:

$$\Delta v_i = \frac{nwv_{rel}}{W + nw}.$$

(ii) If the n men go in succession, calculate the increment in speed imparted by each. For the first man,

$$\Delta v_1 = \frac{w\ v_{rel}}{w + [W + (n - 1)w]},$$

since the original flatcar is now, essentially, replaced by the flatcar plus the (n - 1) men who stay behind. The total change in speed is $\Delta v_{ii} = \Delta v_1 + \Delta v_2 + \ldots + \Delta v_n$, or

$$\Delta v_{ii} = \frac{w\ v_{rel}}{W + nw}\left[1 + \frac{W + nw}{W + (n - 1)w} + \ldots + \frac{W + nw}{W + w}\right].$$

All of the terms inside the square brackets are greater than 1 (except the first which equals one), and there are n of these terms so that the second method yields a greater change in speed than the first.

9-38

Since the coal leaves the slow barge in a direction perpendicular to its velocity, the reaction it imparts is perpendicular to the velocity. Therefore, no additional force is required, although the rudder must be applied to prevent the barge from moving in a circle. The coal arrives on the faster barge with a velocity component parallel to the barge's motion and equal to 20 - 10 = 10 km/h, opposite to the barge's velocity. Hence, the additional force required is

$$F = v_{rel}\frac{dm}{dt} = (10\ \text{km/h})(1000\ \text{kg/min}) = 46.3\ \text{N} \quad \underline{\text{Ans}}.$$

9-39

(a) In ingesting the air the plane experiences a resisting force F_1 equal to

$$F_1 = v_{rel}\frac{dm}{dt} = (600\ \text{ft/s})(4.8\ \text{slug/s}) = 2880\ \text{lb}.$$

The mass ejected each second is the 4.8 slug of air plus 0.2 slug of fuel. The thrust imparted to the plane in ejecting this is

$$F_2 = v_{rel}\frac{dm}{dt} = (1600\ \text{ft/s})(5\ \text{slug/s}) = 8000\ \text{lb}.$$

The net thrust is $F_2 - F_1 = 5120$ lb <u>Ans</u>.

(b) The horsepower delivered is

$$P = (F_2 - F_1)v = \frac{(5120\ \text{lb})\ (600\ \text{ft/s})}{(550\ \text{ft}\cdot\text{lb/s}\cdot\text{hp})} = 5600\ \text{hp}\quad \underline{\text{Ans}}.$$

<u>9-41</u>

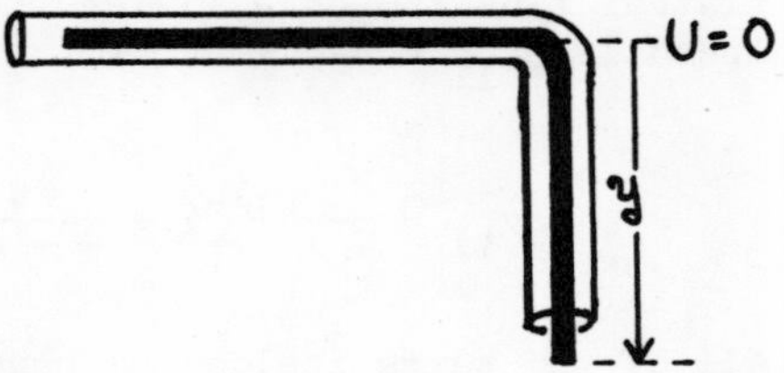

(a) Considering the entire string as the system, its mass is not changing and therefore $v_{rel} = 0$.

(b) Let λ be the mass per unit length of the string. The force accelerating the string is $(\lambda y)g$, but since the entire string, mass $\lambda \ell$, accelerates, the equation of motion will be

$$\lambda y g = (\lambda \ell)\frac{d^2y}{dt^2}.$$

(c) The kinetic and potential energies are

$$K = \tfrac{1}{2}(\lambda \ell)\left(\frac{dy}{dt}\right)^2;\ U = -(\lambda y)g\cdot\tfrac{1}{2}y = -\tfrac{1}{2}\lambda g y^2.$$

With the tube frictionless, $K + U =$ constant; write the constant as $\frac{1}{2}C\lambda$; then,

$$K + U = \tfrac{1}{2}C\lambda,$$

$$\tfrac{1}{2}(\lambda \ell)\left(\frac{dy}{dt}\right)^2 - \tfrac{1}{2}\lambda g y^2 = \tfrac{1}{2}C\lambda,$$

$$\ell\left(\frac{dy}{dt}\right)^2 - g y^2 = C.$$

Differentiating this with respect to time, and noting that $dC/dt = 0$ yields the equation of motion quoted in (b).

(d) The given solution can be verified by direct substitution; it is valid only until $y = \ell$, after which the string undergoes free-fall with acceleration g.

CHAPTER 10

<u>10-7</u>

Since the force increases uniformly,

$$\bar{F} = \tfrac{1}{2}(F_f - F_i) = \tfrac{1}{2}(50) = 25\ \text{N},$$

so that

$$\bar{F}(t_f - t_i) = m(v_f - v_i),$$

$$(25\ \text{N})(4\ \text{s}) = (10\ \text{kg})(v_f - 0),$$

$$v_f = 10\ \text{m/s} \quad \underline{\text{Ans}}.$$

<u>10-10</u>

Let v_{12}, v_{18} be the final speeds. Since the initial speed and hence initial momentum of each spacecraft is zero,

$$m_{12}v_{12} = m_{18}v_{18},$$

$$m_{12}v_{12} + m_{18}v_{18} = 600\ \text{N·s}.$$

Putting $m_{12} = 1200$ kg and $m_{18} = 1800$ kg in the above equations, the final speeds are found to be

$$v_{12} = \frac{1}{4}\ \text{m/s},\ v_{18} = \frac{1}{6}\ \text{m/s}.$$

Since the spacecraft go off in opposite directions, the relative speed is

$$v = v_{12} + v_{18} = \frac{1}{4} + \frac{1}{6} = \frac{5}{12}\ \text{m/s} \quad \underline{\text{Ans}}.$$

10-24

By conservation of momentum and mechanical energy,

$$mv_i = (m + M)V,$$

$$\tfrac{1}{2}mv_i^2 = \tfrac{1}{2}(m + M)V^2 + U_s = \tfrac{1}{2}(m + M)V^2 + f\cdot\tfrac{1}{2}mv_i^2.$$

Here V is the speed of the gun+ball system as the ball sticks, U_s is the potential energy of spring+ball when the ball is stuck and f is the desired fraction. Solve for V in the first equation and substitute into the second; a factor $\frac{1}{2}mv_i^2$ cancels and the rest yields for the fraction,

$$f = \frac{M}{m + M} \quad \underline{\text{Ans}}.$$

10-25

Let v be the speed of the pebbles just before impact; since the collisions are completely inelastic, the pebbles come to rest upon striking the box. Hence, the momentum transferred to the box by each pebble is $p = m(v_f - v_i) = m(0 - v) = -mv$, or $p = mv$ in magnitude. As μ pebbles strike each second, a time interval $T = \mu^{-1}$ elapses between impacts. Thus, the average force $\overline{F}$ imparted to the box by the colliding pebbles is

$$\overline{F} = \frac{p}{T} = \frac{mv}{1/\mu} = m\mu v = m\mu(2gh)^{\frac{1}{2}}.$$

But, after a time t, μt pebbles with total weight $(mg)\mu t$ already reside in the box. Hence, the scale reading R after time t is

$$R = m\mu(2gh)^{\frac{1}{2}} + mg\mu t = mg\mu[(2h/g)^{\frac{1}{2}} + t] \quad \underline{\text{Ans}}.$$

Numerically,

$$R = (0.01)(100)[(2\cdot 25/32)^{\frac{1}{2}} + 10] = 11.25 \text{ lb} \quad \underline{\text{Ans}}.$$

10-28

Let the x-axis be vertical and positive downward, with origin at the top of the shaft. If the ball is dropped at time $t = 0$,

$$x_b = \tfrac{1}{2}gt^2 \; ; \; x_e = h - vt,$$

where v is the speed of the elevator and $h = 60$ ft. If ball and elevator collide at time $t = t'$, then,

$$x_b(t') = x_e(t'),$$

$$\tfrac{1}{2}gt'^2 = h - vt',$$

$$gt' = -v + (v^2 + 2gh)^{\frac{1}{2}},$$

picking, for the last equation, the positive root of the equation for t'. Relative to the elevator, the velocity of the ball is reversed by the collision (elastic collision): just before the collision the relative velocity is $+(v + gt')$ and just after it is $-(v + gt')$. Thus, the velocity of the ball relative to the shaft just after collision is $-(v + gt') - v = -(2v + gt')$. Therefore, the highest point $x = H$ subsequently reached by the ball is given from

$$0^2 = (2v + gt')^2 + 2g[H - x_e(t')],$$

where $x_e(t') = h - vt'$. Putting this in and rearranging gives

$$-4v^2 - 6v(gt') - (gt')^2 = 2g(H - h).$$

Finally, substitute for gt' from above and solve for H:

$$H = -2(\frac{v}{g})(v^2 + 2gh)^{\frac{1}{2}}.$$

(a) Set $v = +6$ ft/s; $H = -23.3$ ft (i.e. above top of shaft) Ans.
(b) Put $v = -6$ ft/s; $H = +23.3$ ft (i.e. below top of shaft) Ans.

10-29

At the instant of maximum compression of the spring, the relative velocity of the two blocks is zero: that is, they move with a common speed V relative to the table. Hence, by conservation of momentum,

$$m_1v_1 + m_2v_2 = (m_1 + m_2)V,$$
$$(2)(10) + (5)(3) = (2 + 5)V,$$
$$V = 5 \text{ m/s}.$$

Then, by conservation of mechanical energy, with x the maximum compression of the spring,

$$\tfrac{1}{2}m_1v_1^2 + \tfrac{1}{2}m_2v_2^2 = \tfrac{1}{2}(m_1 + m_2)V^2 + \tfrac{1}{2}kx^2,$$
$$\tfrac{1}{2}(2)(10)^2 + \tfrac{1}{2}(5)(3)^2 = \tfrac{1}{2}(2 + 5)(5)^2 + \tfrac{1}{2}(1120)x^2,$$
$$x^2 = \frac{70}{1120} = \frac{1}{16},$$
$$x = \tfrac{1}{4} \text{ m} = 25 \text{ cm} \quad \underline{\text{Ans}}.$$

10-30

(a) In parenthesis are indicated the initial and final velocities for the specified mass in the collision under consideration. In the collision of the left (v_0, v') mass with the center (0, u) mass,

$$mv_0 + 0 = mv' + mu; \quad \tfrac{1}{2}mv_0^2 = \tfrac{1}{2}mv'^2 + \tfrac{1}{2}mu^2,$$

giving

$$v' = 0, \quad u = v_0.$$

Thus, the left mass comes to rest. In the subsequent collision between the center (u, u') mass with the right (0, V) mass,

$$mu + 0 = mu' + MV; \quad \tfrac{1}{2}mu^2 = \tfrac{1}{2}mu'^2 + \tfrac{1}{2}MV^2.$$

Solving

$$V = 2\,\frac{m}{m + M}\,v_0; \quad u' = \frac{m - M}{m + M}\,v_0.$$

If $M < m$, $u' > 0$ and the center mass moves to the right with a velocity $u' < V$, so it never catches the right mass and, therefore, there are two collisions.

(b) If $M > m$, the motion of the right mass is as above, but now

$u' < 0$, so that the center mass moves to the left with a speed

$$-u' = \frac{M - m}{M + m} v_0.$$

This mass will collide (the third collision) with the stationary left mass and the two will exchange velocities.

<u>10-31</u>

In the collision between m and m':

$$mv = mv' + m'u, \quad e(v - 0) = u - v';$$

similarly, in the collision between m' and M:

$$m'u = m'u' + MV, \quad e'(u - 0) = V - u',$$

where $0 \leq e, e' \leq 1$ ($e = 0$ for a completely inelastic collision, $e = 1$ for an elastic collision). Eliminating v' from the first pair of equations and u' from the second pair gives

$$u = \frac{mv}{m + m'}(1 + e), \quad V = \frac{m'u}{m' + M}(1 + e').$$

Now eliminate u between these to obtain

$$V = \frac{m\,m'}{(m' + M)(m' + m)} v(1 + e)(1 + e').$$

Therefore, the kinetic energy K of the mass M is

$$K = \tfrac{1}{2}MV^2 = \tfrac{1}{2}\,\frac{Mm^2m'^2}{(m' + M)^2(m' + m)^2} v^2(1 + e)^2(1 + e')^2.$$

Considered as a function of m', find the maximum by the usual method:

$$\frac{dK}{dm'} = \frac{m'Mm^2(1 + e)^2(1 + e')^2v^2}{(m' + M)^3(m' + m)^3}(Mm - m'^2),$$

which is zero for $m' = 0$, $m' = \infty$ and, the desired value,

$$m' = (Mm)^{\frac{1}{2}}.$$

10-39

Since the balls are frictionless, the forces between them at the moment of collision are normal to their surfaces, i.e., along a radius. The geometry at impact is shown in the sketch. Clearly,

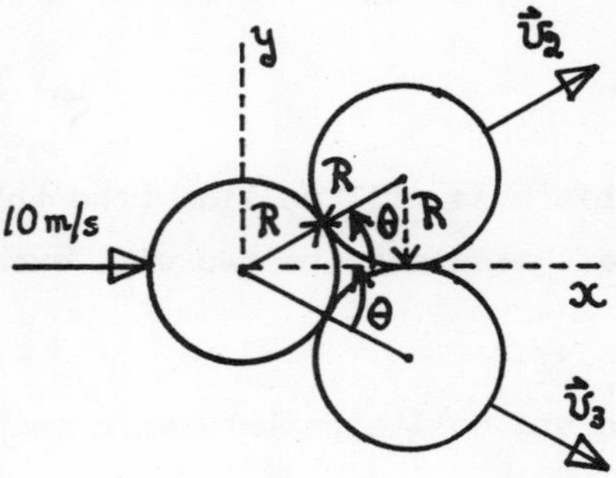

$$\theta = \sin^{-1}\left(\frac{R}{2R}\right) = 30^\circ; \ \cos\theta = \sqrt{3}/2.$$

Also, from the symmetry of the collision, $v_2 = v_3 = v$, say. This follows from conservation of momentum in the y-direction. In addition, as the momentum in the y-direction before impact is zero, it must be zero just after impact. It follows that the velocity of the first ball after collision must be in the x-direction; call this velocity V. By conservation of momentum,

$$m(10) = 2mv\cos\theta + mV,$$

$$10 = \sqrt{3}v + V.$$

By conservation of kinetic energy,

$$\tfrac{1}{2}m(10)^2 = 2\cdot\tfrac{1}{2}mv^2 + \tfrac{1}{2}mV^2,$$

$$100 = 2v^2 + V^2.$$

From the momentum equation,

$$3v^2 = (10 - V)^2 = 100 - 20V + V^2.$$

Substituting this into the equation preceding it yields

$$(5V + 10)(V - 10) = 0,$$

$$V = -2 \text{ m/s} \quad \underline{\text{Ans}}.$$

That is, the ball originally in motion "bounces" and moves off in the direction from whence it came. The other solution of the quadratic equation reproduces the conditions before collision, as expected in elastic collisions. With the result for V above,

$$v = (10 + 2)/\sqrt{3} = 6.93 \text{ m/s} \quad \underline{\text{Ans}}.$$

10-40

Let m be the mass of each object and v their common initial speed. Use the rules for addition of vectors by the geometric method to arrange the initial and final velocities so that the vector sum of the initial momenta equals the momentum of the pair after sticking together in the collision. By conservation of momentum in the x and y-directions,

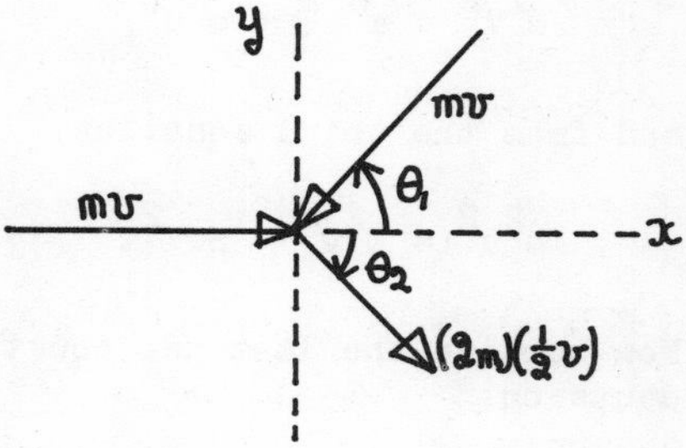

$$mv - mv\cos\theta_1 = (2m)(\tfrac{1}{2}v)\cos\theta_2,$$

$$mv\sin\theta_1 = (2m)(\tfrac{1}{2}v)\sin\theta_2.$$

Dividing out the mv gives

$$1 - \cos\theta_1 = \cos\theta_2,$$

$$\sin\theta_1 = \sin\theta_2.$$

The second equation implies that $\theta_1 = \theta_2$. Using this in the first equation above gives

$$1 - \cos\theta_1 = \cos\theta_1,$$

$$\theta_1 = \cos^{-1}(\tfrac{1}{2}) = \pi/3,$$

so that the angle between the initial velocities ("tail to tail") is $2\pi/3$.

10-41

Conservation of momentum and kinetic energy give

$$mv = MV\cos\theta, \quad mu = MV\sin\theta,$$

$$\tfrac{1}{2}mv^2 = \tfrac{1}{2}mu^2 + \tfrac{1}{2}MV^2,$$

m the mass of the neutron, M the mass of the deuteron. From the first two equations,

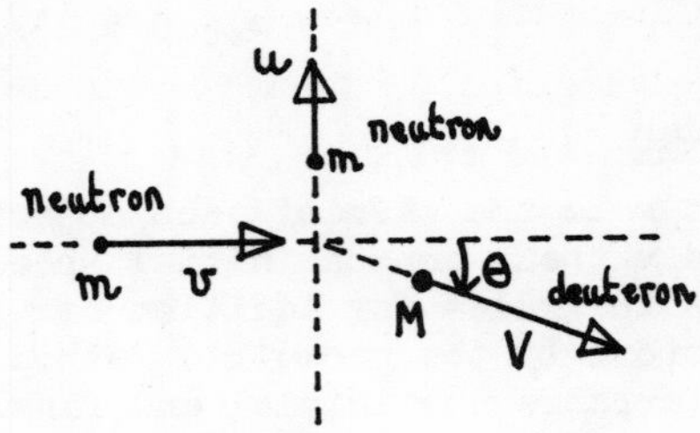

$$M^2V^2 = m^2v^2 + m^2u^2,$$

and from the third equation,

$$m^2u^2 = m^2v^2 - MmV^2.$$

Now combine the last two equations to get the kinetic energy of the deuteron:

$$\tfrac{1}{2}MV^2 = \frac{m^2}{m + M}\,v^2.$$

But M = 2m and therefore

$$\tfrac{1}{2}MV^2 = \frac{m}{m + M}(mv^2) = \frac{1}{3}\,mv^2 = \frac{2}{3}(\tfrac{1}{2}mv^2),$$

establishing the result quoted in the problem.

10-42

By conservation of the two components of momentum and of kinetic energy,

$$m_1v_1 + 0 = m_1u_1\cos\theta_1 + m_2u_2\cos\theta_2,$$
$$0 = m_1u_1\sin\theta_1 + m_2u_2\sin\theta_2,$$
$$\tfrac{1}{2}m_1v_1^2 + 0 = \tfrac{1}{2}m_1u_1^2 + \tfrac{1}{2}m_2u_2^2.$$

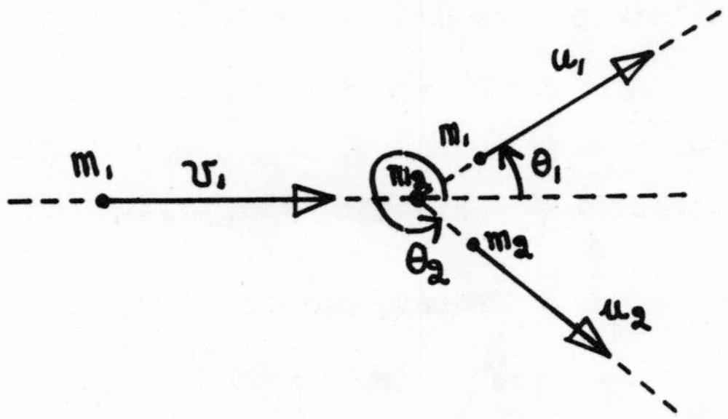

Eliminating u_1 and u_2, the velocities after collision (say, by solving for u_1 from the middle equation, substituting into the first and third equations, then eliminating u_2 between the two resulting equations), yields

$$\tan\theta_1 = \sin 2\theta_2/(\cos 2\theta_2 - m_1/m_2).$$

To find the maximum of $\tan\theta_1$ set $d\tan\theta_1/d\theta_2 = 0$ to obtain

$$\cos 2\theta_2 = m_2/m_1.$$

Substitute this into the equation for $\tan\theta_1$; the resulting $\theta_1 = \theta_m$ is given by

$$\tan^2\theta_m = m_2^2/(m_1^2 - m_2^2).$$

Since

$$\cos^2\theta_m = 1/(1 + \tan^2\theta_m),$$

it follows that

$$\cos^2\theta_m = 1 - m_2^2/m_1^2.$$

(a) For any $m_1 > m_2$, the equation above yields a value for θ_m that, depending on m_2/m_1, will be somewhere between 0 and $\pi/2$.
(b) If $m_1 = m_2$, the equation above cannot be used, since zero denominators would have occured in its derivation. Return to the original equation for $\tan\theta_1$ and set $m_1/m_2 = 1$:

$$\tan\theta_1 = \sin 2\theta_2/(\cos 2\theta_2 - 1) = -\cot\theta_2$$

since $\cos 2\theta = 1 - 2\sin^2\theta$ and $\sin 2\theta = 2\sin\theta\cos\theta$. If θ_2 was taken clockwise instead of counterclockwise, θ_2 above would be replaced with $2\pi - \theta_2$; but $\cot\theta = -\cot(2\pi - \theta)$ so that, for this new θ_2,

$$\tan\theta_1 = \cot\theta_2,$$

$$\theta_1 + \theta_2 = \pi/2.$$

(c) If $m_2 > m_1$, $\cos\theta_2$ can take on the value m_1/m_2 since this is now less than one. But then, from the original equation for $\tan\theta_1$, $\tan\theta_1$ approaches either $+\infty$ or $-\infty$, depending on the sign of $\sin 2\theta_2$. This implies that θ_1 can take on any value.

10-45

The cross section is given from

$$R_x = R_0 N\sigma/A.$$

Here $A = 1\ m^2$. Now 65 g of copper contains $N_0 = 6.02 \times 10^{23}$ copper atoms (Avogadro's number), so that 0.005 g must contain 4.63×10^{19} atoms, by direct proportion. Hence,

$$(4.6 \times 10^{11}\ s^{-1}) = (1.1 \times 10^{18}\ /m^2\cdot s)(4.63 \times 10^{19})\sigma \times 10^{-28}\ m^2,$$

$$\sigma = 90\ b \quad \underline{\text{Ans}}.$$

10-49

(a) Directly from the given information,

$$\Delta(mc^2) = c^2(236 - 132 - 98)u = (6\ u)(1.66 \times 10^{-27}\ kg/u)c^2;$$

Putting $c = 3 \times 10^8$ m/s gives

$$\Delta(mc^2) = 5603\ MeV,$$

since $1\ eV = 1.6 \times 10^{-19}$ J. The energy lost is 5603 - 192 = 5411 MeV.

(b) By conservation of momentum and the definition of Q,

$$m_A v_A = m_B v_B,$$

$$\tfrac{1}{2}m_A v_A^2 + \tfrac{1}{2}m_B v_B^2 - 0 = Q = 192\ MeV = 307 \times 10^{-13}\ J,$$

since the U^{236} nucleus was at rest. With $m_A = 132u$, $m_B = 98u$,

$$v_A = 1.09 \times 10^7 \text{ m/s}, \; v_B = (m_A/m_B)v_A = 1.47 \times 10^7 \text{ m/s} \quad \underline{\text{Ans}}.$$

(c) By straightforward calculation,

$$Q_A = \tfrac{1}{2}m_A v_A^2 = 81.7 \text{ MeV}, \; Q_B = \tfrac{1}{2}m_B v_B^2 = 110 \text{ MeV} \quad \underline{\text{Ans}}.$$

CHAPTER 11

11-3

(a) A wheel with 500 teeth has 500 slots also, so that the angular separation between adjacent slots is $2\pi/500$ rad. Thus the wheel rotates through an angle $\pi/250$ rad in the time t taken for light to travel a distance 2ℓ; this time is $2\ell/c$ and therefore,

$$\omega t = \frac{\pi}{250} \text{ rad},$$

$$t = \frac{2\ell}{c};$$

eliminating t gives

$$\omega = \frac{c\pi}{500\ell} = 3770 \text{ rad/s} \quad \underline{\text{Ans}}.$$

(b) The linear speed at the rim is

$$v = r\omega = (0.05 \text{ m})(3770 \text{ rad/s}) = 188.5 \text{ m/s} \quad \underline{\text{Ans}}.$$

11-8

This situation involves relative angular velocities. Applying the ideas of Section 4-6 to parallel angular, rather than linear, velocities, and noting that P and E revolve in the same direction, the relative angular speed is

$$\omega_S = \omega_E - \omega_P.$$

But $\omega = 2\pi/T$ so that

$$\frac{2\pi}{T_S} = \frac{2\pi}{T_E} - \frac{2\pi}{T_P},$$

$$\frac{1}{T_S} = \frac{1}{T_E} - \frac{1}{T_P}.$$

11-13

(a) The initial and final angular speeds are

$$\omega_i = \frac{(300\text{ rev})(2\pi\text{ rad/rev})}{60\text{ s}} = 10\pi\text{ rad/s},$$

$$\omega_f = \frac{(225\text{ rev})(2\pi\text{ rad/rev})}{60\text{ s}} = 7.5\pi\text{ rad/s}$$

so that

$$\omega_f = \omega_i + \alpha t,$$

$$7.5\pi = 10\pi + \alpha(60),$$

$$\alpha = -\frac{\pi}{24} = -0.131\text{ rad/s}^2 \quad \underline{\text{Ans}}.$$

(b) Measuring the time from the moment of the second observation gives

$$\omega = \omega_0 + \alpha t,$$

$$0 = 7.5\pi + (-\frac{\pi}{24})t,$$

$$t = 180\text{ s} = 3\text{ min} \quad \underline{\text{Ans}},$$

after the second observation, or 4 min after the first.

(c) The angle θ turned through, after the second observation, in coming to rest is

$$\omega^2 = \omega_0^2 + 2\alpha\theta,$$

$$0 = (7.5\pi)^2 + 2(-\frac{\pi}{24})\,\theta,$$

$$\theta = 675\pi\text{ rad}.$$

The number of revolutions is $\theta/2\pi = 337.5$ rev <u>Ans</u>.

11-14

Let ω_1 be the angular velocity at the start of the 4 s interval and θ the angle turned through during the interval. Then,

$$\theta = \tfrac{1}{2}\alpha t^2 + \omega_1 t + \theta_0,$$

$$120 = \tfrac{1}{2}(3)(4)^2 + \omega_1(4),$$

$$\omega_1 = 24 \text{ rad/s}.$$

If T = time the wheel had been rotating before the start of the 4 second interval, and if the wheel started from rest,

$$\omega_1 = \alpha T,$$

$$24 = (3)T,$$

$$T = 8.00 \text{ s} \quad \underline{\text{Ans}}.$$

11-19

If the belt does not slip, the linear speeds at the rims of the two wheels are the same; that is,

$$r_A\omega_A = r_C\omega_C.$$

Differentiating with respect to time gives

$$\alpha_C = (r_A/r_C)\alpha_A.$$

Now $\omega_C = \alpha_C t + \omega_{0C}$ and $\omega_{0C} = 0$. Hence,

$$\omega_C = (r_A/r_C)\alpha_A t.$$

Since ω_C = 100 rev/min corresponds to (100)(2π)/(60) rad/s,

$$\frac{10\pi}{3} = \left(\frac{10}{25}\right)\left(\frac{\pi}{2}\right)t,$$

$$t = \frac{50}{3} = 16.7 \text{ s} \quad \underline{\text{Ans}}.$$

11-20

The angular speed $\omega = 2\pi/\text{period}$; the period of rotation = 24 h = 86,400 s, giving

$$\omega = \frac{2\pi}{86400} = 7.27 \times 10^{-5} \text{ rad/s}$$

everywhere. If r is the radius of the daily circular path, then

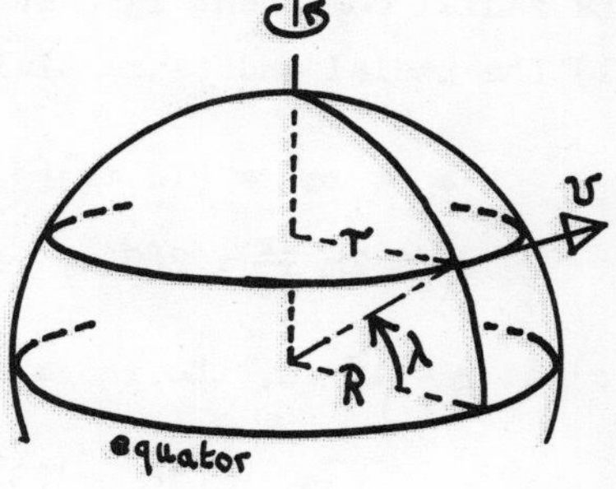

$v = 2\pi r/T$, T the period of rotation. But $r = R\cos\lambda$, R the radius of the earth and λ the latitude. Hence,

$$v = \frac{2\pi R}{T}\cos\lambda.$$

At the equator $v = 2\pi R/T = \omega R = (7.27 \times 10^{-5})(6.37 \times 10^{6}) = 463$ m/s. Since $\cos 45° = \frac{1}{2}\sqrt{2}$, the linear speed v at latitude $\lambda = 45°$ is $v = 327$ m/s.

<u>11-26</u>

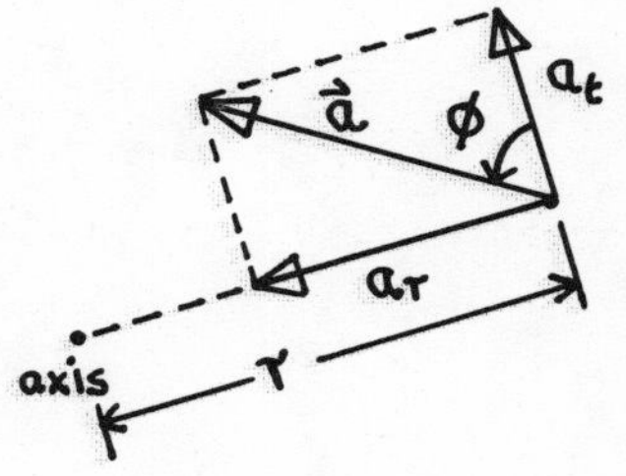

(a) The radial acceleration, assuming the body started from rest at t = 0, is

$$a_r = \omega^2 r = (\alpha t)^2 r = \alpha^2 t^2 r \quad \underline{\text{Ans}}.$$

(b) The tangential acceleration is

$$a_t = \frac{dv}{dt} = \frac{d(\omega r)}{dt} = \alpha r \quad \underline{\text{Ans}}.$$

(c) From the sketch,

$$\tan\phi = a_r/a_t = \frac{\alpha^2 t^2 r}{\alpha r} = \alpha t^2.$$

But the angle turned through is

$$\theta = \tfrac{1}{2}\alpha t^2 = \tfrac{1}{2}\tan\phi = \tfrac{1}{2}\tan 60°,$$

$$\theta = 0.866 \text{ rad} = 49.6° \quad \underline{\text{Ans}}.$$

<u>11-28</u>

(a) The angular speed is $\omega = (33\frac{1}{3})(2\pi)/(60) = 3.49$ rad/s. Hence, the tangential component of the velocity is

$$v_t = r\omega = (12 \text{ cm})(3.49 \text{ rad/s}) = 42 \text{ cm/s} \quad \underline{\text{Ans}};$$

the radial component is just 1.6 cm/s Ans.

(b) The radial and tangential components are

$$a_r = r\omega^2 = (12\text{ cm})(3.49\text{ rad/s})^2 = 146\text{ cm/s}^2 \quad \text{Ans};$$

$$a_t = 2\omega\frac{dr}{dt} = 2(3.49\text{ rad/s})(1.6\text{ cm/s}) = 11\text{ cm/s}^2 \quad \text{Ans}.$$

(c) At r = 16 cm, the radial acceleration is

$$a_r = (16\text{ cm})(3.49\text{ rad/s})^2 = 195\text{ cm/s}^2.$$

The tangential component, independent of r, is unchanged; $a_t = 11$ cm/s^2. Hence, a = 195.3 ≈ 195 cm/s^2; i.e., the tangential component is negligible. By Newton's second law in the inertial system,

$$F = \mu N = \mu mg = ma,$$

$$\mu = \frac{a}{g} = \frac{195}{980} = 0.20 \quad \text{Ans}.$$

12-8

By straightforward calculation,

(a)

$$I = \Sigma m_i r_i^2 = m\ell^2 + m(2\ell)^2 + m(3\ell)^2 = 14m\ell^2 \quad \underline{\text{Ans}}.$$

(b)

$$L_m = I_m\omega = (4m\ell^2)\omega = 4m\ell^2\omega \quad \underline{\text{Ans}}.$$

(c)

$$L = I\omega = 14m\ell^2\omega \quad \underline{\text{Ans}}.$$

12-19

If friction at the end of the stick in contact with the floor can be neglected, then conservation of energy applies. In the vertical position the height h of the center of mass of the stick above the floor is $\frac{1}{2}L$, L the length of the stick (1 m). As it strikes the floor, h = 0. Initially the kinetic energy K = 0; upon striking the floor it is

$$K = \tfrac{1}{2}I\omega^2 = \tfrac{1}{2}(\tfrac{1}{3}\,mL^2)(\tfrac{v}{L})^2 = \tfrac{1}{6}\,mv^2,$$

v the speed at the end of the stick. Hence,

$$mg(\tfrac{1}{2}L) = \tfrac{1}{6}\,mv^2,$$

$$v = (3gL)^{\frac{1}{2}} = 5.42 \text{ m/s} \quad \underline{\text{Ans}}.$$

12-20

(a) The radial acceleration is $a_r = h\omega^2$. By energy conservation,

$$mg(\tfrac{1}{2}h) = \tfrac{1}{2}I\omega^2 + mg(\tfrac{1}{2}h\cos\theta),$$

where $I = mh^2/3$ since the chimney, in effect, is rotating about an axis through its base. Substituting the expression for I yields

$$mgh(1 - \cos\theta) = \frac{1}{3}\, mh^2\omega^2,$$

and therefore

$$a_r = h\omega^2 = 3g(1 - \cos\theta) \quad \underline{\text{Ans}}.$$

(b) The tangential acceleration a_t is

$$a_t = \frac{dv}{dt} = \frac{d}{dt}(h\omega) = h\,\frac{d\omega}{dt}.$$

Using the result from (a),

$$\frac{d\omega}{dt} = \tfrac{1}{2}\omega^{-1}\left(\frac{3g}{h}\sin\theta\right)\frac{d\theta}{dt} = \frac{3g}{2h}\sin\theta$$

since $\omega = \dfrac{d\theta}{dt}$. Thus the tangential acceleration is

$$a_t = \frac{3g}{2}\sin\theta \quad \underline{\text{Ans}}.$$

(c) The total linear acceleration a is

$$a = (a_r^2 + a_t^2)^{\frac{1}{2}} = \frac{3g}{2}[(1 - \cos\theta)(5 - 3\cos\theta)]^{\frac{1}{2}}.$$

This is zero at $\theta = 0°$ and increases as θ increases, passing g near $\theta = 34°$. It is easily seen that $a > g$ near $\theta = 90°$, for then $\cos\theta \approx 0$. Hence, the answer is YES.

(d) As the chimney tips over, the weight of the upper part acts transverse to the column. Chimneys standing upright are not subject to this force and are not specifically designed to withstand it.

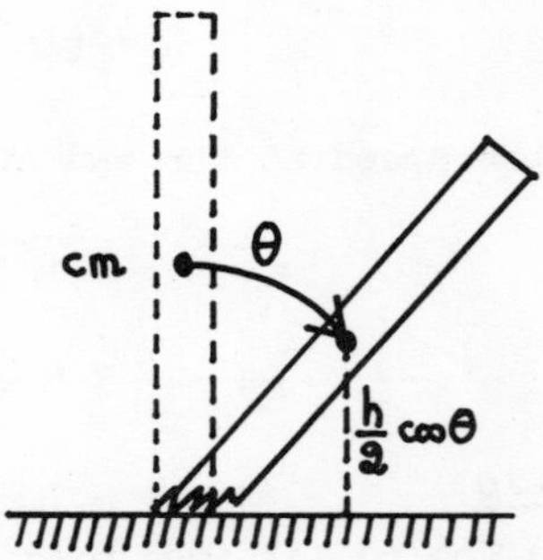

12-28

Use energy methods. Let the original level of the small object m be the zero level of gravitational potential energy U, so that the initial potential energy $U_i = 0$. After the object has fallen a distance h, its potential energy is $U_f = -mgh$. As the system is released from rest its initial kinetic energy $K_i = 0$. After falling a distance h the object has kinetic energy $\frac{1}{2}mv^2$. If the string does not slip, the tangential linear speeds of the rim of the pulley and the equator of the sphere each are v also; the respective angular speeds are $\frac{v}{r}$ and $\frac{v}{R}$. The rotational inertias are I for the pulley and $\frac{2}{3}MR^2$ for the shell. By conservation of energy,

$$K_i + U_i = K_f + U_f,$$

$$0 + 0 = \tfrac{1}{2}mv^2 + \tfrac{1}{2}I(\tfrac{v}{r})^2 + \tfrac{1}{2}(\tfrac{2}{3}MR^2)(\tfrac{v}{R})^2 - mgh,$$

$$v = \left(\frac{2gh}{1 + 2M/3m + I/mr^2}\right)^{\frac{1}{2}} \quad \underline{\text{Ans.}}$$

12-29

(a)

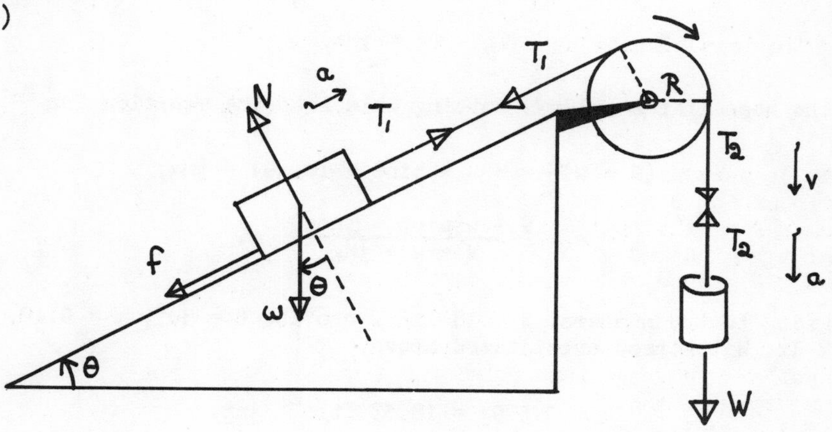

Assume that W accelerates and moves down, so that w slides up the plane, and the pulley rotates clockwise. Neither the weight Mg of the pulley nor the force exerted by the pulley's supports exert a torque on the pulley, so these forces are not shown in the diagram. The equations of motion (Newton's second law) for w,W are, with μ the coefficient of friction between w and the inclined plane, and the positive directions in the direction of the accelerations,

$$T_1 - w\sin\theta - f = \frac{w}{g}\cdot a,$$

$$N - w\cos\theta = 0,$$

$$f = \mu N;$$

$$W - T_2 = \frac{W}{g}\cdot a.$$

The middle two equations give $f = \mu w\cos\theta$; substituting this into the first equation and then adding the first and fourth equations yields

$$(W + w)\frac{a}{g} = W - w(\sin\theta + \mu\cos\theta) - (T_2 - T_1).$$

The rotation equation for the pulley is

$$\tau = I\alpha,$$

$$(T_2 - T_1)R = I\alpha = (\tfrac{1}{2}MR^2)(\tfrac{a}{R}) = \tfrac{1}{2}MRa,$$

$$T_2 - T_1 = \tfrac{1}{2}Ma,$$

M the mass of the pulley. Putting this into the equation for $\frac{a}{g}$,

$$(W + w)\frac{a}{g} = W - w(\sin\theta + \mu\cos\theta) - \tfrac{1}{2}Ma,$$

$$a = \frac{W - w(\sin\theta + \mu\cos\theta)}{W + w + \tfrac{1}{2}Mg}\,g.$$

Turning to the numbers, $W = 18$ lb, $w = 6$ lb, $\theta = 30°$, $\mu = 0.10$, $Mg = 2$ lb. With these substituted above,

$$a = 0.579g = 18.53 \text{ ft/s}^2 \quad \underline{\text{Ans}}.$$

(b) The tension T_2 follows from the fourth equation:

$$T_2 = W(1 - \frac{a}{g}) = (18)(1 - 0.579) = 7.58 \text{ lb} \quad \underline{\text{Ans}}.$$

The other tension T_1 is

$$T_1 = T_2 - \tfrac{1}{2}Ma = T_2 - \tfrac{1}{2}(Mg)\frac{a}{g},$$

$$T_1 = 7.58 - \tfrac{1}{2}(2)(0.579) = 7.00 \text{ lb} \quad \underline{\text{Ans}}.$$

12-33

The walls constrain the initial velocities to be parallel to the walls. If the ladder remains in contact with both surfaces as it slips, then

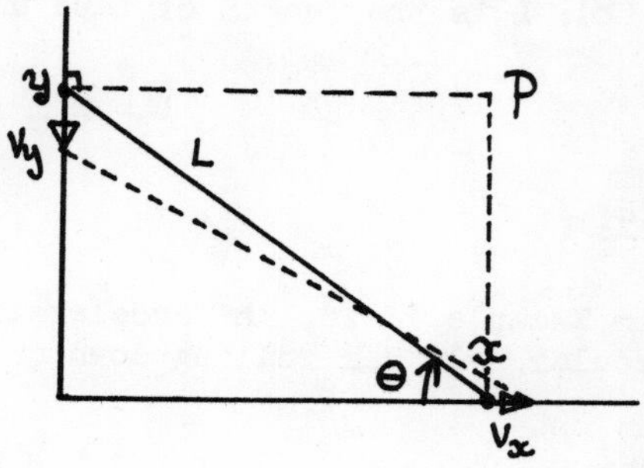

$$x^2 + y^2 = L^2,$$

$$2xv_x + 2yv_y = 0,$$

$$v_x = -\frac{y}{x}v_y.$$

If P is the instantaneous axis, P constructed as shown,

$$v_x = y\omega; \; v_y = -x\omega.$$

But, upon eliminating ω, these equations give

$$v_x = -\frac{y}{x}v_y,$$

the same equation required by the geometric constraints. Hence, the axis is indeed at P, a distance $x = L\cos\theta = (10 \text{ ft})\cos 60^\circ = 5$ ft from the wall, and a distance $y = L\sin\theta = (10 \text{ ft})\sin 60^\circ = 8.66$ ft from the floor.

12-38

(a) With the linear acceleration zero,

$$T - Mg = 0,$$

$$T = Mg \quad \underline{\text{Ans}}.$$

(b) By the work-energy theorem,

$$W = \Delta K = \tfrac{1}{2}I\omega^2 - 0 = \tfrac{1}{2}(\tfrac{1}{2}MR^2)\omega^2 = \tfrac{1}{4}MR^2\omega^2 \quad \underline{\text{Ans}}.$$

(c) If θ is the angle turned through,

$$W = \tau\theta = (TR)(\tfrac{L}{R}) = TL = MgL = \tfrac{1}{4}MR^2\omega^2,$$

by (b); L is the length of tape unwound. Thus,

$$L = R^2\omega^2/4g \quad \underline{\text{Ans}}.$$

12-41

From Example 12-10, the acceleration of the center of mass of a circular cylinder rolling down an inclined plane is

$$a = \frac{2}{3}\, g\sin\theta,$$

independent of the radius of the cylinder. Thus, provided the wound tape remains circular, the acceleration is constant, and the time for the tape to unwind, and hence for the center of mass to move a distance ℓ the length of the tape, is given from

$$\ell = \tfrac{1}{2}aT^2 = \tfrac{1}{2}\left(\frac{2}{3}\, g\sin\theta\right)T^2,$$

$$T = \left[\frac{3\ell}{g\sin\theta}\right]^{\frac{1}{2}} \quad \underline{\text{Ans}}.$$

12-42

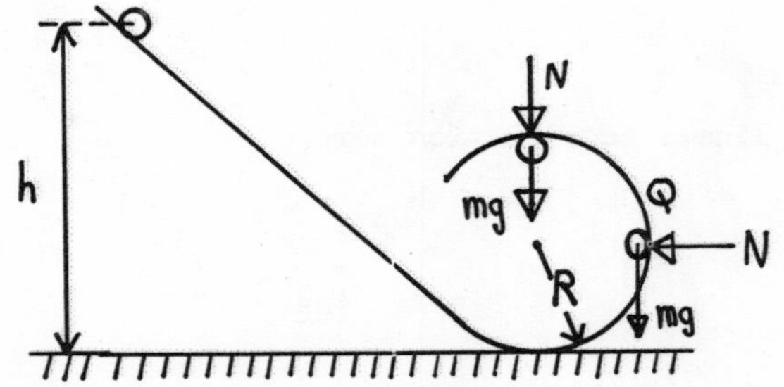

(a) The kinetic energy of the rolling marble is $\frac{1}{2}I(v/r)^2$, where I is the rotational inertia of the marble about the point of contact with the track and v/r is the angular speed of rotation. For I, use the parallel-axis theorem to get

$$I = I_{cm} + mr^2 = \frac{2}{5}mr^2 + mr^2 = \frac{7}{5}mr^2.$$

The kinetic energy of rolling, then, is

$$K = \tfrac{1}{2}(\tfrac{7}{5}mr^2)(\tfrac{v}{r})^2 = \frac{7}{10}\,mr^2.$$

Applying energy conservation,

$$mgh = mg(2R - r) + \frac{7}{10}mv^2.$$

Newton's second law implies that

$$N + mg = m\,\frac{v^2}{R - r}.$$

Eliminating v^2 between these equations yields

$$N = \frac{mg}{7}\left(\frac{10h - 27R + 17r}{R - r}\right).$$

As the marble loses contact, N vanishes, so that

$$10h - 27R + 17r = 0.$$

For $r \ll R$, this becomes

$$10h - 27R = 0,$$

$$h = 2.7R \quad \underline{\text{Ans}}.$$

(b) The horizontal force at Q is

$$N = \frac{mv^2}{R - r}.$$

By energy conservation,

$$mg(6R) = mgR + \frac{7}{10}mv^2.$$

Again, eliminate v^2 to obtain, for $r \ll R$,

$$N = \frac{50}{7}\,mg \quad \underline{\text{Ans}}.$$

12-43

The equations of motion are

$$mg - T = ma,$$

$$\tau = I\alpha = Tr = I\cdot\frac{a}{r}.$$

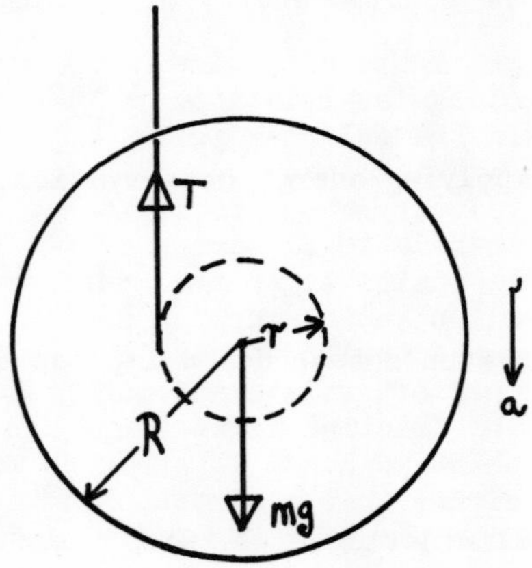

Note that the acceleration is directed down during both the descent and ascent; at the bottom point the velocity is reversed, giving the yo-yo an initial upward velocity. Put $I = \frac{1}{2}mR^2$, ignoring the contribution of the shaft, which should be small if $r \ll R$.

(a) Solving the equations for the tension T, for both up and down motion, gives

$$T = mg\left(\frac{1}{1 + 2r^2/R^2}\right) \quad \underline{\text{Ans}}.$$

(b) Let t be the desired time. If the length of the string is L + R the center of the yo-yo will descend a distance L. Therefore, since

$$t = 2\left(\frac{2L}{a}\right)^{\frac{1}{2}},$$

and, from the equations of motion,

$$a = g\left(\frac{1}{1 + R^2/2r^2}\right),$$

the time is

$$t = 2\left[\frac{2L}{g}(1 + R^2/2r^2)\right]^{\frac{1}{2}} \quad \underline{\text{Ans}}.$$

Generally, for real yo-yos, the "1" in the last answer can be neglected, so that

$$t \approx \frac{2R}{r}\left(\frac{L}{g}\right)^{\frac{1}{2}}.$$

12-44

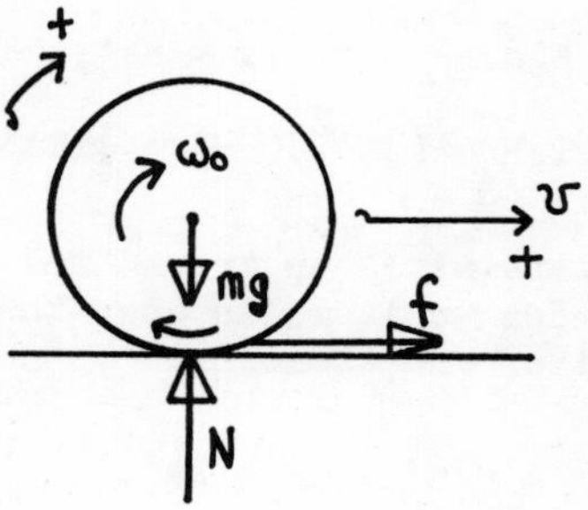

Let ω_0 be clockwise and let clockwise rotations be positive. If the cylinder rolls, a clockwise rotation leads to a linear velocity v to the right, so let translational motion to the right be positive. As the cylinder slips, points on its rim in contact with the table move to the left relative to the table; the friction force f opposes this and hence f points to the right, in the direction of motion; note that $f = \mu N$, but it is not necessary to use this relation. For translational motion,

$$f = ma = m\frac{dv}{dt},$$

$$v = \frac{ft}{m} + v_0 = \frac{ft}{m},$$

since $v_0 = 0$, the cylinder being dropped vertically. For rotation, noting that f will give rise to counterclockwise rotation,

$$\tau = -fR = I\alpha = (\tfrac{1}{2}mR^2)(\frac{d\omega}{dt}),$$

$$\omega = -\frac{2ft}{mR} + \omega_0.$$

When at time $t = t^*$ rolling sets in, $v^* = R\omega^*$;

$$\frac{ft^*}{m} = -\frac{2ft^*}{m} + R\omega_0,$$

$$t^* = Rm\omega_0/3f.$$

Hence, at this instant,

$$v^* = \frac{ft^*}{m} = \frac{1}{3}R\omega_0 \quad \underline{\text{Ans}}.$$

12-45

Apparently the pulley is to be considered massless, so that the tension T in the tape is the same throughout. With f the force of rolling friction ($f \neq \mu N$), the equations of motion, translation and rotation, are

$$T - mg = ma,$$

$$Mg\sin\theta - T - f = Ma',$$

$$(f - T)R = (\tfrac{1}{2}MR^2)(\frac{a'}{R}) = \tfrac{1}{2}MRa'.$$

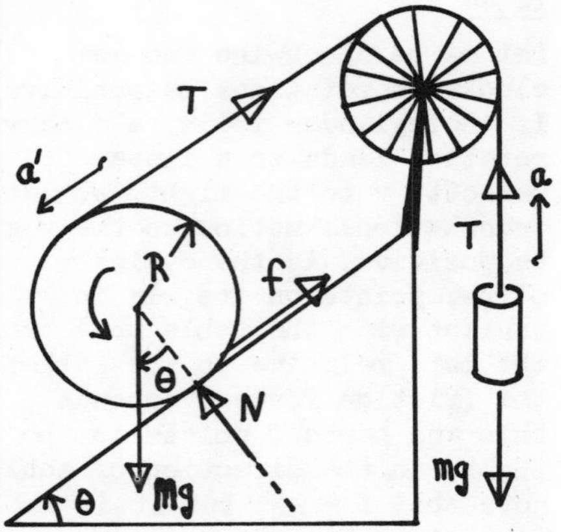

The tape winds up around the cylinder as it rolls down the incline, and therefore

$$a = a' + \alpha R = 2a'.$$

Using this equation to eliminate a, and eliminating f between the second and third equations gives

(a)
$$a' = 2g\,\frac{M\sin\theta - 2m}{3M + 8m} = g/21 = 0.467 \text{ m/s}^2 \quad \underline{\text{Ans}};$$

(b)
$$T = m(g + a) = m(g + 2a') = \frac{23}{21}\,mg = 48.3 \text{ N} \quad \underline{\text{Ans}},$$

since M = 23 kg, m = 4.5 kg, $\theta = 30°$. The results are independent of the radius R of the cylinder.

<u>12-46</u>

Since the impulse is central, the initial motion of the ball is pure sliding. If v_0 is to the right (taken as the positive direction), f points to the left. Let clockwise rotations be positive. Then,

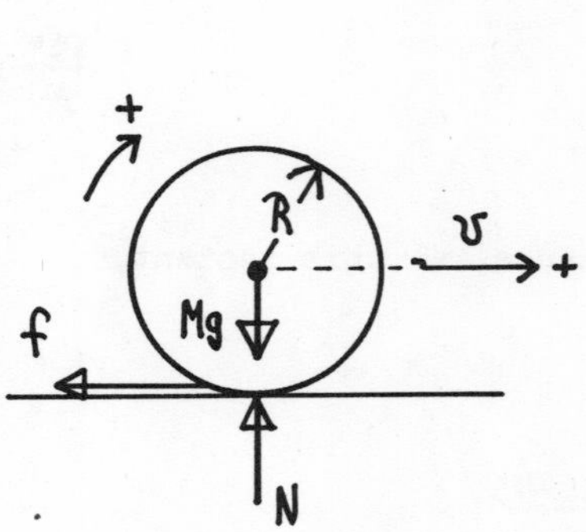

$$-f = +Ma = M\,\frac{dv}{dt},$$

$$v = -\frac{ft}{M} + v_0,$$

$$x = -\tfrac{1}{2}(\frac{f}{M})t^2 + v_0 t.$$

The rotation equation is

$$\tau = fR = I\alpha = (\frac{2}{5}MR^2)\frac{d\omega}{dt},$$

$$\omega = \frac{5ft}{2MR},$$

the initial angular speed being zero. When, at $t = t^*$, rolling sets in, then $v^* = R\omega^*$, so that

$$-\frac{ft^*}{M} + v_0 = \frac{5ft^*}{2M},$$

$$t^* = 2Mv_0/7f.$$

Thus, the distance x^* traveled is

$$x^* = -\tfrac{1}{2}(\frac{f}{M})t^{*2} + v_0t^* = \frac{12}{49}Mv_0^2/f.$$

Finally, $f = \mu N = \mu Mg$, so that

$$x^* = \frac{12}{49}v_0^2/\mu g \quad \underline{\text{Ans}}.$$

CHAPTER 13

13-4

The precession frequency is

$$\omega_p = MgR/L,$$

where M is the total mass of the top or gyroscope, R the distance from support to center of mass and L is the spin angular momentum. Neglecting the mass of the axle, $M = m + m^*$, m the mass of the disc. Adding m^* to the end of the axle shifts the center of mass farther out, a distance R from the support given by

$$R = \frac{m(\frac{1}{2}\ell) + m^*(\ell)}{m + m^*} = \frac{m + 2m^*}{m + m^*}(\tfrac{1}{2}\ell),$$

where ℓ is the length of the axle. Since m^* is located on the axle, it does not contribute to the rotational inertia of the gyroscope about the axle. Hence,

$$L = I\omega = \tfrac{1}{2}ma^2\omega,$$

a the radius of the disc. Substituting these expressions into the equation for the precession frequency gives

$$\omega_p = (m + m^*)g\left[\frac{m + 2m^*}{m + m^*}(\tfrac{1}{2}\ell)\right]/(\tfrac{1}{2}ma^2\omega) = (1 + 2\,\frac{m^*}{m})(g\ell/a^2\omega).$$

The second factor is the precession frequency without m^*; from Problem 13-3 this is 43 rev/min. Thus, for this situation,

$$\omega_p = 43(1 + 2r) \text{ rev/min} \quad \underline{\text{Ans}},$$

where $r = m^*/m$.

13-17

Let $\vec{F}(t)$ be the force exerted by the disc on the cylinder as the disc slips by; by Newton's third law, the cylinder exerts a force $-\vec{F}(t)$ on the disc. Integrate over the time interval during which the disc slips and apply the equation of angular impulse to the cylinder and linear impulse to the disc to obtain

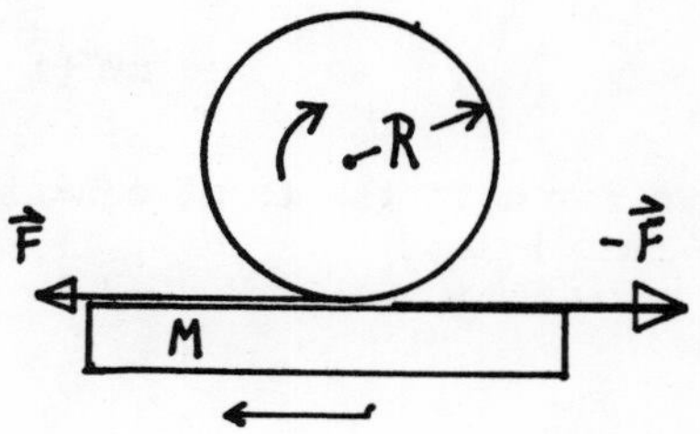

$$R\int F(t)dt = I\omega,$$

$$-\int F(t)dt = M(v_2 - v_1),$$

ω being the final angular velocity of the cylinder. When slipping ceases, then $v_2 = R\omega$. Invoking this, and eliminating $\int Fdt$ from the two equations, and solving for v_2 gives

$$M(v_1 - v_2) = \frac{I}{R}\,\omega = \frac{I}{R}(v_2/R),$$

$$v_2 = \frac{M\,v_1}{M + I/R^2} = \frac{v_1}{1 + I/MR^2} \quad \underline{\text{Ans}}.$$

13-18

(a) Conserved quantities are:

angular momentum (no torques exerted on puck+stick system);
linear momentum (conserved in collisions);
kinetic energy (elastic collision).

(b) Applying the conservation principles above, in order:

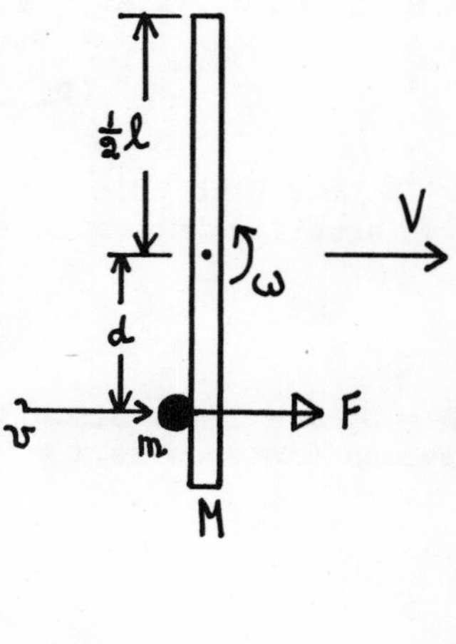

$$mvd = I\omega;$$

$$mv = MV;$$

$$\tfrac{1}{2}mv^2 = \tfrac{1}{2}MV^2 + \tfrac{1}{2}I\omega^2.$$

Use the middle equation to eliminate V from the third, to obtain

$$mv^2(1 - \frac{m}{M}) = I\omega^2.$$

Solve for ω in the first equation and substitute into the above; the result is,

$$1 - \frac{m}{M} = \frac{M - m}{M} = \frac{md^2}{I}.$$

But $I = \frac{1}{12}M\ell^2$ so that

$$M - m = 12md^2/\ell^2$$

$$m = M(1 + 12d^2/\ell^2)^{-1} \quad \underline{\text{Ans}}.$$

<u>13-19</u>

Let R be the average reaction force, and v^* and ω^* the velocity of the center of mass and the angular velocity about the suspension S, immediately after impact. The impulse of the striking blow is $F\Delta t$, so that

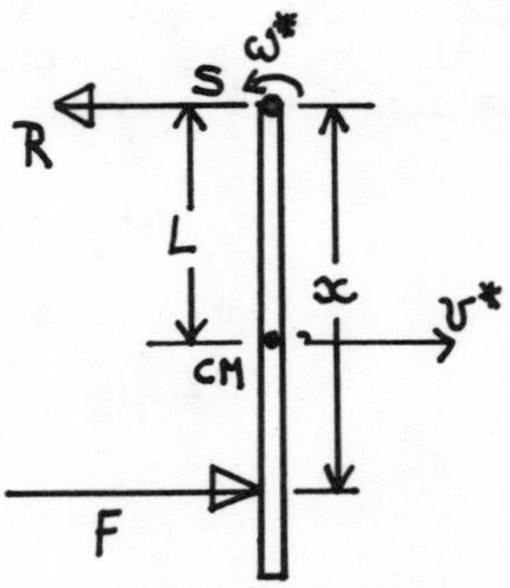

$$(F - R)\Delta t = \Delta(mv) = mv^*.$$

The angular impulse, calculated about S is $\tau\Delta t = (Fx)\Delta t$, so that

$$(Fx)\Delta t = \Delta(I\omega) = I\omega^*.$$

But $v^* = L\omega^*$; with this, v^* and ω^* can be eliminated by dividing the two equations above to yield

$$R = F(1 - \frac{mLx}{I}).$$

For $R = 0$, $x = I/mL$. Since $I = 4mL^2/3$ about the point of suspension the needed distance is $x = 4L/3$ <u>Ans</u>.

13-20

(a) The friction forces f on each wheel must be equal in magnitude and opposite in direction, by Newton's third law. Let ω_1, ω_2 be the angular speeds when slipping ceases. The equations of angular impulse, applied to each wheel, are

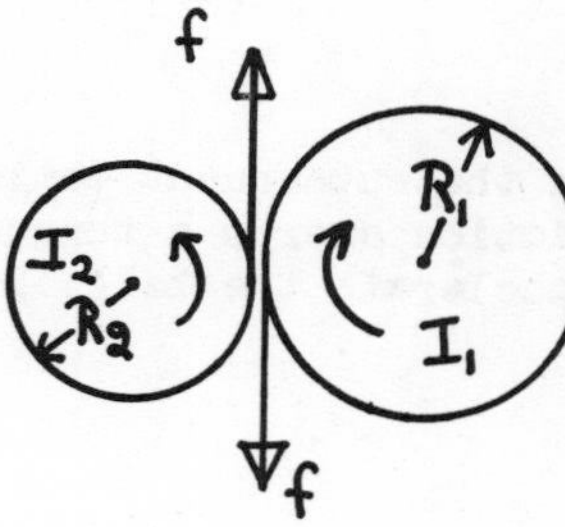

$$R_2\int f dt = I_2(\omega_2 - 0),$$
$$-R_1\int f dt = I_1(\omega_1 - \omega_0).$$

Dividing these equations, to eliminate the impulse, gives

$$-\frac{R_2}{R_1} = \frac{I_2}{I_1}\frac{\omega_2}{\omega_1 - \omega_0}.$$

When slipping ceases, the linear speeds at the rims must be equal;

$$R_1\omega_1 = R_2\omega_2.$$

Using this to eliminate ω_1 yields

$$\omega_2 = \omega_0\left[\frac{R_1 I_2}{R_2 I_1} + \frac{R_2}{R_1}\right]^{-1} \quad \underline{\text{Ans}}.$$

(b) Angular momentum is not conserved since external torques must be applied to keep the wheels from "climbing over" each other.

13-21

The force of friction is f. Positive directions for rotation and translation are shown on the diagram. The impulse and angular impulse equations, assuming that $F \gg f$, are

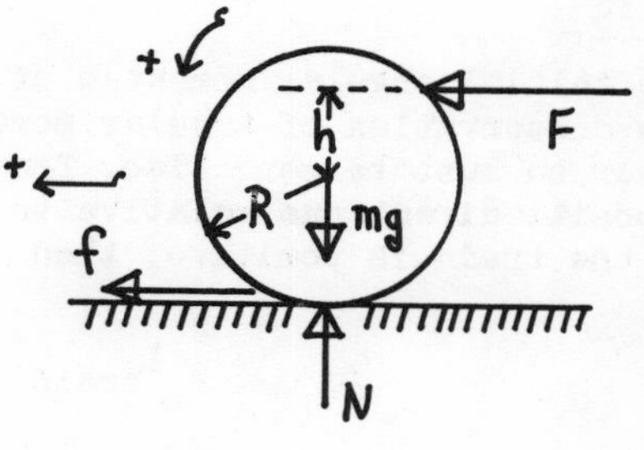

$$\int F dt = mv_0,$$
$$h\int F dt = I\omega_0,$$

where ω_0 is the initial angular velocity. Solving for this,

$$\omega_0 = hmv_0/I.$$

During the subsequent motion, until slipping ceases, the force f of friction exerts a torque tending to decrease the angular speed and accelerate the ball. Hence,

$$-fR = I\alpha = \frac{2}{5}mR^2 \frac{d\omega}{dt},$$

$$ft = -\frac{2}{5}mR(\omega - \omega_0),$$

$$ft = -\frac{2}{5}mR(\omega - hmv_0/I),$$

using the expression for ω_0 derived above. In addition,

$$f = m\frac{dv}{dt},$$

$$v = \frac{ft}{m} + v_0.$$

Thus,

$$m(v - v_0) = -\frac{2}{5}mR(\omega - hmv_0/I).$$

When rolling sets in, $v = R\omega = 9v_0/7$, the last by supposition. Replacing v with this results in

$$\frac{2}{7}mv_0 = -\frac{2}{5}mR\left[\frac{9v_0}{7R} - \frac{hmv_0}{\frac{2}{5}mR^2}\right],$$

$$h = \frac{4}{5}R \quad \underline{\text{Ans}}.$$

<u>13-28</u>

The initial angular momentum of the system wheel+train is zero; by the conservation of angular momentum, the angular momentum with the power on must be zero also. Thus, the wheel and train rotate in opposite directions relative to the earth. If the sense of rotation of the train is positive, then for the magnitudes of the momenta,

$$L_{train} - L_{wheel} = 0.$$

Since the rotational inertia of the wheel is $I = MR^2$,

$$L_{wheel} = MR^2\omega.$$

The speed V of the train relative to the earth is

$$V = v - R\omega$$

since $R\omega$ is the speed of the track. Thus,

$$MR^2\omega = mVR = m(v - R\omega)R,$$

$$\omega = \frac{m}{M + m}\left(\frac{v}{R}\right) \quad \underline{\text{Ans}}.$$

<u>13-30</u>

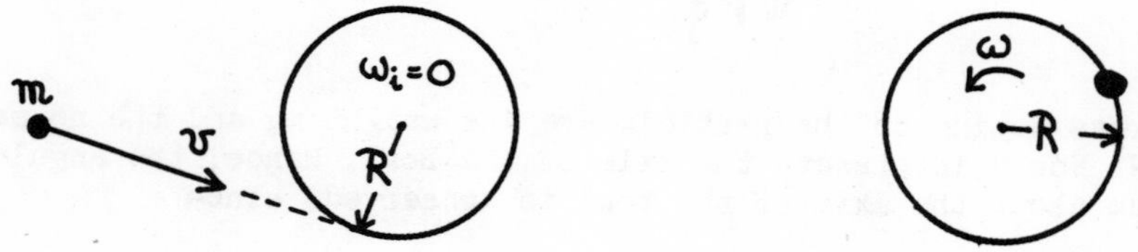

The angular momentum of the system child+merry-go-round about the axis of the merry-go-round is conserved. If m and v are the mass and speed of the child before jumping on, $I = Mk^2$ the rotational inertia of the merry-go-round (k is the radius of gyration), then since the merry-go-round is motionless until the child arrives, angular momentum conservation gives

$$mvR = (I + mR^2)\omega = (Mk^2 + mR^2)\omega,$$

treating the child as a point-mass. Therefore, using "English" data

$$\omega = \frac{mvR}{Mk^2 + mR^2} = \frac{(3)(10)(4)}{12(3)^2 + 3(4)^2} = 0.77 \text{ rad/s} \quad \underline{\text{Ans}}.$$

13-34

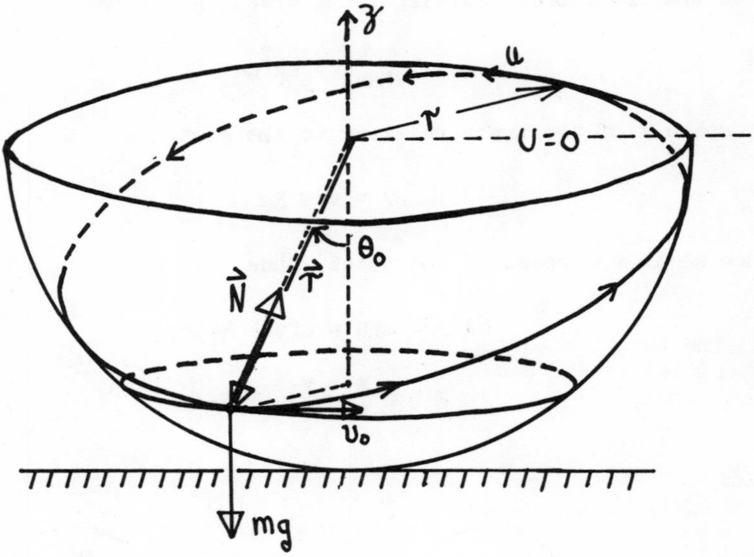

The forces acting on the particle are its weight mg and the normal force N. Now N intersects the axis of the bowl. Hence, the angular momentum about the axis of the bowl is conserved, since

$$\vec{\tau} = \vec{r} \times \vec{N} = 0.$$

(The torque due to mg has no component parallel to the axis of the bowl since mg is parallel to the axis.) If the particle just reaches the top of the bowl, its speed u there is directed at a tangent to the bowl's top. Hence, conservation of angular momentum gives

$$mv_0(r\sin\theta_0) = mur,$$

$$v_0\sin\theta_0 = u.$$

Conservation of energy requires that

$$\tfrac{1}{2}mv_0^2 - mgr\cos\theta_0 = \tfrac{1}{2}mu^2.$$

Substituting for u from the preceding equation yields

$$\tfrac{1}{2}mv_0^2 - mgr\cos\theta_0 = \tfrac{1}{2}mv_0^2\sin^2\theta_0,$$

$$\tfrac{1}{2}v_0^2(1 - \sin^2\theta_0) = gr\cos\theta_0,$$

$$\tfrac{1}{2}v_0^2\cos^2\theta_0 = gr\cos\theta_0,$$

$$v_0 = (2gr\sec\theta_0)^{\frac{1}{2}} \quad \underline{\text{Ans}}.$$

13-35

(a) The total angular momentum is initially zero and must remain zero; thus,

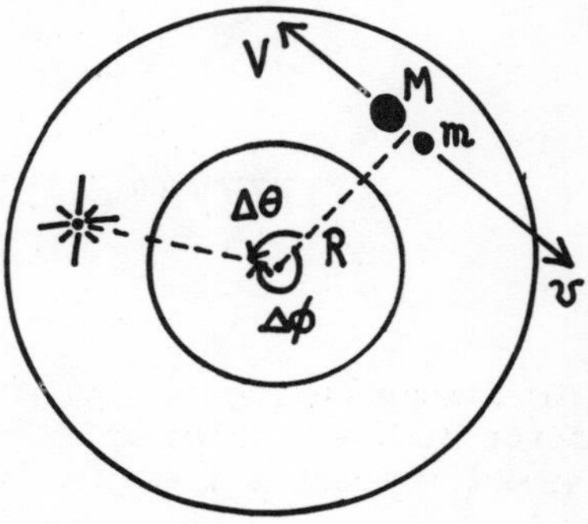

$$MVR = mvR.$$

If $\omega_\theta = d\theta/dt$ and $\omega_\phi = d\phi/dt$, then,

$$M(R\omega_\theta)R = m(R\omega_\phi)R,$$

$$M\omega_\theta = m\omega_\phi,$$

$$M\Delta\theta = m\Delta\phi.$$

Now at collision $\Delta\theta + \Delta\phi = 2\pi$, so that

$$M\Delta\theta = m(2\pi - \Delta\theta),$$

$$\Delta\theta = \frac{2\pi m}{m + M} \quad \underline{\text{Ans}}.$$

(b) By conservation of energy,

$$\tfrac{1}{2}MV^2 + \tfrac{1}{2}mv^2 = U_0,$$

$$\tfrac{1}{2}MR^2\omega_\theta^2 + \tfrac{1}{2}mR^2\omega_\phi^2 = U_0,$$

while from (a)

$$\omega_\phi = \frac{M}{m}\,\omega_\theta,$$

so that, since $\theta(0) = 0$,

$$\omega_\theta = [2U_0 m/MR^2(m + M)]^{\frac{1}{2}},$$

$$\theta = [2U_0 m/MR^2(m + M)]^{\frac{1}{2}} t.$$

Let the time of the collision be $t = t_c$; $\theta(t_c) = \Delta\theta = 2\pi m/(m + M)$, from (a). Using this gives

$$\frac{2\pi m}{m + M} = [2U_0 m/MR^2(m + M)]^{\frac{1}{2}} t_c,$$

$$t_c = \left[\frac{2\pi^2 mMR^2}{(m + M)U_0}\right]^{\frac{1}{2}} \quad \underline{\text{Ans}}.$$

14-5

Isolate half the chain. Since the half is in equilibrium,

$$T\cos\theta = T_0,$$

$$T\sin\theta = \tfrac{1}{2}W.$$

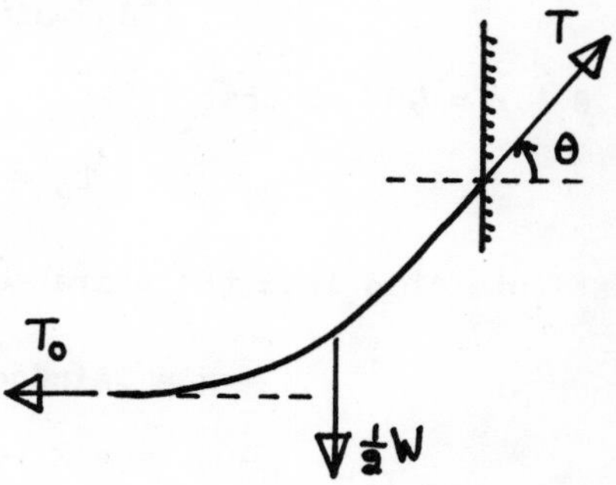

Divide the equations to obtain,

$$\frac{\cos\theta}{\sin\theta} = \cot\theta = T_0/\tfrac{1}{2}W,$$

$$T_0 = \tfrac{1}{2}W\cot\theta \quad \underline{\text{Ans}}.$$

Now substitute T_0 into the very first equation to obtain

$$T = T_0/\cos\theta = \tfrac{1}{2}W\csc\theta \quad \underline{\text{Ans}};$$

this is tangent to the chain. By Newton's third law, the chain exerts an equal and opposite force on the support.

14-6

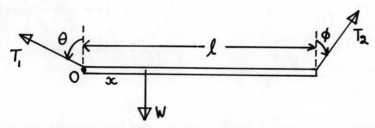

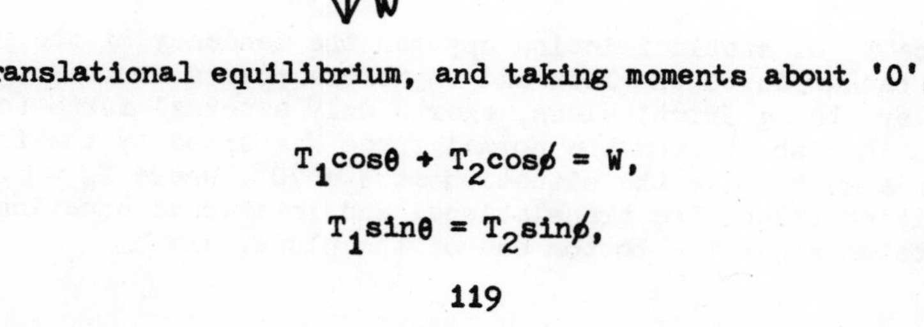

For translational equilibrium, and taking moments about 'O',

$$T_1\cos\theta + T_2\cos\phi = W,$$

$$T_1\sin\theta = T_2\sin\phi,$$

$$Wx = (T_2 \cos\phi)\ell.$$

Multiply the first equation by $\sin\theta$, the second by $\cos\theta$ and subtract, to obtain for T_2,

$$T_2 = \frac{W \sin\theta}{\sin(\theta + \phi)}.$$

But $\theta + \phi = 90°$ so that

$$T_2 = W\sin\theta.$$

Substitute this into the third equation to find

$$x = \ell\sin\theta\cos\phi = \ell\sin^2\theta,$$

$$x = 2.20 \text{ m} \quad \underline{\text{Ans}}.$$

<u>14-11</u>

The normal force vanishes as the wheel lifts. Taking moments about '0' as this happens,

$$Wx = F(r - h).$$

But,

$$x^2 = r^2 - (r - h)^2,$$

$$x = (2rh - h^2)^{\frac{1}{2}},$$

so that

$$F = W\,\frac{(2rh - h^2)^{\frac{1}{2}}}{r - h} \quad \underline{\text{Ans}}.$$

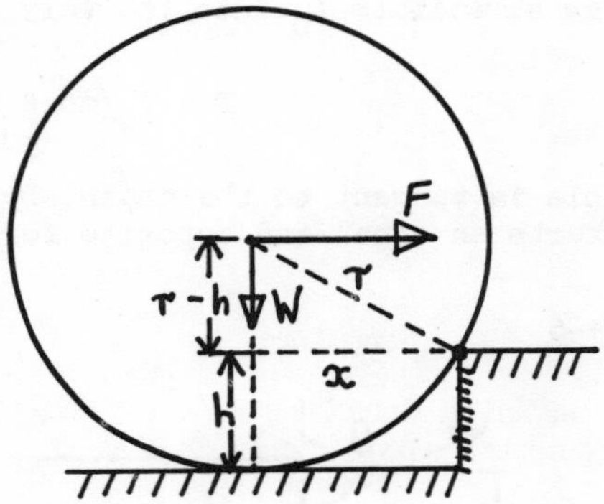

<u>14-20</u>

The force f_s of static friction opposes the tendency of the plank to slip backwards to the left and therefore points to the right. The roller, being frictionless, exerts only a normal force (called R to distinguish it from the normal force N exerted by the floor) on the plank. Examine the situation at $\theta = 70°$, where $f_s = \mu_s N$, its limiting value. The translational and rotational equations, the latter taken about the bottom end of the plank, are

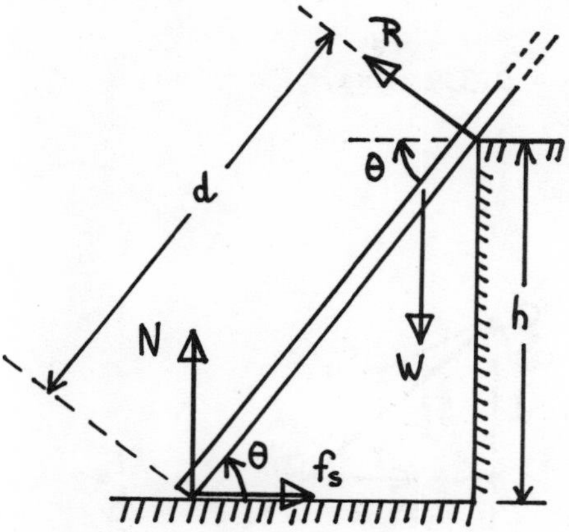

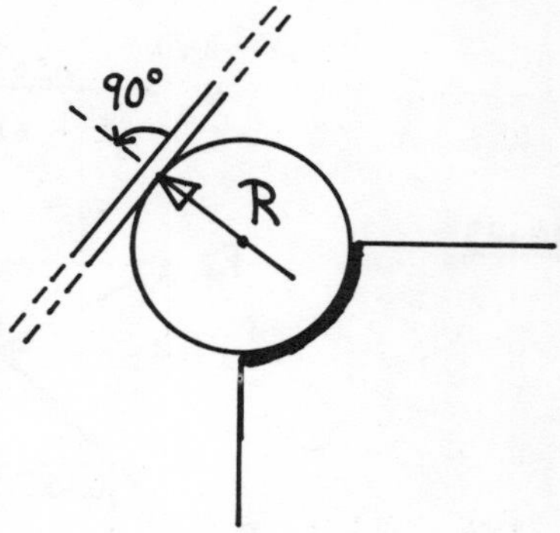

$$R\cos\theta + N - W = 0,$$

$$R\sin\theta - \mu_s N = 0,$$

$$Rd - W\frac{\ell}{2}\cos\theta = 0.$$

Multiply the first equation by μ_s and add to the second to obtain

$$R(\sin\theta + \mu_s\cos\theta) = \mu_s W.$$

The third equation gives $R = (W\ell/2d)\cos\theta$; substitute this into the equation above to find that

$$\sin\theta\cos\theta + \mu_s\cos^2\theta = \frac{2d}{\ell}\mu_s.$$

But $d = h/\sin\theta$; therefore

$$\sin^2\theta\cos\theta + \mu_s\sin\theta\cos^2\theta = \frac{2h}{\ell}\mu_s,$$

$$\mu_s = \frac{\sin^2\theta\cos\theta}{2h - \ell\sin\theta\cos^2\theta}\,\ell.$$

Numerically, $2h = \ell$ and $\theta = 70^\circ$ giving

$$\mu_s = \frac{\sin^2\theta\cos\theta}{1 - \sin\theta\cos^2\theta} = 0.339 \quad \underline{\text{Ans}}.$$

14-23

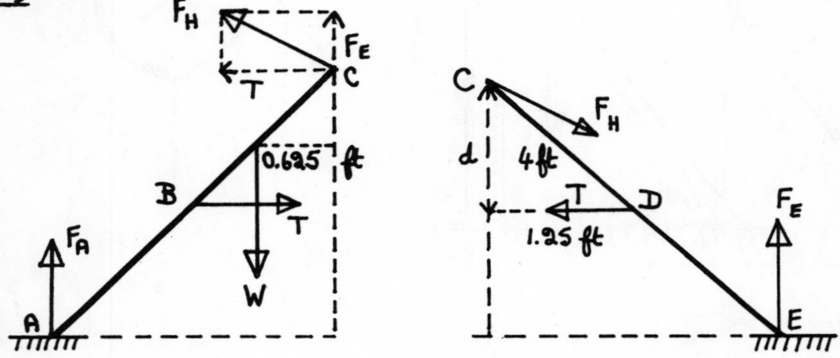

The distance d is

$$d = (4^2 - 1.25^2)^{\frac{1}{2}} = 3.80 \text{ ft}.$$

Note that the vertical component of the force F_H exerted by the hinge on the left leg equals the force F_E exerted by the ground on the right leg, for the equations for translational equilibrium for the left leg and for the ladder as a whole are identical:

$$F_A + F_E - W = 0.$$

Taking moments about C gives, first for the right leg,

$$2.5F_E - 3.8T = 0,$$

and for the left leg,

$$0.625W + 3.8T - 2.5F_A = 0.$$

Set W = 192 lb; solution of the three equations gives

(a) $$T = 47 \text{ lb} \quad \underline{\text{Ans}};$$

(b) $$F_A = 120 \text{ lb}, \; F_E = 72 \text{ lb} \quad \underline{\text{Ans}}.$$

14-24

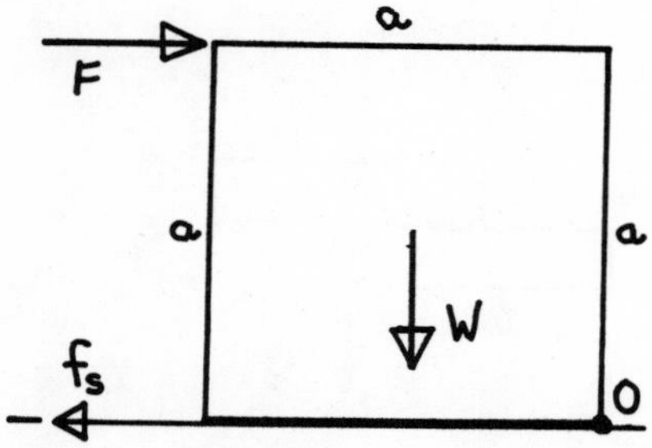

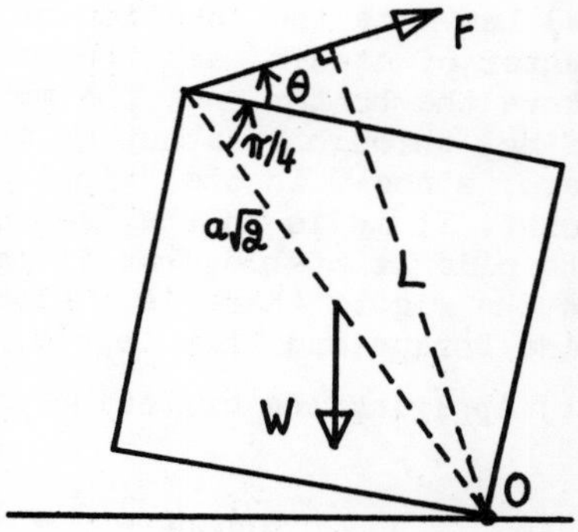

(a) The normal force vanishes as the box begins to roll. Taking moments about O,

$$W\frac{a}{2} = Fa,$$

$$F = \frac{W}{2} = 445 \text{ N} \quad \underline{\text{Ans}}.$$

(b) For the minimum force, assume the box was on the verge of slipping, so that $f_s = \mu_s N$. Translational equilibrium requires that $f_s = F$; thus

$$F = \mu_s N = \mu_s W,$$

$$\mu_s = \frac{F}{W} = 0.5 \quad \underline{\text{Ans}}.$$

(c) If F is applied at an angle θ above the horizontal at the top of the box (for the greatest moment arm) then, again taking moments about O,

$$F(\sqrt{2}a)\sin(\theta + \tfrac{1}{4}\pi) = W(\tfrac{1}{2}a),$$

$$F\sin(\theta + \tfrac{1}{4}\pi) = \tfrac{1}{4}W\sqrt{2}.$$

For F to be a minimum, set $\sin(\theta + \frac{1}{4}\pi) = 1$, its maximum possible value. Hence, $\theta = \frac{1}{4}\pi$ and

$$F = \tfrac{1}{4}(890)\sqrt{2} = 315\ \text{N} \quad \underline{\text{Ans}}.$$

<u>14-26</u>

(a) Let X be the location of the center of mass of all the bricks above the bottom one. The moment of mg, through X, about O, is zero, since O is the "tipping" point. If mg is to the left of O the pile is stable, but if mg is to the right, there is a clockwise torque and they topple.

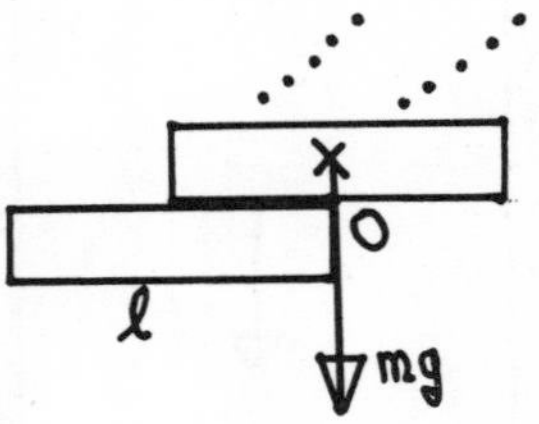

(b) Applying the criterion of (a)

$$\text{overhang} = \frac{\ell}{2} + \frac{\ell}{4} + \frac{\ell}{6} + \dots = \frac{\ell}{2}\cdot\sum_{1}^{\infty}\left(\frac{1}{n}\right) = \infty.$$

(c) Suppose there are N bricks. The center of gravity of those above the first, or lowest, brick is at a distance x from the end of the lowest brick away from the overhang, where x satisfies the relation defining the location of the center of mass:

$$(N - 1)mx = m\left[\frac{\ell}{2} + \frac{\ell}{n}\right] + m\left[\frac{\ell}{2} + \frac{2\ell}{n}\right] + \dots + m\left[\frac{\ell}{2} + \frac{(N-1)\ell}{n}\right].$$

The maximum allowed $x = \ell$; this gives

$$(N - 1) = \frac{N-1}{2} + \frac{1}{n}[1 + 2 + 3 + \dots + (N - 1)],$$

$$N - 1 = \frac{N-1}{2} + \frac{1}{n}\,\frac{(N-1)N}{2},$$

$$n = N \quad \underline{\text{Ans}}.$$

<u>14-28</u>

Check for stability. In the original position the height of the cube's center above O, the center of the cylinder, is $r + \frac{1}{2}a$. If the cube is tipped by rolling (so that P'R = arcSR) a small angle θ, then P and P' nearly coincide, so that, in order that the cube's center rise on rolling,

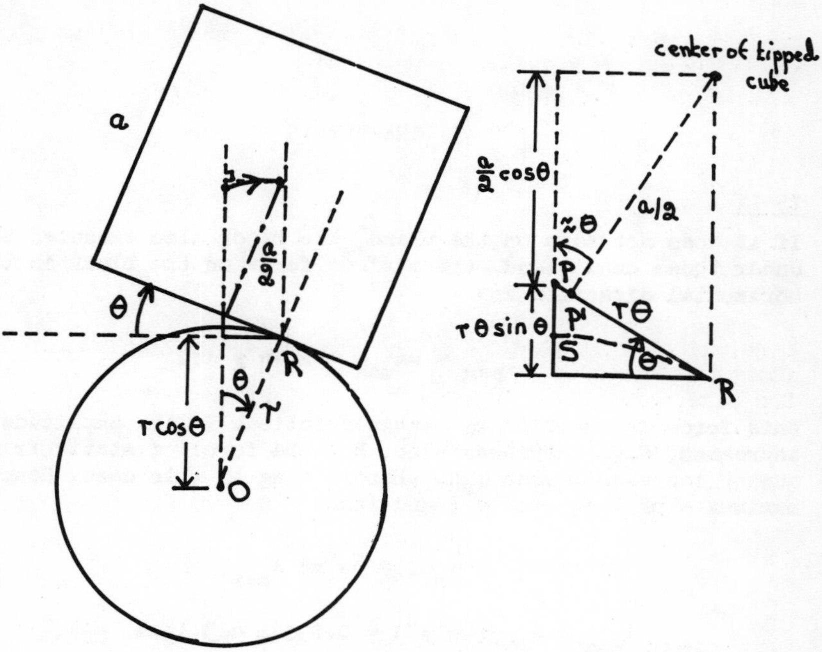

$$r\cos\theta + r\theta\sin\theta + \tfrac{1}{2}a\cos\theta > r + \tfrac{1}{2}a,$$

$$r(1 - \tfrac{1}{2}\theta^2) + r\theta(\theta) + \tfrac{1}{2}a(1 - \tfrac{1}{2}\theta^2) > r + \tfrac{1}{2}a,$$

$$(r + \tfrac{1}{2}a) + \theta^2(\tfrac{1}{2}r - \tfrac{1}{4}a) > r + \tfrac{1}{2}a,$$

$$\tfrac{1}{2}r - \tfrac{1}{4}a > 0,$$

$$r > a/2.$$

CHAPTER 15

15-11

If it does not slip on the plane, the block also executes SHM. Under these conditions, the maximum force on the block in the horizontal direction is

$$F_{max} = ma_{max} = m(4\pi^2\nu^2 A).$$

This force is supplied by static friction. If the amplitude A increases, F_{max} increases also. But the force of static friction cannot increase beyond $\mu_s N$, where N = mg in this case. Hence, the maximum amplitude can be found from

$$\mu_s mg = 4\pi^2 m\nu^2 A_{max},$$

$$A_{max} = \mu_s g/(4\pi^2\nu^2) = 0.031 \text{ m} = 3.1 \text{ cm} \quad \underline{\text{Ans}},$$

using $\mu_s = 0.5$ and $\nu = 2$ Hz.

15-16

(a) The amplitude of each is $\frac{1}{2}A$ and their periods are T = 1.5 s. If their displacements be x_1 and x_2, then,

$$x_1 = \tfrac{1}{2}A\cos(\frac{2\pi}{T}t),$$

$$x_2 = \tfrac{1}{2}A\cos(\frac{2\pi}{T}t - \frac{\pi}{6}),$$

so that particle 1 is at one end of the line at time t = 0 and particle 2 is at this same point at time

$$t = \frac{\pi/6}{2\pi/T} = \frac{1}{8} \text{ s},$$

since T = 1.5 s. Hence, particle 1 leads particle 2. It is required to find their distance apart at time t = 0.125 + 0.5 =

0.625 s. Using the first two equations, it is found that at this time,

$$x_1 = \tfrac{1}{2}A\cos(5\pi/6) = -0.4330A;$$

$$x_2 = \tfrac{1}{2}A\cos(2\pi/3) = -0.2500A.$$

Hence, their distance apart, a positive number, is

$$s = x_2 - x_1 = 0.1830A \quad \underline{\text{Ans}}.$$

(b) To establish the directions of motion, examine the velocities at $t = 0.625$ s. Since $v = dx/dt$, these involve the sines of the angles encountered in (a). But $\sin(5\pi/6)$ and $\sin(2\pi/3)$ both are positive, indicating that the particles are moving in the same direction at this time.

15-18

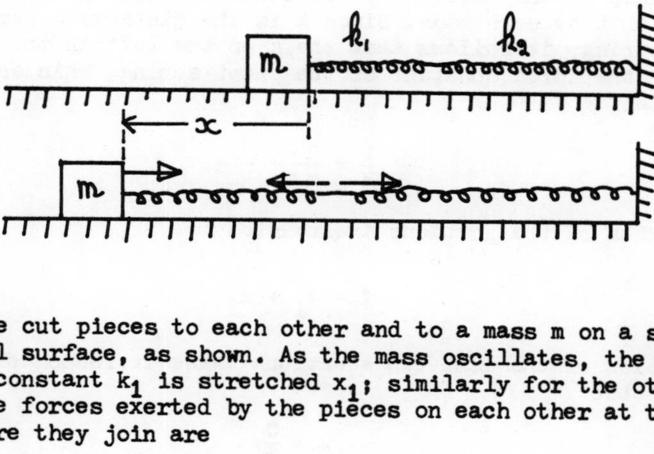

Attach the cut pieces to each other and to a mass m on a smooth horizontal surface, as shown. As the mass oscillates, the piece of force constant k_1 is stretched x_1; similarly for the other piece. The forces exerted by the pieces on each other at the point where they join are

$$F_1 = k_1x_1,\ F_2 = k_2x_2.$$

By Newton's third law $F_1 = F_2$, so that

$$k_1x_1 = k_2x_2.$$

Now

$$x = x_1 + x_2;$$

hence,

$$x_1 = x - x_2 = x - \frac{k_1}{k_2} x_1,$$

$$x_1 = \frac{k_2}{k_1 + k_2} x.$$

Now, the force on the mass m is $F_1 = k_1 x_1$; in terms of its displacement x this is

$$F_1 = \frac{k_1 \cdot k_2}{k_1 + k_2} x.$$

Thus, the pieces joined in this way act like a spring with a force constant

$$k = k_1 k_2/(k_1 + k_2).$$

But, joined as they are, they are indistinguishable from the spring from which they were made. Since x is the distance m stretches the entire spring, it follows that the k on the left in the equation above is the force constant of the uncut spring. This equation can be written

$$\frac{1}{k} = \frac{1}{k_1} + \frac{1}{k_2}.$$

The lengths of the portions are related by

$$\ell = \ell_1 + \ell_2.$$

Examination of the last two equations makes it reasonable to suppose that

$$k = \frac{C}{\ell},$$

that is, with C a constant, the force constant of a spring is inversely proportional to its length. Thus,

$$k_1 = \frac{C}{\ell_1}, \; k_2 = \frac{C}{\ell_2}$$

with $C = k\ell$. If $\ell_1 = n\ell_2$, then

$$\ell_1 + \ell_2 = n\ell_2 + \ell_2 = \ell,$$

$$\ell_2 = \frac{\ell}{n+1},\ \ell_1 = \frac{n}{n+1}\ell.$$

Therefore,

$$k_1 = \frac{n+1}{n}k,\ k_2 = (n+1)k \quad \underline{\text{Ans}}.$$

15-19

If the block is displaced x to the right, say, from the position of equilibrium, then the right-hand spring is compressed by x and exerts a force k_2x on the block directed to the left (push); the left-hand spring is stretched the same distance x and pulls on the block with a force k_1x, also to the left. Thus, the resultant force F on the block is $F = (k_1 + k_2)x = Kx$, K the effective force constant, with $K = k_1 + k_2$. The frequency, then, is

$$\nu = \frac{1}{2\pi}\left(\frac{K}{m}\right)^{\frac{1}{2}} = \frac{1}{2\pi}\left(\frac{k_1 + k_2}{m}\right)^{\frac{1}{2}} \quad \underline{\text{Ans}}.$$

15-22

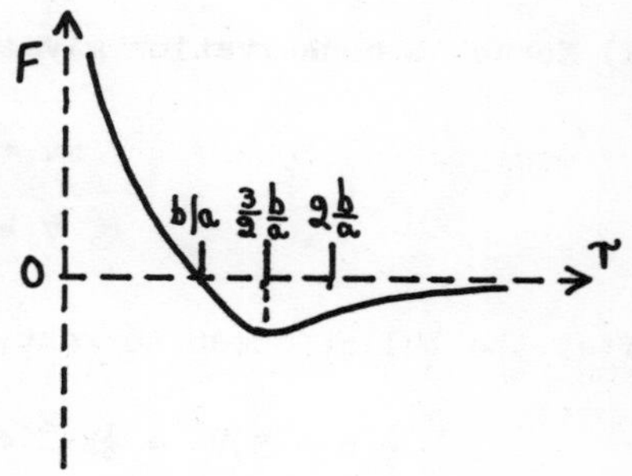

(a) The separation at equilibrium is at $r = b/a$, since $F(r = b/a) = 0$, as can be verified by direct substituting; the position where no net force acts on the particle is defined as the equilibrium position.

(b) Consider points only a very small distance $x = r - b/a$ from the equilibrium point b/a. The force at these points, from the equation for F, is

$$F = -\frac{a}{(x + b/a)^2} + \frac{b}{(x + b/a)^3} = -\frac{a^3/b^2}{(1 + ax/b)^2} + \frac{a^3/b^2}{(1 + ax/b)^3}.$$

By assumption, $x/(\frac{b}{a}) = ax/b \ll 1$. Expanding in powers of ax/b and retaining only linear terms gives

$$F \approx -(a^3/b^2)(1 - 2\,\frac{ax}{b}) + (a^3/b^2)(1 - 3\,\frac{ax}{b}) = -\frac{a^4}{b^3}\,x.$$

If this is written as $F = -kx$, the force constant $k = a^4/b^3$ Ans.

(c) The period T of simple harmonic motion is

$$T = 2\pi[m/k]^{\frac{1}{2}} = 2\pi[mb^3/a^4]^{\frac{1}{2}} \quad \text{Ans.}$$

15-28

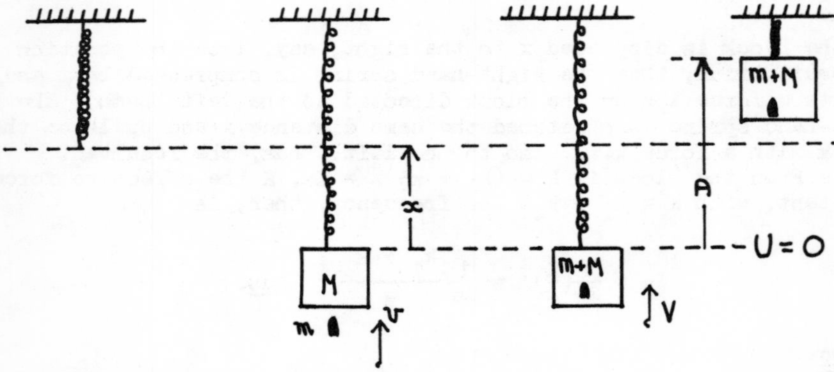

(a) Momentum conservation gives

$$mv = (m + M)V,$$

$$V = \frac{m}{m + M}\,v.$$

After the bullet comes to rest, energy conservation implies that

$$\tfrac{1}{2}(m + M)V^2 + \tfrac{1}{2}kx^2 = \tfrac{1}{2}k(A - x)^2 + (m + M)gA.$$

The equilibrium position of the block before the bullet is fired is taken as the zero-level of gravitational potential energy; x is the distance the block stretches the spring before the bullet is fired. The difference in equilibrium positions for the block (mass M) before the bullet is fired and the block+bullet (mass m + M) after the bullet is fired is ignored. This last also indicates that

$$(M + m)g = kx.$$

Substituting this into the previous equation gives

$$\tfrac{1}{2}(m + M)V^2 = \tfrac{1}{2}kA^2,$$

that is, the gravitational field appears to "drop out." Putting in the expression found for V and solving for A results in

$$A = \frac{mv}{[k(m + M)]^{\frac{1}{2}}} = 0.518 \text{ ft} \quad \underline{\text{Ans}},$$

using k = 3 lb/in. = 36 lb/ft, m = 0.1/32 = 3.125 X 10^{-3} slug, M = 8/32 = 0.25 slug, V = 500 ft/s.

(b) The desired fraction is

$$\frac{\frac{1}{2}kA^2}{\frac{1}{2}mv^2} = \frac{\frac{1}{2}(m + M)V^2}{\frac{1}{2}mv^2} = \frac{m}{m + M} = \frac{w}{w + W} = \frac{1}{81},$$

or 1.23%. The remaining 98.77% is dissipated as heat, vibration, sound,etc.

15-30

The translational kinetic energy is

$$K_t = \tfrac{1}{2}Mv^2;$$

since the rotational inertia of a cylinder about its axis is $\frac{1}{2}MR^2$ and the cylinder is rolling, the rotational energy is

$$K_r = \tfrac{1}{2}I\omega^2 = \tfrac{1}{2}(\tfrac{1}{2}MR^2)(\tfrac{v}{R})^2 = \tfrac{1}{4}Mv^2;$$

finally, the potential energy is $\frac{1}{2}kx^2$. Hence, the total energy E is

$$E = \frac{3}{4} Mv^2 + \tfrac{1}{2}kx^2.$$

Evaluate E at the moment of release; as the cylinder was released from rest,

$$E = \tfrac{1}{2}kx_i^2 = \tfrac{1}{2}(3)(0.25)^2 = \frac{3}{32} \text{ J}.$$

At equilibrium $x = 0$ and $v = v_{max}$; so, by energy conservation,

$$E = \frac{3}{4} Mv_{max}^2 = \frac{3}{32},$$

$$Mv_{max}^2 = \frac{1}{8},$$

and therefore (a) $K_t = \frac{1}{16}$ J and (b) $K_r = \frac{1}{32}$ J.

(c) As before,

$$E = \frac{3}{4} Mv^2 + \tfrac{1}{2}kx^2.$$

Differentiate with respect to time; note that $dE/dt = 0$ and then factor out $2dx/dt = 2v$ to get

$$\frac{d^2x}{dt^2} + \frac{2k}{3M} x = 0.$$

This is the differential equation for SHM: the coefficient of x is the square of the angular speed; hence,

$$T = 2\pi\left(\frac{3M}{2k}\right)^{\frac{1}{2}}.$$

15-31

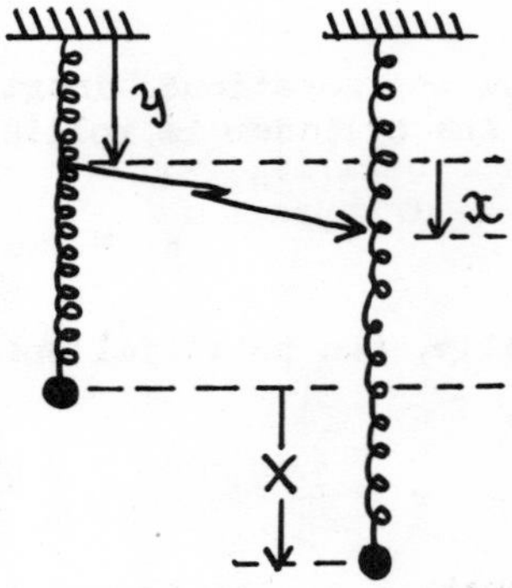

Use the following notation:

x = extension of an element (speed v) of the spring located a distance y from the fixed end if unstretched;

X = extension of the moving end, where speed is V;

L = unstretched length of spring.

The potential energy of the system is, since $F = -kX$,

$$V = \tfrac{1}{2}kX^2;$$

the kinetic energy can be written

$$T = \tfrac{1}{2}mV^2 + \tfrac{1}{2}\int v^2 dm_s = \tfrac{1}{2}mV^2 + \tfrac{1}{2}\int v^2 (m_s/L)dy.$$

By assumption, $x/X = y/L$, so that $v = (\frac{y}{L})V$; using this in the above yields

$$T = \tfrac{1}{2}mV^2 + \tfrac{1}{2}(m_s/L)(V/L)^2\int_0^L y^2 dy = \tfrac{1}{2}(m + m_s/3)V^2.$$

Now invoke conservation of energy and differentiate with respect to time:

$$\tfrac{1}{2}kX^2 + \tfrac{1}{2}(m + m_s/3)V^2 = E,$$

$$kXV + (m + m_s/3)V\frac{dV}{dt} = 0,$$

$$\frac{d^2X}{dt^2} + \frac{k}{m + m_s/3}X = 0,$$

$$T = \frac{2\pi}{\omega} = 2\pi\left(\frac{m + m_s/3}{k}\right)^{\frac{1}{2}}.$$

15-35

(a) For a physical pendulum,

$$\nu = \frac{1}{2\pi}\left(\frac{MgR}{I}\right)^{\frac{1}{2}}.$$

By the parallel-axis theorem the rotational inertia about the point of support is

$$I = I_{cm} + Mh^2 = MR^2 + MR^2 = 2MR^2.$$

Therefore,

$$\nu = \frac{1}{2\pi}\left(\frac{MgR}{2MR^2}\right)^{\frac{1}{2}} = \frac{1}{2\pi}\left(\frac{g}{2R}\right)^{\frac{1}{2}} = \frac{1}{2\pi}\left(\frac{32}{4}\right)^{\frac{1}{2}} = 0.450 \text{ Hz} \quad \underline{\text{Ans}}.$$

(b) The frequency of a simple pendulum is $\nu_s = \frac{1}{2\pi}(\frac{g}{\ell})^{\frac{1}{2}}$. If $\nu = \nu_s$, then $\ell = 2R = 4$ ft Ans.

<u>15-37</u>

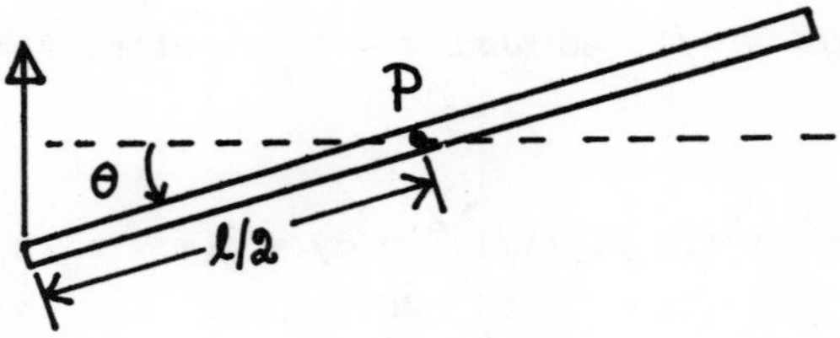

The torque exerted by the spring force F about the pivot P is

$$\tau = F\cdot\tfrac{1}{2}\ell = (-kx)(\tfrac{1}{2}\ell) = -k(x)(\tfrac{1}{2}\ell) \approx -k(\tfrac{1}{2}\ell\cdot\theta)(\tfrac{1}{2}\ell) = I\alpha,$$

$$-\tfrac{1}{4}k\ell^2 = I\,\frac{d^2\theta}{dt^2},$$

if θ is small. Since $I = m\ell^2/12$,

$$-\tfrac{1}{4}k\ell^2\theta = \frac{1}{12}\,m\ell^2\,\frac{d^2\theta}{dt^2},$$

$$\frac{d^2\theta}{dt^2} + \frac{3k}{m}\,\theta = 0.$$

This is the standard equation for SHM; identifying the coefficient of θ with ω^2 gives for the period T,

$$T = 2\pi\left(\frac{m}{3k}\right)^{\frac{1}{2}} \quad \underline{\text{Ans}}.$$

<u>15-38</u>

(a) Since the period T = 0.5 s, the frequency $\nu = 1/T = 2$ Hz; combining this with the given amplitude of π rad gives

$$\theta = \pi\cos(4\pi t + \phi),$$

for the angular displacement as a function of time. The angular velocity is

$$\frac{d\theta}{dt} = -4\pi^2\sin(4\pi t + \phi).$$

Since the sine varies between -1 and +1, the maximum value of $\frac{d\theta}{dt}$ is $4\pi^2 = 39.5$ rad/s Ans.

(b) When the displacement $\theta = \pi/2$, then $\cos(4\pi t + \phi) = \frac{1}{2}$, by the first equation. Hence $\sin(4\pi t + \phi) = \pm\sqrt{3}/2$, so that $\frac{d\theta}{dt} = \pm 2\pi^2\sqrt{3} = \pm 34.2$ rad/s at this instant. In either case, the angular speed, as opposed to angular velocity, is 34.2 rad/s Ans.

(c) When $\theta = \pi/4$, $\cos(4\pi t + \phi) = \frac{1}{4}$. But the angular acceleration is

$$\frac{d^2\theta}{dt^2} = -16\pi^3\cos(4\pi t + \phi),$$

giving, at the instant when $\theta = \pi/4$, an angular acceleration of $-4\pi^3 = -124$ rad/s^2 Ans.

15-45

To obtain the path, i.e., y vs. x, it is necessary to eliminate the time t. To do this, write

$$y = A\cos(\omega t + \phi_y) = A(\cos\omega t\cdot\cos\phi_y - \sin\omega t\cdot\sin\phi_y),$$

$$y = A[\tfrac{x}{A}\cdot\cos\phi_y - (1 - x^2/A^2)^{\frac{1}{2}}\sin\phi_y].$$

Rearrange and square:

$$y - x\cos\phi_y = -A(1 - x^2/A^2)^{\frac{1}{2}}\sin\phi_y,$$

$$y^2 - 2xy\cos\phi_y + x^2\cos^2\phi_y = A^2(1 - x^2/A^2)\sin^2\phi_y,$$

$$y^2 - 2xy\cos\phi_y + x^2 = A^2\sin^2\phi_y,$$

the last equation being for the path. It is the equation of an ellipse and, since it remains unchanged if x and y are exchanged, the axes of the ellipse must bisect the x and y-coordinate axes.

(a) If $\phi_y = 0$, the equation reduces to $x = y$, and the ellipse to a straight line.

(b) With $\phi_y = 30^\circ$, the full ellipse is present; the angle ϕ_y determines the elongation of the ellipse.

(c) For $\phi_y = 90^\circ$, the ellipse becomes the circle $x^2 + y^2 = A^2$.

<u>15-48</u>

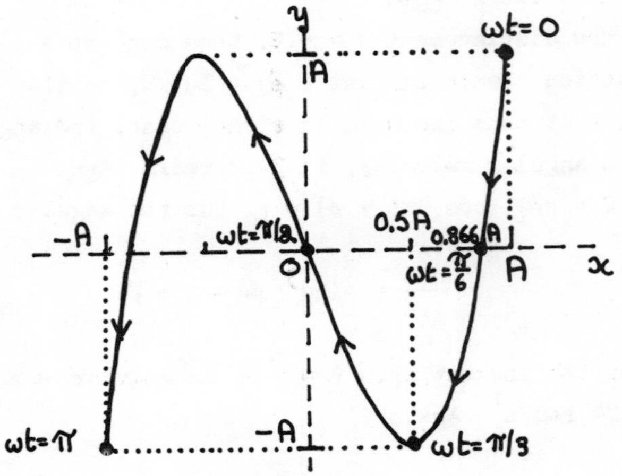

(a) The trajectory is shown in the sketch above.

(b) Since $\vec{r} = A(\vec{i}\cos\omega t + \vec{j}\cos 3\omega t)$,

$$\vec{v} = \frac{d\vec{r}}{dt} = -\omega A(\vec{i}\sin\omega t + \vec{j}3\sin 3\omega t),$$

and therefore $\vec{L} = m\vec{r} \times \vec{v}$ is just

$$\vec{L} = -mA^2\omega(3\sin 3\omega t\cos\omega t - \sin\omega t\cos 3\omega t)\vec{k} \quad \underline{\text{Ans}}.$$

(c) Since $\vec{F} = m\vec{a}$ and $\vec{a} = \frac{d\vec{v}}{dt}$,

$$\vec{F} = -m\omega^2 A(\vec{i}\cos\omega t + \vec{j}9\cos 3\omega t) \quad \underline{\text{Ans}}.$$

(d) For this particle $x = A\cos\omega t$ and $y = A\cos 3\omega t$. Hence, from (a)

$$F_x = -m\omega^2 x = -\frac{\partial U}{\partial x}; \quad U = \tfrac{1}{2}m\omega^2 x^2 + f(y);$$

$$F_y = -9m\omega^2 y = -\frac{\partial U}{\partial y}; \quad U = \frac{9}{2}m\omega^2 y^2 + g(x).$$

These two expressions for U imply that

$$U = \tfrac{1}{2}m\omega^2(x^2 + 9y^2) = \tfrac{1}{2}mA^2\omega^2(\cos^2\omega t + 9\cos^2 3\omega t) \quad \underline{\text{Ans}}.$$

(e) The kinetic energy is simply

$$K = \tfrac{1}{2}mv^2 = \tfrac{1}{2}m\omega^2A^2(\sin^2\omega t + 9\sin^2 3\omega t).$$

Since the total energy $E = K + U$ and $\sin^2\theta + \cos^2\theta = 1$,

$$E = 5m\omega^2A^2 \quad \underline{\text{Ans}}.$$

(f) The motion is periodic with period $T = 2\pi/\omega$ <u>Ans</u>.

<u>15-53</u>

By conservation of momentum,

$$m_1v_1 + m_2v_2 = 0,$$

provided the center of mass is at rest; if it is not, use center of mass coordinates. If $v = v_1 - v_2$ is the relative velocity,

$$v_1 = \frac{m_2v}{m_1 + m_2}, \quad v_2 = \frac{m_1v}{m_1 + m_2}.$$

Using these the kinetic energy $K = \tfrac{1}{2}m_1v_1^2 + \tfrac{1}{2}m_2v_2^2$ becomes

$$K = \tfrac{1}{2}m_1m_2^2v^2/(m_1 + m_2)^2 + \tfrac{1}{2}m_2m_1^2v^2(m_1 + m_2)^2,$$

$$K = \frac{1}{2}\,\frac{m_1\,m_2}{m_1 + m_2}\,v^2 = \tfrac{1}{2}\mu v^2.$$

<u>15-54</u>

The displacement x of the block is given by

$$x = Ae^{-bt/2m}\cos(\omega' t + \delta).$$

Now let t be the time required for the amplitude to fall to one-third of its initial value, which is A; then,

$$\frac{1}{3} A = Ae^{-bt/2m},$$

$$t = \frac{2m}{b} \ln 3.$$

The number N of oscillations completed in this time is

$$N = \nu' t = \frac{\omega'}{2\pi} \frac{2m}{b} \ln 3 = \frac{\omega' m}{\pi b} \ln 3.$$

Evaluating ω':

$$\omega' = (k/m - b^2/4m^2)^{\frac{1}{2}} = 2.31 \text{ rad/s}.$$

Using this, the preceding equation gives N = 5.26 Ans.

CHAPTER 16

16-8

(a) If R is the radius of the earth, the difference in weight is

$$W' - W = GMm/R^2 - GMm/(R + h)^2.$$

Usually $h/R \ll 1$ and therefore this can be written

$$W' - W \approx \frac{GMm}{R^2} - \frac{GMm}{R^2}(1 - 2\,\frac{h}{R}) = 2GMmh/R^3 = \frac{8\pi}{3}\,G\rho mh \quad \underline{\text{Ans}},$$

since $M = \frac{4\pi}{3}\rho R^3$.

(b) Solving (a) for h,

$$h = (W' - W)/(8\pi G\rho m/3) = \frac{W' - W}{W}\,\frac{3g}{8\pi G\rho}.$$

Putting $(W' - W)/W = 10^{-6}$ and substituting the numerical values $g = 980\ \text{cm/s}^2$, $G = 6.67 \times 10^{-8}\ \text{dyne}\cdot\text{cm}^2/\text{g}^2$ and $\rho = 5.5\ \text{g/cm}^3$ yields $h = 319\ \text{cm} = 3.19\ \text{m}$ Ans.

16-10

(a) Let the suspended body have mass m, the spring a force constant k. The forces on m in suspension are its true weight mg directed down and the spring force kx directed up. This last is equal in magnitude to the force exerted by m on the spring and equals the scale reading. The mass m is moving on the equator (a circle) with linear speed V relative to space; hence,

$$mg - kx = m\,\frac{V^2}{R}.$$

R is the radius of the earth. The speed V is made up of the speed $R\omega$ due to the earth's rotation and the speed v of the ship relative to the earth; that is

$$V = R\omega \pm v,$$

the + sign if the ship is sailing in the direction of the earth's rotation, i.e., to the east, and the - sign if in the opposite direction against the rotation of the earth. The scale reading, then, is

$$kx = mg - m\frac{(R\omega \pm v)^2}{R},$$

$$kx = mg - mR\omega^2 \pm 2m\omega v - m\frac{v^2}{R},$$

the + sign now for a ship sailing west. Now, $v \ll R\omega$, that is, the speed of the ship relative to the earth is, very likely, much smaller than the speed due to the earth's rotation (about 1000 mph at the equator). This means that the last term in the equation above numerically is much smaller than the others and can be dropped with little error. When the ship is at rest, the scale reading is W_0 given by

$$W_0 = mg - mR\omega^2$$

so that

$$W = W_0 \pm 2m\omega v = W_0\left(1 \pm \frac{2m\omega v}{W_0}\right).$$

Finally, in the last term on the right above,

$$\frac{2m\omega v}{W_0} = \frac{2m\omega v}{mg - mR\omega^2} = \frac{2\omega v}{g}\left(1 + \frac{R\omega^2}{g} + \ldots\right).$$

$R\omega^2$ is the acceleration imparted by the rotation of the earth and is much less than g, the acceleration due to gravity. Thus, to a good approximation,

$$W = W_0(1 \pm 2\omega v/g).$$

(b) Since the term mv^2/R was transposed during the analysis, in the final equation above the - sign applies when sailing east, the + sign when sailing west along the equator.

16-13

For a circular orbit and assuming that the stars of the galaxy are distributed in space in a spherically symmetric manner (which they are not), then

$$GM_{gal}M_s/R^2 = M_s v^2/R,$$

where v is the speed of the sun in its orbit, M_s is the mass of the sun and R is the distance of the sun from the galactic center. The period T of revolution is

$$T = \frac{2\pi R}{v} = 2\pi(R^3/GM_{gal})^{\frac{1}{2}},$$

$$M_{gal} = \frac{4\pi^2}{G}\frac{R^3}{T^2},$$

(Kepler's third law on a grand scale)

$$M_{gal} = \frac{4\pi^2}{6.67 \times 10^{-11}\ \mathrm{N{\cdot}m^2/kg^2}} \cdot \frac{(2.4 \times 10^{20}\ \mathrm{m})^3}{(7.9 \times 10^{15}\ \mathrm{s})^2} = 1.3 \times 10^{41}\ \mathrm{kg}.$$

The mass of the sun is 2×10^{30} kg and if this is typical of the stars in the galaxy their number will be approximately

$$N = (1.3 \times 10^{41})/(2 \times 10^{30}) = 6.5 \times 10^{10} \quad \underline{\text{Ans.}}$$

Strictly speaking, this gives an estimate only of the number of stars in the galaxy lying exterior to the orbit of the sun.

16-14

Let m, M refer to the falling object and the earth. Then, in a fixed inertial frame centered at the center of mass of earth+object

$$G\frac{mM}{(r_m + r_M)^2} = ma_m;\ G\frac{mM}{(r_m + r_M)^2} = Ma_M.$$

These expressions indicate that a_m is independent of m but not of M and that a_M is independent of M but not of m. However,

$$a_m + a_M = GM/(r_m + r_M)^2 + Gm/(r_m + r_M)^2,$$

$$a_m + a_M = G(m + M)/(r_m + r_M)^2,$$

so that $a_m + a_M$, the relative acceleration, depends on $m + M$. Therefore, the statement that g is the same regardless of the mass of the falling body means g relative to the center of mass, the source mass M being kept constant. The earth is not a fixed, inertial reference frame. However, the Galilean conclusion that all bodies fall at the same rate is almost exactly right for $m \ll M$ but is fundamentally right vis-a-vis Aristotle; this would clearly be shown by dropping equal masses m (successively increasing m in successive trials) toward the earth from opposite sides of the earth (the earth is then at the center of mass and still $m \ll M$).

16-17

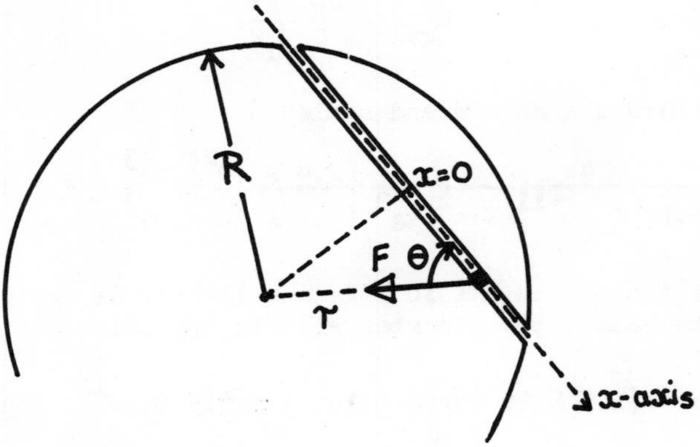

(a) The component of the gravitational force F on the object along the chord is $F\cos\theta = F(x/r)$, toward $x = 0$, r the distance of the object from the center of the earth. If M^* is the mass of that part of the earth interior to r and ρ is the density of the earth (assumed uniform), then

$$F = GM^*m/r^2 = G(\frac{4}{3}\pi r^3\rho)m/r^2 = \frac{4}{3}\pi G\rho mr.$$

Therefore,

$$F\cos\theta = (\frac{4}{3}\pi G\rho m)x = -m\frac{d^2x}{dt^2},$$

$$\frac{d^2x}{dt^2} + \frac{G\rho}{3\pi}(2\pi)^2x = 0,$$

the minus sign in the preceding equation since the force always is directed toward $x = 0$. The last equation is one describing simple harmonic motion with a frequency ν given by

$$\nu^2 = \frac{G\rho}{3\pi}$$

(b) By comparison with the case of a chute dug along a diameter, the frequencies are seen to be equal (see Example 16-3); hence, the periods $T = 1/\nu = 84.2$ min are the same also.

(c) In simple harmonic motion, $v_{max} = 2\pi A/T$. In the present case the amplitude $A = \frac{1}{2}$(chord length) $< R$, the amplitude for a chute along a diameter. Since the periods T are the same along chord and diameter, v_{max} will be smaller for the former situation.

<u>16-18</u>

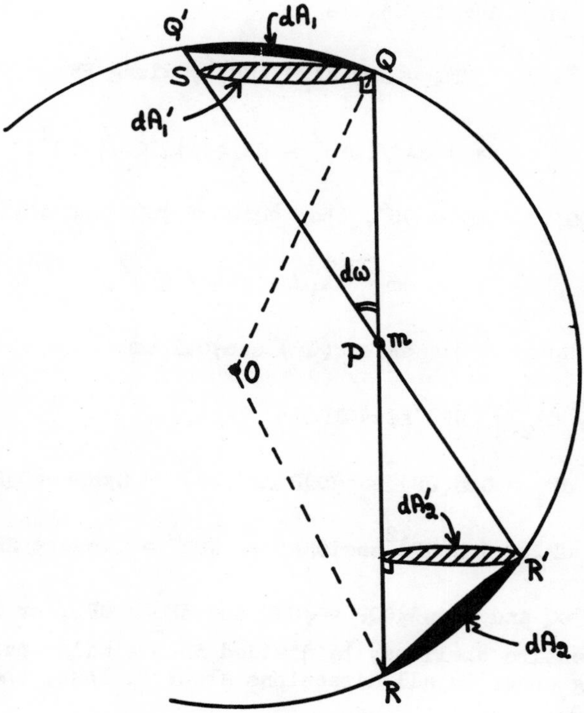

(a) Let the mass of the particle at P be m, and $d\omega$ the solid angle of the narrow double cone constructed with apex at P. The cone intercepts areas dA_1 and dA_2 on the surface of the shell (which is of uniform density σ). These areas are not perpendicular to the generators of the cone, so construct areas dA_1' and dA_2' that are perpendicular. The areas dA_1 and dA_2 are so small that Newton's law of universal gravitation can be applied to them. The force on m due to dA_1 is

$$dF_1 = Gm\sigma(dA_1)/(QP)^2, \text{ along PQ.}$$

The force on m due to dA_2 is

$$dF_2 = Gm\sigma(dA_2)/(RP)^2, \text{ along PR.}$$

But,

$$d\omega = dA_1'/(QP)^2 = dA_1\cos\angle Q'QS/(QP)^2.$$

Since $\angle OQQ' = \angle SQP = 90^\circ$, then $\angle Q'QS = \angle OQP$ and therefore

$$d\omega = dA_1\cos\angle OQP/(QP)^2,$$

$$dA_1 = (QP)^2\sec\angle OQP\, d\omega.$$

Similarly $dA_2 = (RP)^2\sec\angle ORP\, d\omega$. Thus,

$$dF_1 = Gm\sigma(QP)^2\sec\angle OQP\, d\omega/(QP)^2 = Gm\sigma\sec\angle OQP\, d\omega,$$

$$dF_2 = Gm\sigma(RP)^2\sec\angle ORP\, d\omega/(RP)^2 = Gm\sigma\sec\angle ORP\, d\omega.$$

But OR = OQ and thus $\angle OQP = \angle ORP$ and $dF_1 = dF_2$, or $d\vec{F}_1 + d\vec{F}_2 = 0$.
(b) The entire shell may be divided into similar pairs of areas by taking cones in all directions about P. Thus, the net force on m is zero.

16-25

Start with the second law and substitute in Kepler's third law $T^2 = kR^3$:

$$F = m\frac{v^2}{R} = m\frac{(2\pi R/T)^2}{R} = 4\pi^2 m\frac{R}{T^2} = 4\pi^2 m\frac{R}{kR^3},$$

$$F = (4\pi^2 m/k)R^{-2}.$$

Newton, of course, was able to derive an expression for the constant k from his law of gravitation.

16-27

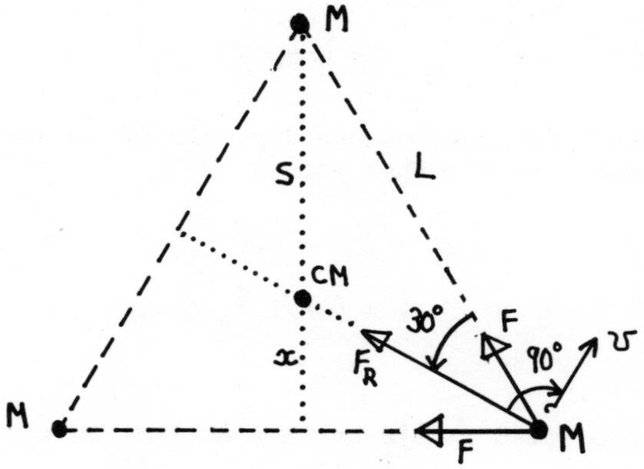

The motion to be described is about the center of mass of the system and, therefore, the center of mass must first be located. Clearly, it lies along the perpendicular bisectors drawn from the masses, at their intersection. The length of a perpendicular bisector is $L\sin 60^\circ = \sqrt{3}L/2$. Hence, the center of mass is at a distance x along a bisector, as shown, where x is found from

$$(3M)x = M(\sqrt{3}L/2) + 2M(0),$$

$$x = \sqrt{3}L/6$$

by the definition of center of mass. The distance s of any mass from the center of mass is

$$s = \sqrt{3}L/2 - x = \sqrt{3}L/3.$$

Now s is the radius of the circular orbit of each mass; for

uniform circular motion,

$$F_R = Mv^2/s,$$

F_R the resultant force on each mass M. But,

$$F_R = 2F\cos 30° = \sqrt{3}F,$$

F the gravitational attraction between any pair of masses. By the law of gravitation,

$$F = GM^2/L^2,$$

since L is the distance between any pair of masses. Putting the last four equations together yields

$$\sqrt{3}GM^2/L^2 = Mv^2/(\sqrt{3}L/3),$$

$$v = (GM/L)^{\frac{1}{2}} \quad \underline{\text{Ans}}.$$

<u>16-34</u>

(a) The escape speed at the earth's surface is $v_e = (2GM_e/R_e)^{\frac{1}{2}}$. Using $g = GM_e/R_e^2$ this becomes $v_e = (2gR_e)^{\frac{1}{2}}$. Since the speed of the rocket was $v = 2(gR_e)^{\frac{1}{2}} > v_e$, the rocket escapes.

(b) By conservation of energy,

$$\tfrac{1}{2}mv^2 - GM_em/R_e = \tfrac{1}{2}mV^2 - 0,$$

the potential energy being nearly zero when the rocket is very far from the earth, its speed out there being V. Solving for V gives

$$V^2 = v^2 - 2GM_e/R_e = 4gR_e - 2gR_e = 2gR_e,$$

$$V = (2gR_e)^{\frac{1}{2}} \quad \underline{\text{Ans}}.$$

Hence, in this situation, the rocket's "final" speed is just the escape speed at the earth's surface!

16-37

Let i,f refer to the initial configuration (masses far apart) and to the configuration when the separation of the particles is d. Then,

$$\Delta U = U_f - U_i = -\frac{GMm}{d} - 0,$$

$$\Delta K = K_f - K_i = \tfrac{1}{2}mv^2 + \tfrac{1}{2}MV^2 - 0,$$

where v,V are the speeds of the particles relative to a fixed coordinate system when their separation is d. By the conservation of energy, $\Delta U + \Delta K = 0$ so that

$$\frac{GMm}{d} = \tfrac{1}{2}mv^2 + \tfrac{1}{2}MV^2.$$

The initial momentum was zero; thus, to conserve momentum

$$mv + M(-V) = 0,$$

$$mv = MV.$$

Using this, the energy equation becomes

$$\frac{GMm}{d} = \tfrac{1}{2}mv(v + V).$$

Let the relative speed of the two oncoming particles be u; using the momentum equation, u can be expressed in terms of v:

$$u = v + V = v + \frac{m}{M}v = \frac{m + M}{M}v.$$

Therefore,

$$\frac{GMm}{d} = \tfrac{1}{2}m\left(\frac{Mu}{m + M}\right)u = \frac{1}{2}\frac{Mm}{m + M}u^2,$$

$$u = [2G(m + M)/d]^{\frac{1}{2}} \quad \underline{\text{Ans}}.$$

16-38

Let a = radius of the earth's orbit, assumed to be a circle, and T = time for the comet to move from D to B. If dA/dt = rate at which the line from sun to comet sweeps out area, and A* = area of sector BSD of the parabola on which the comet moves, then by conservation of angular momentum

$$\frac{A^*}{T} = \frac{dA}{dt}.$$

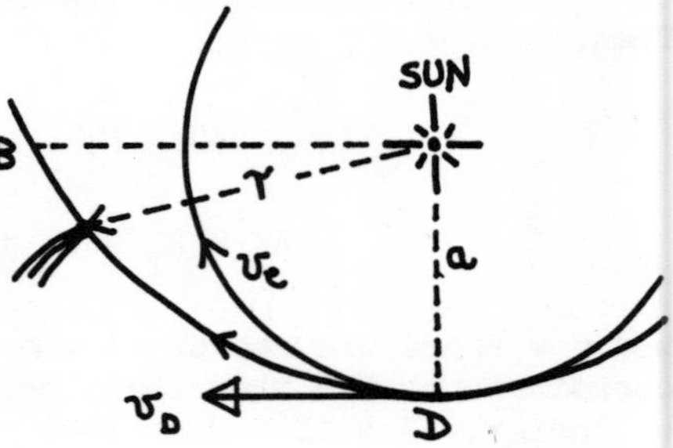

Since dA/dt is a constant, it can be evaluated at any point on the comet's orbit. It is easiest to compute dA/dt when the comet is at D, its point of closest approach to the sun. Suppose that the speed of the comet at D is v_D; then,

$$\frac{dA}{dt} = \tfrac{1}{2}\omega r^2 = \tfrac{1}{2}(v_D/r_D)r_D^2 = \tfrac{1}{2}r_D v_D = \tfrac{1}{2}av_D,$$

since r is perpendicular to v at point D. Now, by Problem 16-32,

$$v_D = \sqrt{2}v_e,$$

where v_e is the speed of the earth in its orbit, so that

$$\frac{dA}{dt} = \tfrac{1}{2}a(\sqrt{2}v_e) = \sqrt{2}(\tfrac{1}{2}av_e).$$

But the earth's radius vector sweeps out area at a rate

$$dA_e/dt = \pi a^2/P = \frac{\pi a^2}{(2\pi a/v_e)} = \tfrac{1}{2}av_e,$$

P the period of the earth's revolution about the sun. Hence,

$$\frac{dA}{dt} = \sqrt{2}(\pi a^2/P),$$

and therefore, for the comet,

$$T = A^*/\frac{dA}{dt} = A^*/(\sqrt{2}\pi a^2/P) = \frac{A^*P}{\sqrt{2}\pi a^2}.$$

The area $A^* = \frac{4}{3}a^2$ (see a calculus text; the sun is at the focus of the parabola); thus,

$$T = \frac{4P}{3\sqrt{2}\pi},$$

and since P = 1 year,

$$T = 4/3\sqrt{2}\pi \approx 0.300 \text{ years} \quad \underline{\text{Ans}}.$$

<u>16-48</u>

(a) Use the notation R, h, M, m for the radius of the earth, height of the circular orbit above the earth's surface, mass of earth and of satellite. For circular orbits,

$$\frac{GMm}{(R + h)^2} = m\,\frac{v^2}{R + h},$$

$$v = \left(\frac{GM}{R + h}\right)^{\frac{1}{2}} = 7.54 \text{ km/s} \quad \underline{\text{Ans}},$$

using $M = 5.98 \times 10^{24}$ kg, $R = 6.37 \times 10^6$ m.

(b) The period T is given by

$$T = \frac{2\pi(R + h)}{v} = \frac{2\pi\,(7.01 \times 10^6 \text{ m})}{7.543 \times 10^3 \text{ m/s}},$$

$$T = 5.84 \times 10^3 \text{ s} = 97.3 \text{ min} \quad \underline{\text{Ans}}.$$

(c) The mechanical energy is

$$E = K + U = \tfrac{1}{2}mv^2 - \frac{GMm}{R + h} = -\,\tfrac{1}{2}\,\frac{GMm}{R + h}.$$

As the orbit lowers h diminishes and therefore E changes at a rate

$$\frac{dE}{dt} = \frac{1}{2}\,\frac{GMm}{(R + h)^2}\,\frac{dh}{dt}$$

assuming that the orbit remains nearly circular. Numerically,

$$\frac{GMm}{(R + h)^2} = 1.7857 \times 10^3 \text{ N},$$

and therefore

$$-1.4 \times 10^5 \text{ J/rev} = \tfrac{1}{2}(1.7857 \times 10^3 \text{ N})\frac{dh}{dt},$$

$$\frac{dh}{dt} = -156.8 \text{ m/rev},$$

the negative sign indicating that h is diminishing, just as the minus sign for dE/dt (implied) indicates a loss of energy. After 1500 revolutions, the orbital altitude drops by (156.8)(1500) = 235 km. Thus, the new altitude is 640 - 235 = 405 km. The equations in (a) and (b) now give v = 7.68 km/s and T = 92.3 min.

(d) Let F be the force; the rate of energy loss (power) is

$$\frac{dE}{dt} = \tau\omega = F(R + h)\cdot\frac{2\pi}{T}.$$

The period T is

$$T = 2\pi\,\frac{(R + h)^{3/2}}{(GM)^{\frac{1}{2}}}$$

(Kepler's third law), so that

$$F = \left(\frac{R + h}{GM}\right)^{\frac{1}{2}}\frac{dE}{dt}.$$

For the highest (initial) orbit, $dE/dt = 1.4 \times 10^5$ J/5.84×10^3 s = 24 J/s in absolute value; for this orbit $F = 3.18 \times 10^{-3}$ N.

(e) The resistive force exerts a torque on the satellite and its orbital angular momentum decreases. If all influences originating outside the earth-satellite system (sun, moon, etc.) are ignored, the system is isolated and its total angular momentum must remain constant.

16-49

(a) At the extreme positions, denoted 1 and 2, the velocity is perpendicular to the radius vector. Let the given data refer to the latter position. By conservation of angular momentum,

$$mr_1v_1 = mr_2v_2,$$

$$r_1v_1 = a(k/2ma)^{\frac{1}{2}}.$$

Conservation of energy gives

$$\tfrac{1}{2}mv_1^2 - \frac{k}{r_1} = \tfrac{1}{2}mv_2^2 - \frac{k}{r_2} = \tfrac{1}{2}m\left(\frac{k}{2ma}\right) - \frac{k}{a} = -\frac{3}{4}\frac{k}{a},$$

$$\tfrac{1}{2}mv_1^2 = \frac{k}{r_1} - \frac{3}{4}\frac{k}{a}.$$

But $r_1^2 v_1^2 = ka/2m$. Thus

$$r_1^2 = \frac{ka/2m}{\frac{2}{m}\left(\frac{k}{r_1} - \frac{3}{4}\frac{k}{a}\right)},$$

$$r_1^2 - \frac{4a}{3}r_1 + \frac{a^2}{3} = 0$$

$$(r_1 - a)(r_1 - \frac{a}{3}) = 0,$$

$$r_1 = \frac{a}{3} \quad \underline{\text{Ans}}.$$

(b) From the momentum equation,

$$v_1 = a(k/2ma)^{\frac{1}{2}}/\tfrac{a}{3} = 3(k/2ma)^{\frac{1}{2}} \quad \underline{\text{Ans}}.$$

<u>16-50</u>

(a) The northern edge of the floor of a room in the northern hemisphere is nearer the axis of rotation of the earth than is the southern edge, and therefore its linear velocity due to the earth's rotation is less than for the southern edge since the circumference of its path is smaller. Thus the floor twists around continually, everywhere except at the earth's equator. The pendulum is constrained to follow the changes in the direction of g, i.e., the vertical, but is otherwise free. Its plane of oscillation, then, will appear to rotate in the opposite direction to the twisting of the ground, and at the same rate.

(b) The vertical component of ω at latitude λ is $\omega\sin\lambda$. Thus the earth appears to rotate beneath the pendulum with a period

$$T = \frac{2\pi}{\omega\sin\lambda} = \frac{24\text{ h}}{\sin\lambda} \quad \underline{\text{Ans}},$$

or, as seen from the earth, the pendulum appears to rotate in this time.

(c) At the poles the plane of oscillation is fixed and the earth rotates beneath it: thus, T = 24 h. At the equator, the plane of oscillation gets carried around with the rotation of the earth, following the changing direction of g, but its orientation with respect to the ground does not change. Hence, at this location, $T = \infty$.

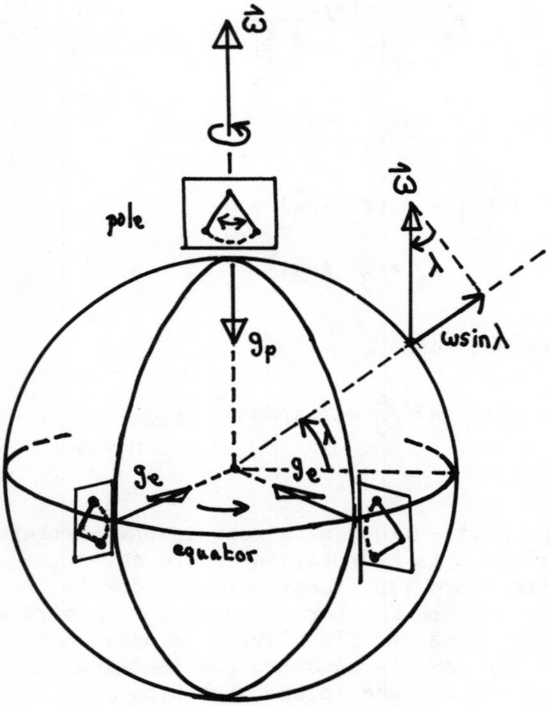

CHAPTER 17

17-2

Let p be the pressure inside the box. The air remaining in the box exerts a force pA against the lid from the inside, in the same direction as the force F required to pull off the lid. The force tending to keep the lid on is PA, where P is the outside air pressure. If F is the minimum force needed, then,

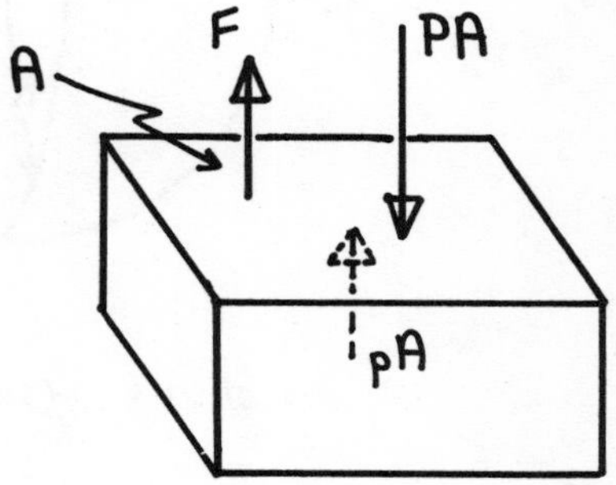

$$pA + F = PA,$$

$$p = P - \frac{F}{A} = 15 - \frac{108}{12},$$

$$p = 6\ \text{lb/in.}^2 \quad \underline{\text{Ans}}.$$

17-3

(a) Each horse must exert a force F equal, in magnitude, to the resultant of the forces due to the air inside and outside the hemisphere. The vertical components due to the air vanish; it is the horizontal components that must be calculated. Hence,

$$F = \int (dF_{out} - dF_{in})\cos\theta = \int (P_{out} - P_{in})\cos\theta\, dA,$$

$$F = \int P\cos\theta\, dA = P\int \cos\theta\, dA.$$

In spherical coordinates $dA = 2\pi R^2 \sin\theta d\theta$ and therefore

$$F = 2\pi PR^2 \int_0^{\frac{1}{2}\pi} \sin\theta\cos\theta d\theta = \pi R^2 P.$$

(b) Since atmospheric pressure is 14.7 lb/in.^{2},

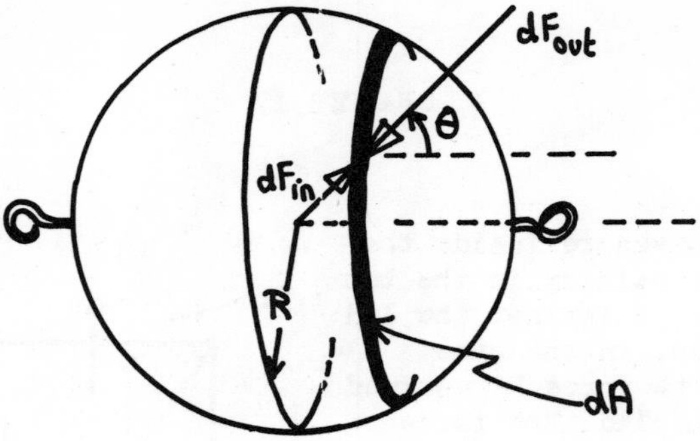

$$P = 14.7 - \frac{1}{10}(14.7) = 13.23\ \text{lb/in.}^2,$$

giving

$$F = \pi(1\ \text{ft})^2\ \frac{13.23\ \text{lb}}{(1/12\ \text{ft})^2} = 6000\ \text{lb}\quad \underline{\text{Ans}}.$$

(c) Two teams of horses look more impressive than one, but one could be used if the other hemisphere is hooked to the side of a building, say.

17-12

(a) The force dF exerted by the water against the shaded area of the dam face is

$$dF = pdA = (\rho gx)(Wdx).$$

The force F against the entire face of the dam is

$$F = \int dF = \rho gW\int_0^D xdx = \tfrac{1}{2}\rho gWD^2\quad \underline{\text{Ans}}.$$

(b) The horizontal component of the torque due to dF is $d\tau = dF\cdot(D - x)$; the total horizontal torque becomes

$$\tau = \int_0^D \rho gWx(D - x)dx = \frac{1}{6}\rho gWD^3\quad \underline{\text{Ans}}.$$

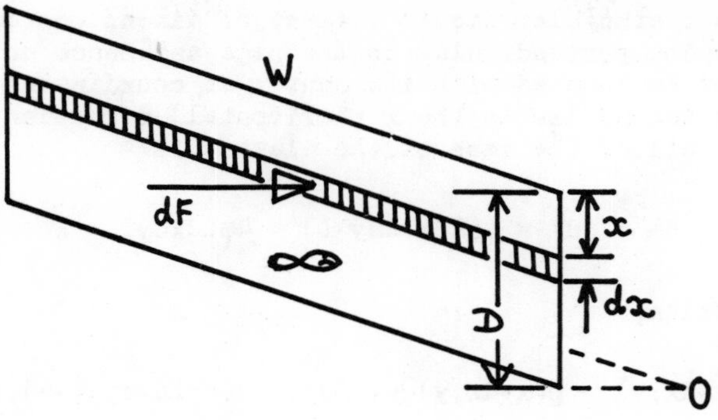

(c) The line of action of F in (a) to yield the torque in (b) must be at a distance d above the bottom of the dam, where

$$(\tfrac{1}{2}\rho gWD^2)d = \tfrac{1}{6}\rho gWD^3,$$

$$d = \frac{D}{3} \quad \underline{\text{Ans.}}$$

<u>17-16</u>

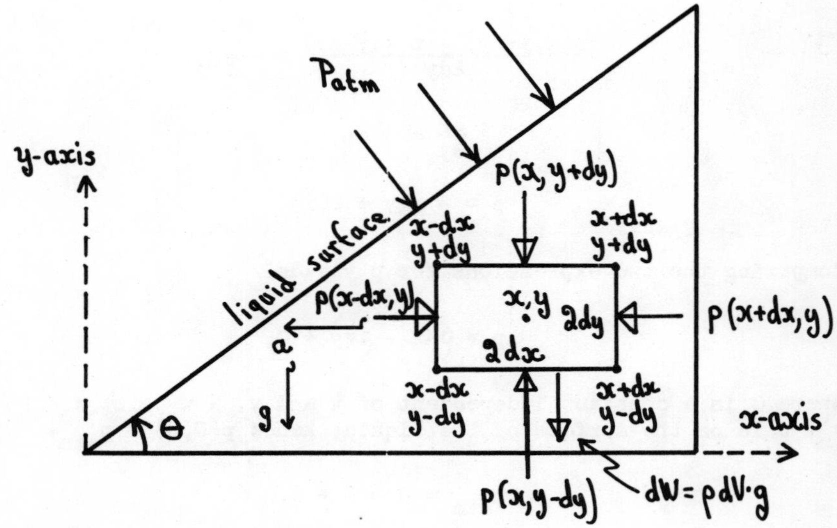

(a) Consider a rectangular liquid element of dimensions 2dx, 2dy, L the last dimension perpendicular to the page and hence not shown. Let the element be located with its center at coordinates x,y. Apply Newton's second law in the x (horizontal) direction and y (vertical) direction. The mass of the element is

$$dm = \rho dV = \rho(2dx \cdot 2dy \cdot L) = 4\rho L dx dy.$$

In the x-direction,

$$[p(x-dx,y) - p(x+dx,y)](L \cdot 2dy) = (4\rho L dx dy)(-a),$$

$$\frac{p(x+dx,y) - p(x-dx,y)}{2dx} = \rho a,$$

$$\frac{\partial p}{\partial x} = \rho a,$$

$$p = \rho a x + f(y).$$

Similarly, in the y-direction,

$$[p(x,y-dy) - p(x,y+dy)](L \cdot 2dx) - (4\rho L dx dy)g = 0,$$

$$\frac{p(x,y+dy) - p(x,y-dy)}{2dy} = -\rho g,$$

$$\frac{\partial p}{\partial y} = -g\rho,$$

$$p = -\rho g y + e(x).$$

Comparing the two expressions for p yields

$$p = \rho a x - \rho g y + C,$$

where C is a constant independent of x and y. Now point x = 0, y = 0 is on the surface of the liquid; hence $p(0,0) = p_{atm}$:

$$p_{atm} = 0 - 0 + C,$$

$$C = p_{atm},$$

so that

$$p = \rho ax - \rho gy + p_{atm}.$$

But the pressure at any point on the surface must be p_{atm}. The surface is described by the equation $\tan\theta = y/x$; hence, if $y = x\tan\theta$, then $p = p_{atm}$:

$$p_{atm} = \rho ax - \rho g(x\tan\theta) + p_{atm},$$

$$\tan\theta = \frac{a}{g}.$$

(b) The vertical depth h increases downward, y increases upward; thus, from (a),

$$\frac{\partial p}{\partial y} = -\frac{\partial p}{\partial h} = -\rho g,$$

$$\frac{\partial p}{\partial h} = \rho g.$$

17-18

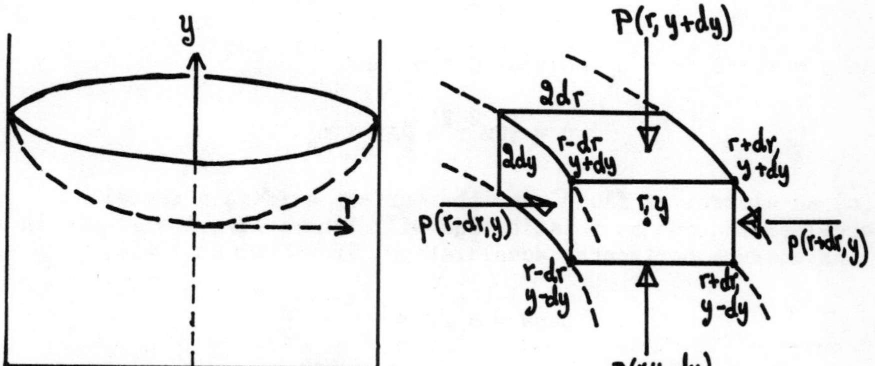

(a) Consider a complete 360° ring-shaped element of fluid (a part of the element is shown above); the element contains a mass

$$dm = \rho dV = \rho(2\pi r \cdot 2dr \cdot 2dy) = 8\pi\rho r dr dy.$$

Applying Newton's second law in the radial (r) direction,

$$[p(r-dr,y) - p(r+dr,y)](2\pi r\cdot 2dy) = -(8\pi\rho r dr dy) r\omega^2,$$

$$\frac{p(r+dr,y) - p(r-dr,y)}{2dr} = +\rho r\omega^2,$$

$$\frac{\partial p}{\partial r} = \rho r\omega^2.$$

(b) In the vertical (y) direction,

$$[-p(r,y+dy) + p(r,y-dy)](2\pi r\cdot 2dr) - 8\pi\rho g r dr dy = 0,$$

$$\frac{p(r,y+dy) - p(r,y-dy)}{2dy} = -\rho g,$$

$$\frac{\partial p}{\partial y} = -\rho g.$$

Integrating this and the expression derived in (a) gives

$$p = \tfrac{1}{2}\rho\omega^2 r^2 - \rho g y + C.$$

At $y = r = 0$, $p = p_c$ so that $C = p_c$ and

$$p = \tfrac{1}{2}\rho\omega^2 r^2 - \rho g y + p_c.$$

(c) An element of fluid near the surface suffers a radial acceleration due to rotation and will behave like the liquid in a tank under a horizontal acceleration. From Problem 17-16,

$$\tan\theta = a_r/g = \frac{\omega^2 r}{g} = \frac{dy}{dr},$$

$$y = \tfrac{1}{2}\omega^2 r^2/g.$$

(d) From (b), with h increasing down and y increasing up,

$$\frac{\partial p}{\partial h} = -\frac{\partial p}{\partial y} = \rho g.$$

17-21

(a) For the minimum area A of ice, let the ice sink 1 ft so that the top surface of the ice is at the water line. The buoyant force must support the weight of car + ice. The ice weighs

$$(0.92)(62.4\ \mathrm{lb/ft^3})(A)(1\ \mathrm{ft})$$

and the car weighs 2500 lb. The buoyant force is the weight of the water displaced and so equals

$$(62.4\ \mathrm{lb/ft^3})(A)(1\ \mathrm{ft}).$$

For equilibrium,

$$62.4A = 2500 + (0.92)(62.4)A,$$

$$A = 501\ \mathrm{ft}^2 \quad \underline{\text{Ans}}.$$

(b) Place the car in the center of the ice, otherwise the ice will tilt and the resulting buoyant force will diminish, since there is less ice in the water and therefore less water displaced.

17-28

(a) Let the cube have volume V and weight W. Then, floating in the mercury alone,

$$W = \tfrac{1}{4}V\rho_m g.$$

In the mercury plus water,

$$fV\rho_m g + (1 - f)V\rho_w g = W = \tfrac{1}{4}V\rho_m g,$$

where f = fraction of volume of cube in mercury, ρ_m = density of mercury = $1.36 \times 10^3\ \mathrm{kg/m^3}$, ρ_w = density of water = $10^3\ \mathrm{kg/m^3}$. Put the numbers into the equation and solve for f to obtain

$$f = 0.19 \quad \underline{\text{Ans}}.$$

(b) Only the volume V and weight W of the object enter the analysis in (a); hence the shape is of no importance and the answer is NO. It is assumed that the object is everywhere convex as seen from the outside.

17-30

(a) The forces acting on the log are its weight W and the buoyant force F(x), x measuring the vertical displacement from equilibrium. If a length ℓ of the log is submerged when in equilibrium, then

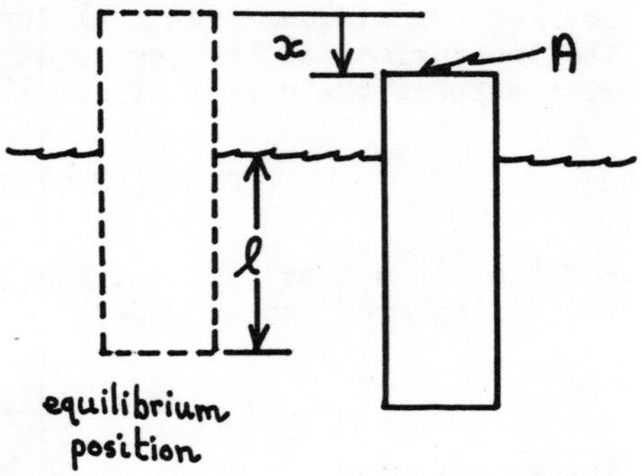

$$W = \rho g A \ell,$$

ρ the density of water and A the cross-sectional area of the log. When the log is displaced down a distance x, x positive downward, Newton's second law gives

$$W - F = ma = \frac{W}{g} a,$$

$$\rho g A \ell - \rho g A(\ell + x) = \frac{\rho g A \ell}{g} \frac{d^2x}{dt^2},$$

$$\frac{d^2x}{dt^2} + \frac{g}{\ell} x = 0.$$

The last equation describes simple harmonic motion (e.g., of the simple pendulum).

(b) The period of oscillation is $T = 2\pi(\frac{\ell}{g})^{\frac{1}{2}} = 3.14$ s Ans.

17-32

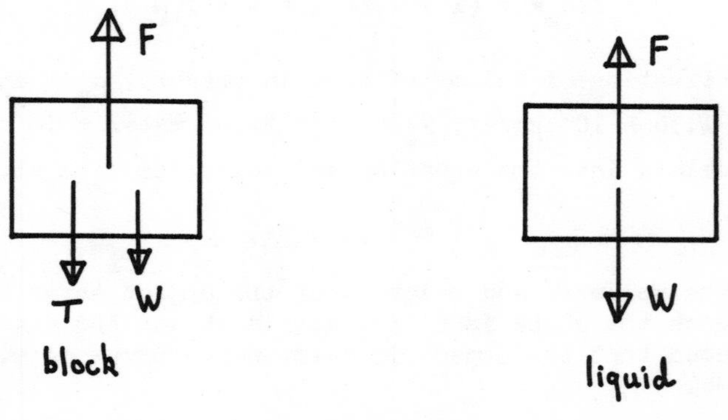

For the block, Newton's second law gives

$$F - W - T = ma = \frac{W}{g}a,$$

$$T = F - W(1 + \frac{a}{g}),$$

m,W the mass and weight of the block, F the buoyant force. Also, Newton's second law applied to an equal volume of liquid yields

$$F - \rho gV = (\rho V)a,$$

V the volume of block and liquid (which is of density ρ). The buoyant force F is the same for block and liquid since the buoyant force depends only on the volume of liquid displaced, and the volumes have been chosen equal above. The previous equation can be rewritten as

$$F = \rho gV(1 + \frac{a}{g}) = F_0(1 + \frac{a}{g}),$$

F_0 the buoyant force for zero acceleration, in which case the equation of motion of the block becomes

$$F_0 - W - T_0 = 0,$$

$$F_0 = W + T_0.$$

Finally, the second, fourth and sixth equations combined show that

$$T = (W + T_0)(1 + \frac{a}{g}) - W(1 + \frac{a}{g}),$$

$$T = T_0(1 + \frac{a}{g}).$$

CHAPTER 18

18-4

The work W done by the pump on a mass m of water is

$$W = mgh + \tfrac{1}{2}mv^2.$$

Thus the power P supplied by the pump is

$$P = \frac{dW}{dt} = \frac{dm}{dt}(gh + \tfrac{1}{2}v^2).$$

But the mass flow rate is

$$\frac{dm}{dt} = Av\rho,$$

so that

$$P = Av\rho(gh + \tfrac{1}{2}v^2),$$

$$P = 65.8\ \text{W} \quad \underline{\text{Ans}}.$$

18-11

Let ρ be the density of air and ρ^* the density of water. By Bernoulli's equation,

$$p + \tfrac{1}{2}\rho v^2 = p_0,$$

the streamlines being horizontal. Thus, the net pressure force on the water column, which must equal the weight of "excess" water for equilibrium, is

$$F = (p_0 - p)A = (hA)\rho^* g.$$

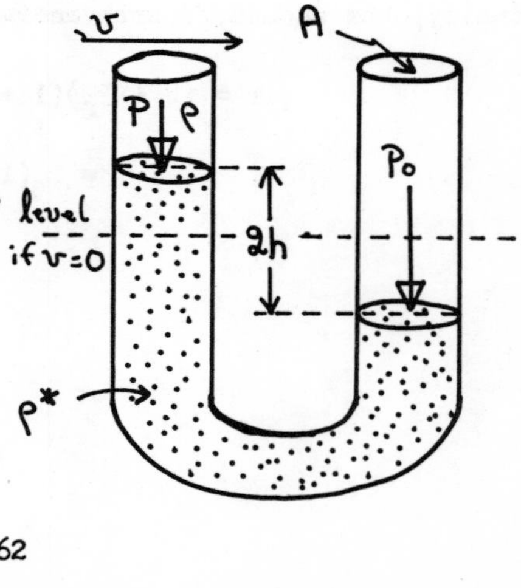

Solving for h,

$$h = \tfrac{1}{2}\rho v^2 A/\rho^* g A,$$

$$h = \frac{(\frac{1}{2})(1.2\ \text{kg/m}^3)(15\ \text{m/s})^2}{(10^3\ \text{kg/m}^3)(9.8\ \text{m/s}^2)} = 0.014\ \text{m} = 1.4\ \text{cm}\quad \underline{\text{Ans}}.$$

<u>18-12</u>

(a) Assume slow removal, so that $v_D = 0$ approximately. If the container is open, $p_D = p_{atm}$. Also, $p_C = p_{atm}$ since the liquid emerges into the atmosphere at C. Thus, Bernoulli's equation applied between points D and C gives

$$p_{atm} = p_{atm} + \tfrac{1}{2}\rho v^2 - \rho g(h_2 + d),$$

v the speed of efflux at C. Hence,

$$v = [2g(h_2 + d)]^{\frac{1}{2}}\quad \underline{\text{Ans}}.$$

(b) If the tube has a uniform cross-sectional area, then $v_B = v_C = v$, by the equation of continuity. Applying Bernoulli's equation between points D and B gives

$$p_{atm} = p_B + \tfrac{1}{2}\rho v^2 + \rho g h_1,$$

$$p_B = p_{atm} - \rho g(h_2 + d + h_1)\quad \underline{\text{Ans}},$$

using (a).

(c) The minimum $p_B = 0$, in which event

$$p_{atm} = \rho g(h_2 + d + h_1).$$

For maximum h_1, arrange things so that $h_2 + d = 0$ to obtain

$$h_{1,max} = p_{atm}/\rho g = 39\ \text{ft} = 10.3\ \text{m}\quad \underline{\text{Ans}}.$$

18-20

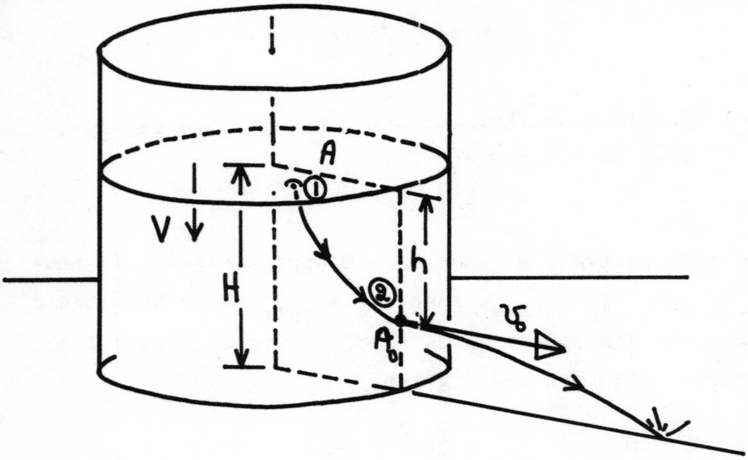

(a) Bernoulli's equation applied at points 1 and 2 gives

$$p_1 + \tfrac{1}{2}\rho v^2 + \rho g H = p_2 + \tfrac{1}{2}\rho v_0^2 + \rho g(H - h),$$

v the speed at which the liquid surface is falling, v_0 the speed of efflux. If the tank is open, the pressures at the surface and at the hole are atmospheric pressure (at the hole by definition of hole); that is, $p_1 = p_2 = p_{atm}$. Hence,

$$\tfrac{1}{2}\rho v^2 = \tfrac{1}{2}\rho v_0^2 - \rho g h,$$

$$v_0^2 = v^2 + 2gh.$$

(b) The equation of continuity states that

$$Av = A_0 v_0.$$

Using this to eliminate v yields

$$v_0^2 = (A_0 v_0/A)^2 + 2gh,$$

$$v_0 = [2gh/(1 - A_0^2/A^2)]^{\frac{1}{2}}.$$

(c) If $A_0 \ll A$, then $A_0^2/A^2 \ll 1$. Using the expansion

$$\frac{1}{(1 - x)^{\frac{1}{2}}} \approx 1 + \tfrac{1}{2}x,$$

with $x = A_0^2/A^2$ gives

$$v_0 \approx (2gh)^{\frac{1}{2}}(1 + A_0^2/2A^2).$$

<u>18-21</u>

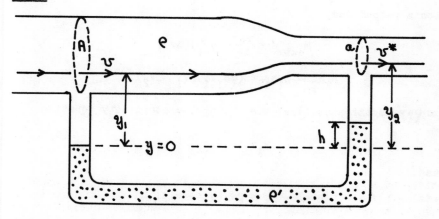

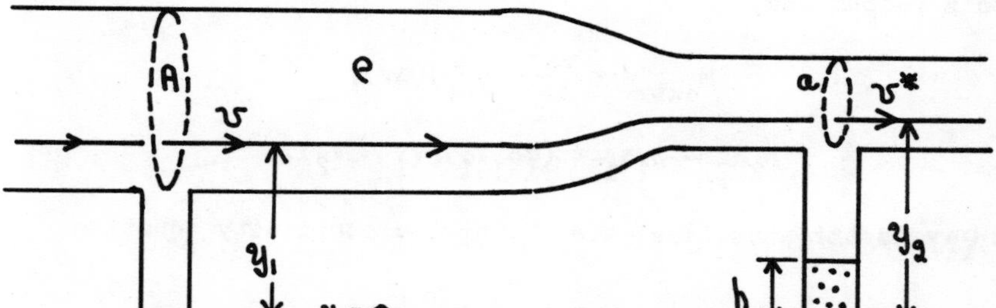

Apply Bernoulli's equation to points 1 and 2 along a streamline:

$$p_1 + \rho g y_1 + \tfrac{1}{2}\rho v^2 = p_2 + \rho g y_2 + \tfrac{1}{2}\rho v^{*2}.$$

By the continuity equation

$$Av = av^*$$

so that

$$p_1 + \rho g y_1 + \tfrac{1}{2}\rho v^2 = p_2 + \rho g y_2 + \tfrac{1}{2}\rho(A^2/a^2)v^2.$$

Equate expressions for the pressures at $y = 0$ in the left and right hand tubes of the U-tube:

$$p_1 + \rho g y_1 = p_2 + \rho g(y_1 - h) + \rho' g h.$$

Solve each of the two preceding equations for $p_1 - p_2 + \rho g(y_1 - y_2)$ and equate the results; then solve for v:

$$v = a\left[\frac{2(\rho' - \rho)gh}{\rho(A^2 - a^2)}\right]^{\frac{1}{2}}.$$

18-24

(a) Since the enlargement is abrupt, the pressure p_1 still acts over an area a_2 in the wide portion close to the enlargement. By Newton's second law,

$$F_{ext} = m\cdot\frac{\Delta v}{\Delta t} = \left(\frac{m}{\Delta t}\right)(\Delta v),$$

$$p_2a_2 - p_1a_1 = (\rho a_1v_1)(v_1 - v_2),$$

since ρav is the mass flow rate. Using the continuity equation

$$v_1a_1 = v_2a_2$$

gives

$$p_2 - p_1 = \rho v_2(v_1 - v_2).$$

(b) If the pipe widened gradually (no turbulence), Bernoulli's equation can be applied; since the pipe is horizontal,

$$p_1 + \tfrac{1}{2}\rho v_1^2 + \rho gy_1 = p_2' + \tfrac{1}{2}\rho v_2^2 + \rho gy_2,$$

$$p_2' - p_2 = \tfrac{1}{2}\rho(v_1^2 - v_2^2).$$

(c) From (a) and (b) the loss in pressure due to the abrupt enlargement is

$$p_2' - p_2 = \tfrac{1}{2}\rho(v_1^2 - v_2^2) - \rho v_2v_1 + \rho v_2^2 = \tfrac{1}{2}\rho(v_1 - v_2)^2 \quad \underline{\text{Ans}}.$$

18-28

(a) Let the outward direction be positive. Newton's second law gives

$$-(p + dp)A + pA = -(\rho A dr)\cdot\frac{v^2}{r},$$

$$\frac{dp}{dr} = \rho\frac{v^2}{r}.$$

(b) By Bernoulli's equation, assuming the same constant for nearby streamlines,

$$p + \tfrac{1}{2}\rho v^2 = (p + dp) + \tfrac{1}{2}\rho(v + dv)^2.$$

Ignoring the products of small quantities (products of differentials), this becomes

$$dp = -\rho v dv.$$

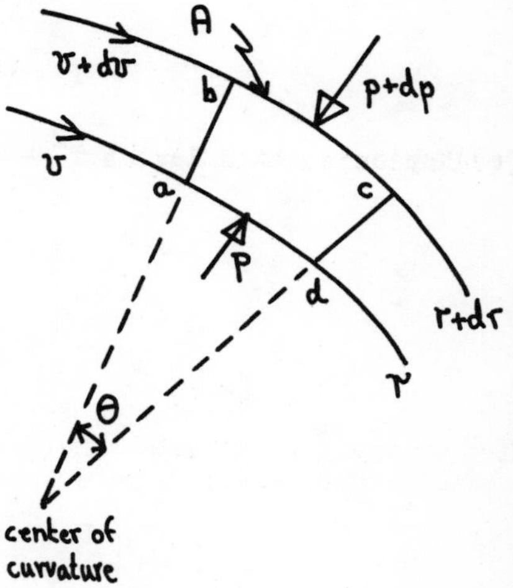

Invoking (a) gives

$$\rho\frac{v^2}{r}\,dr = -\rho v dv,$$

$$\frac{dr}{r} = -\frac{dv}{v}.$$

Integrating,

$$\ln r = -\ln v + C,$$

$$rv = \text{constant}.$$

(c) From Problem 18-26, compute $\int\vec{v}\cdot d\vec{s}$, choosing the path abcd. Note that $\vec{v}$ is perpendicular to $\vec{ds}$ along ab and cd; then

$$\oint\vec{v}\cdot\vec{ds} = \int_b^c\vec{v}\cdot\vec{ds} + \int_d^a\vec{v}\cdot\vec{ds} = (v + dv)(r + dr)\theta - rv\theta.$$

Use the result from (b) to obtain

$$\oint\vec{v}\cdot\vec{ds} = \theta(vr + rdv + vdr - vr) = \theta(rdv + vdr) = 0.$$

In obtaining the last result the product of differentials has been dropped. Since any closed path in the fluid can be constructed from many small paths like those used above, it is concluded that the flow is irrotational.

18-29

(a) From Problem 18-28(b),

$$v = \frac{\text{constant}}{r} = \frac{K}{r}.$$

(b) For circular orbits,

$$v = \frac{2\pi r}{T} = \frac{K}{r},$$

$$T = \left(\frac{2\pi}{K}\right) r^2.$$

(c) Kepler's third law is $T^2 = kr^3$, clearly different from (b)

CHAPTER 19

19-5

(a) At some instant t,

$$kx_1 - \omega t = kx_2 - \omega t + \frac{\pi}{3}.$$

Thus, the distance between the points is

$$x_1 - x_2 = \frac{\pi}{3k} = \frac{\pi}{3(2\pi/\lambda)} = \frac{\lambda}{6}.$$

To find the wavelength λ, set

$$v = \nu\lambda,$$

$$350 = 500\lambda,$$

$$\lambda = 0.70 \text{ m}.$$

Hence,

$$x_1 - x_2 = \frac{0.7}{6} = 0.117 \text{ m} = 11.7 \text{ cm} \quad \underline{\text{Ans}}.$$

(b) In this case,

$$\Delta(\text{phase}) = (kx - \omega t_1) - (kx - \omega t_2),$$

$$\Delta(\text{phase}) = \omega(t_2 - t_1) = 2\pi\nu(t_2 - t_1),$$

$$\Delta(\text{phase}) = 2\pi(500)(0.001) = \pi \text{ rad} \quad \underline{\text{Ans}}.$$

19-14

(a) If T is the tension in the rope,

$$v(y) = [T(y)/\mu]^{\frac{1}{2}}$$

where $\mu = m/L$. Consider a small portion of the rope a distance y from the lower end (which is at y = 0). The tension in this portion is the weight of that part of the rope hanging beneath it;

$$T = (\mu y)g.$$

Therefore,

$$v = \left(\frac{\mu g y}{\mu}\right)^{\frac{1}{2}} = (yg)^{\frac{1}{2}}.$$

(b) Since $v = dy/dt$, the result of (a) can be written as

$$y^{-\frac{1}{2}}dy = g^{\frac{1}{2}}dt.$$

Upon integrating, this last becomes

$$2y^{\frac{1}{2}} = g^{\frac{1}{2}}t + C,$$

C being the constant of integration. To evaluate C, let the wave start at the bottom ($y = 0$) at time $t = 0$;

$$2(0)^{\frac{1}{2}} = g^{\frac{1}{2}}(0) + C,$$

$$C = 0.$$

Hence,

$$2y^{\frac{1}{2}} = g^{\frac{1}{2}}t.$$

If the wave reaches the top ($y = L$) of the rope at time $t = t^*$,

$$2L^{\frac{1}{2}} = g^{\frac{1}{2}}t^*,$$

$$t^* = 2\left(\frac{L}{g}\right)^{\frac{1}{2}}.$$

(c) Provided that the rope has a non-zero mass so that a tension will appear in the rope, the results in (a) and (b) are independent of the numerical value of the mass.

19-18

(a) If the particle displacement is $y = y_m \sin(kx \pm \omega t)$, the particle velocity will be

$$\frac{\partial y}{\partial t} = u = \pm \omega y_m \cos(kx \pm \omega t),$$

$$u_{max} = \omega y_m.$$

(b) The energy of an element of length Δx is

$$\tfrac{1}{2}(\mu\Delta x)\cdot u_m^2 = \tfrac{1}{2}\mu\Delta x\omega^2 y_m^2 = 2\pi^2\mu\Delta x\nu^2 y_m^2.$$

The energy per unit length is just this energy divided by the length Δx of the element:

$$E_1 = 2\pi^2\mu\nu^2 y_m^2.$$

(c) The average power P is

$$P = \frac{\text{energy/unit length}}{\text{time for wave to cross element}},$$

the time in the denominator being $\Delta x/v$, so that

$$P = (2\pi^2\mu\Delta x\nu^2 y_m^2)/(\Delta x/v) = 2\pi^2\mu\nu^2 y_m^2 v = E_1 v.$$

19-19

(a) Assuming no absorption, the power crossing a sphere centered at the source is independent of the sphere's radius. By definition of intensity this means that

$$4\pi r_1^2 I_1 = 4\pi r_2^2 I_2 = \text{constant}.$$

But the intensity I is proportional to the square of the amplitude A of the wave; thus also

$$4\pi r_1^2 A_1^2 = 4\pi r_2^2 A_2^2 = \text{constant},$$

where the constants in the two equations are different. In general (i.e., for any location) then,

$$A = \frac{\text{constant}}{r} = \frac{Y}{r}$$

say. Let the source be at $r = 0$, as implied above. The sinusoidal part of the particle displacement is

$$\sin(kr - \omega t) = \sin k(r - \frac{\omega}{k} t) = \sin k(r - vt),$$

giving

$$y = \frac{Y}{r}\sin k(r - vt),$$

when combined with the previous result.

(b) The sine factor in the above is dimensionless. Since both y and r have dimensions of length, Y must have dimensions of the square of length.

19-24

Waves having the same period have the same frequency. Therefore, with $\theta = kx - \omega t$, the waves can be written as

$$y_1 = A\sin\theta,$$

$$y_2 = \frac{A}{2}\sin(\theta + \frac{\pi}{2}) = \frac{A}{2}\cos\theta,$$

$$y_3 = \frac{A}{3}\sin(\theta + \pi) = -\frac{A}{3}\sin\theta.$$

The resultant $y_T = y_1 + y_2 + y_3$ becomes

$$y_T = \frac{2A}{3}\sin\theta + \frac{A}{2}\cos\theta = \frac{2A}{3}(\sin\theta + \frac{3}{4}\cos\theta),$$

$$y_T = \frac{2A}{3}[(\alpha\sin\beta)\sin\theta + (\alpha\cos\beta)\cos\theta],$$

$$y_T = \frac{2\alpha A}{3}\cos(\theta - \beta).$$

This requires that

$$\alpha\sin\beta = 1,$$

$$\alpha\cos\beta = \frac{3}{4}.$$

and therefore,

$$\alpha = \frac{5}{4};\ \beta = \tan^{-1}(\frac{4}{3}).$$

Thus the amplitude of y_T is $2\alpha A/3 = 5A/6$. At $\theta = 0$, $y_T = \frac{5A}{6}\cos(-\beta) = \frac{5A}{6}\cos\beta = \frac{5A}{6}\cos[\tan^{-1}(4/3)] = \frac{1}{2}A$. The resultant wave,

$$y_T = \frac{5A}{6}\cos[\theta - \tan^{-1}(\frac{4}{3})]$$

is sketched on the following page.

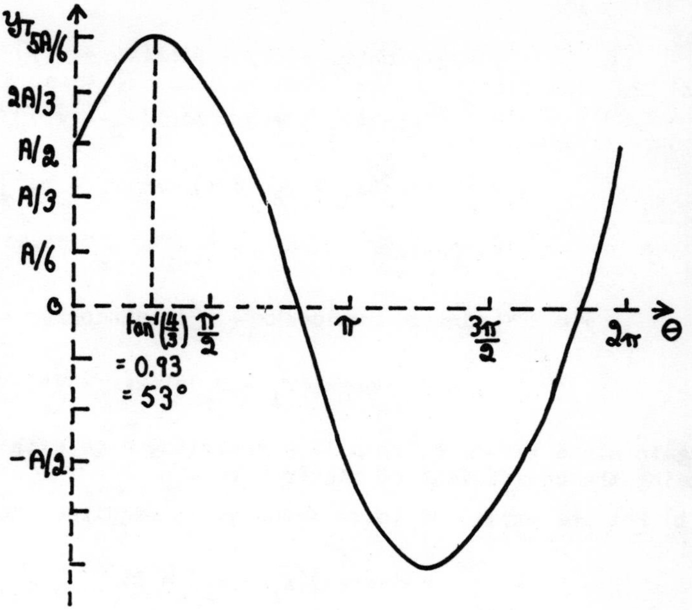

<u>19-25</u>

(a) The two spherical waves are

$$y_1 = \frac{Y}{r_1}\sin k(r_1 - vt), \; y_2 = \frac{Y}{r_2}\sin k(r_2 - vt).$$

Now let

$$r_1 = r - x = r(1 - \frac{x}{r}),$$

$$r_2 = r + x = r(1 + \frac{x}{r}),$$

so that

$$r = \tfrac{1}{2}(r_2 + r_1), \; x = \tfrac{1}{2}(r_2 - r_1).$$

If $r_1 \approx r_2$, then $x/r \ll 1$. Under these conditions, if $y = y_1 + y_2$,

$$y \approx \frac{Y}{r}[(1 + \frac{x}{r})\sin k(r_1 - vt) + (1 - \frac{x}{r})\sin k(r_2 - vt)],$$

$$y = \frac{Y}{r}[\sin k(r_1 - vt) + \sin k(r_2 - vt)]$$

$$+ \frac{Y}{r}\cdot\frac{x}{r}[\sin k(r_1 - vt) - \sin k(r_2 - vt)],$$

$$y = \frac{Y}{r}[2\sin\frac{k}{2}(r_1 + r_2 - 2vt)\cos\frac{k}{2}(r_1 - r_2)]$$

$$+ \frac{Y}{r}\cdot\frac{x}{r}[2\sin\frac{k}{2}(r_1 - r_2)\cos\frac{k}{2}(r_1 + r_2 - 2vt)],$$

$$y = 2\,\frac{Y}{r}[\sin k(r - vt)\cos kx - (\frac{x}{r})\sin kx\cos k(r - vt)],$$

$$y \approx 2\,\frac{Y}{r}\,\cos k\tfrac{1}{2}(r_1 - r_2)\sin k(r - vt)$$

again since $x/r \ll 1$. This is a traveling wave with the amplitude being the coefficient of $\sin k(r - vt)$.

(b) For the amplitude to be zero, it is required that

$$\cos k\tfrac{1}{2}(r_1 - r_2) = 0.$$

But $k = 2\pi/\lambda$, so this becomes

$$\frac{\pi}{\lambda}(r_1 - r_2) = (n + \tfrac{1}{2})\pi, \quad n = 0, 1, \ldots,$$

$$r_1 - r_2 = (n + \tfrac{1}{2})\lambda.$$

For total reinforcement,

$$\cos k\tfrac{1}{2}(r_1 - r_2) = 1,$$

$$\frac{\pi}{\lambda}(r_1 - r_2) = n\pi,$$

$$r_1 - r_2 = n\lambda,$$

where again n is zero or a positive integer.

<u>19-27</u>

(a) If $y_1 = y_m\sin(kx - \omega t)$ and $y_2 = y_m\sin(kx + \omega t)$, then the sum $y = y_1 + y_2$ is given by

$$y = 2y_m\sin(kx)\cos(\omega t) = 0.5\sin(\frac{\pi x}{3})\cos(40\pi t).$$

Therefore the amplitudes are $y_m = \frac{1}{2}(0.5) = 0.25$ cm. The wave speed v is

$$v = \frac{\omega}{k} = \frac{40\pi}{\pi/3} = 120 \text{ cm/s} \quad \underline{\text{Ans}}.$$

(b) The desired distance is $\frac{1}{2}\lambda$. But

$$\lambda = \frac{2\pi}{k} = \frac{2\pi}{\pi/3} = 6 \text{ cm},$$

and therefore the distance between adjacent nodes is 3 cm.

(c) The particle velocity u is

$$u = \frac{\partial y}{\partial t} = -20\pi\sin(\frac{\pi x}{3})\sin(40\pi t),$$

$$u = -20\pi\sin(\frac{\pi}{2})\sin(45\pi) = -20\pi(+1)(0) = 0 \quad \underline{\text{Ans}}.$$

<u>19-31</u>

Associated with a length dx is kinetic energy dK and potential energy dU. The first of these is

$$dK = \frac{1}{2}(dm)u^2 = \frac{1}{2}(\mu dx)u^2$$

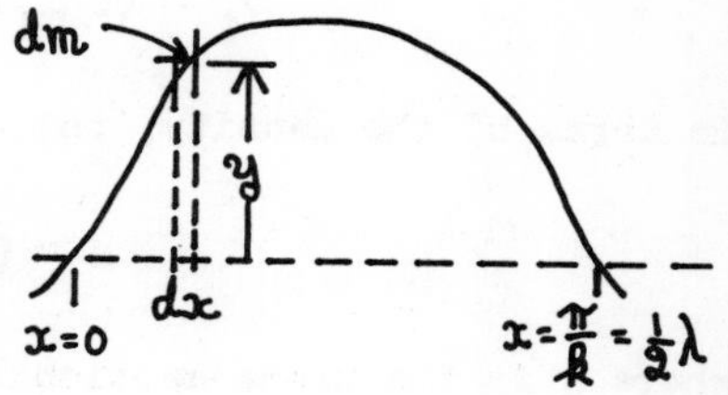

where u is the speed of the element. From the equation of a standing wave this speed is

$$u = \frac{\partial y}{\partial t} = -2y_m\omega\sin(kx)\sin(\omega t)$$

so that

$$dK = 2\mu\omega^2y_m^2\sin^2kx\cdot\sin^2\omega t\ dx.$$

The potential energy is that of a simple harmonic oscillator of mass dm; if 'a' is the force constant,

$$dU = \frac{1}{2}ay^2$$

where

$$a = (dm)\omega^2 = \mu\omega^2 dx.$$

Hence,

$$dU = \tfrac{1}{2}\mu\omega^2 y^2 dx = 2\mu\omega^2 y_m^2 \sin^2 kx \cdot \cos^2 \omega t dx.$$

Adding dU and dK to get the total energy dE,

$$dE = 2\mu\omega^2 y_m^2 \sin^2 kx dx,$$

and integrating over one loop, substituting $\omega = vk = 2\pi\nu$,

$$E = 2\mu\omega^2 y_m^2 \int_0^{\pi/k} \sin^2 kx \, dx = 2\pi^2 \mu y_m^2 \nu v \quad \underline{\text{Ans}}.$$

<u>19-34</u>

(a) From the tensions T in the wires,

$$T = mg = (10 \text{ kg})(9.8 \text{ m/s}^2) = 98 \text{ N},$$

the wave speeds in the two wires can be found from

$$v_1 = (T/\mu_1)^{\frac{1}{2}}, \ v_2 = (T/\mu_2)^{\frac{1}{2}}.$$

In terms of the density, the mass per unit length is

$$\mu = \rho A\ell/\ell = \rho A,$$

where A is the cross-sectional area of the wires and ρ is the density of the wire. Numerically,

$$\mu_1 = (2600 \text{ kg/m}^3)(10^{-6} \text{ m}^2) = 2.6 \text{ X } 10^{-3} \text{ kg/m},$$

$$\mu_2 = 7.8 \text{ X } 10^{-3} \text{ kg/m}.$$

These give for the wave speeds v_1 = 194.1 m/s and v_2 = 112.1 m/s. The distance between adjacent nodes is $\lambda/2$ so that if it is required that the joint be a node,

$$n_1\lambda_1/2 = \ell_1,\ n_2\lambda_2/2 = \ell_2.$$

It is given that $\ell_1 = 0.6$ m and $\ell_2 = 0.866$ m and therefore, since $v = \lambda\nu$,

$$n_1 = 0.00618\nu_1,\ n_2 = 0.0155\nu_2.$$

It is found by trial and error that the smallest integers n_1 and n_2 giving $\nu_1 = \nu_2$ are $n_1 = 2$ and $n_2 = 5$. The frequency is $2/0.00618 = 5/0.0155 = 323$ Hz <u>Ans</u>.

(b) There are $5 + 2 = 7$ loops or 8 nodes in all. If the two at the ends are not counted, the number of nodes is $8 - 2 = 6$ <u>Ans</u>.

20-9

The time t_1 required for a stone to fall to the bottom of a well of depth d is

$$t_1 = (2d/g)^{\frac{1}{2}},$$

and the time t_2 needed for sound, traveling at speed v, to cover this same distance d is

$$t_2 = d/v.$$

The total time t that elapses between dropping the stone and hearing the splash is $t_1 + t_2$ given by

$$t = (2d/g)^{\frac{1}{2}} + d/v.$$

Rearranging and squaring the above equation gives a quadratic equation for d in terms of t:

$$gd^2 - d[(2v)(gt + v)] + (vt)^2 g = 0,$$

$$d = v[(\frac{v}{g} + t) \pm \sqrt{(\frac{v}{g})(\frac{v}{g} + 2t)}].$$

The negative sign is appropriate since t = 0 should indicate d = 0. With $g = 9.8\ m/s^2$, v = 331 m/s, a time of t = 3 s gives d = 40.6 m.

20-12

(a) The path difference between the two waves going via SBD in the two positions must be half a wavelength:

$$\tfrac{1}{2}\lambda = 2(1.65\ cm),$$

$$\lambda = 0.066\ m.$$

The frequency ν must be

$$\nu = \frac{v}{\lambda} = \frac{331\ \text{m/s}}{0.066\ \text{m}} = 5015\ \text{Hz}\quad \underline{\text{Ans}}.$$

(b) Let A be the amplitude of the wave when at D that went via route SAD and B the amplitude at D of the wave that traversed SBD in each of the two positions. Since the intensity of a wave is proportional to the square of the resultant amplitude,

$$(A + B)^2 = 900,$$

$$(A - B)^2 = 100.$$

Thus,

$$\frac{B}{A} = 0.5\quad \underline{\text{Ans}}.$$

(c) The waves going by SAD and SBD travel different distances and therefore lose different amounts of energy by, for example, gas friction with the walls of the tubes.

<u>20-13</u>

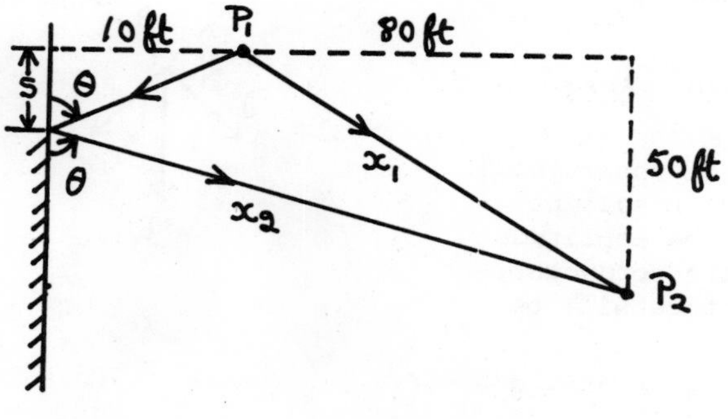

Assuming that the angle of incidence equals the angle of reflection,

$$\tan\theta = \frac{90}{50 - s} = \frac{10}{s},$$

$$s = 5\ \text{ft}.$$

The total path lengths are

$$x_1 = (80^2 + 50^2)^{\frac{1}{2}} = 94.34\ \text{ft},$$

$$x_2 = (10^2 + 5^2)^{\frac{1}{2}} + (90^2 + 45^2)^{\frac{1}{2}} = 111.80 \text{ ft.}$$

The path difference is 17.46 ft. For constructive interference this must be an integral number of wavelengths, there being no phase change on reflection; that is,

$$n\lambda = 17.46 \text{ ft.}$$

(i) $n = 1$: $\lambda = 17.46$ ft; frequency $\nu = v/\lambda = (1100 \text{ ft/s})/(17.46 \text{ ft}) = 63$ Hz.

(ii) $n = 2$: $\lambda = 8.73$ ft; $\nu = 126$ Hz.

20-19

(a) The intensity of either wave at a distance of 5 m is

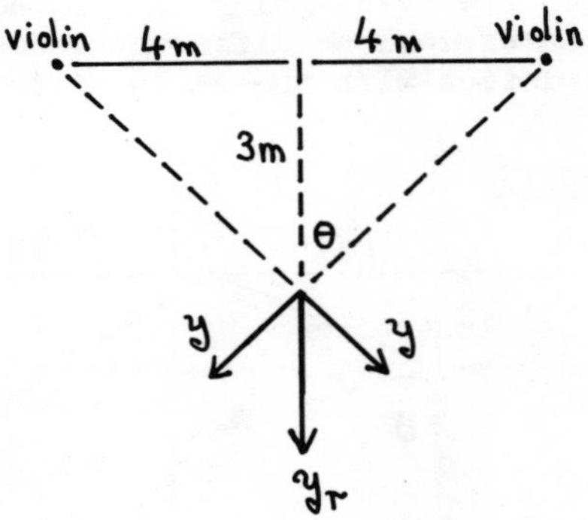

$$I = \frac{S}{4\pi r^2} = \frac{10^{-4} \text{ W}}{4\pi(5 \text{ m})^2},$$

$$I = 3.2 \times 10^{-7} \text{ W/m}^2 \quad \underline{\text{Ans}}.$$

(b) The intensity is proportional to the square of the resultant amplitude. If y is the amplitude of either wave at the conductor, the resultant amplitude will be

$$y_r = 2y\cos\theta = 2y(\tfrac{3}{5}) = \tfrac{6}{5}\, y.$$

Squaring gives for the resultant intensity

$$I_r = \frac{36}{25} I = 4.6 \times 10^{-7} \text{ W/m}^2 \quad \underline{\text{Ans}}.$$

It has been assumed that the violinists are playing in phase. Also, diffraction of the sound waves and the complex receiving characteristics of the human ear have been ignored.

<u>20-26</u>

(a) In the fundamental mode,

$$L = \lambda_0/2 = v/2\nu_0.$$

If ℓ is the length by which the string is shortened,

$$L - \ell = v(2\cdot r\nu_0).$$

Eliminating ν_0 between these equations gives

$$\ell = L(1 - \frac{1}{r}) \quad \underline{\text{Ans}}.$$

(b) For L = 80 cm, successive substitution of r = 6/5, 5/4, 4/3, 3/2 gives ℓ = 13.3, 16, 20, 26.7 cm.

<u>20-30</u>

(a) For spherically symmetric pulsations, the center of the star must be a displacement node.

(b) In the fundamental mode of oscillation, antinodes exist at the surface and nowhere else. Since the distance between antinodes is one-half the wavelength and $\lambda = v/\nu$, v the sound speed,

$$\tfrac{1}{2}v/\nu = 2R,$$

$$\frac{1}{\nu} = T = 4R/v \quad \underline{\text{Ans}}.$$

(c) The speed of sound is

$$v = (\gamma P/\rho)^{\frac{1}{2}} = [(\tfrac{4}{3})(10^{22})/(10^{10})]^{\frac{1}{2}} = 1.1547 \times 10^6 \text{ m/s}.$$

The radius R of the star is $(0.009)(7 \times 10^8 \text{ m}) = 6.3 \times 10^6$ m. Thus,

$$T = \frac{(4)(6.3 \times 10^6)}{1.1547 \times 10^6} = 22 \text{ s} \quad \underline{\text{Ans}}.$$

20-32

Possible frequencies are

$$\frac{v}{2L}, \frac{v}{L}, \frac{3v}{2L} \text{ etc.,}$$

among which are 880, 1320 Hz as consecutive frequencies. But 880/1320 = 2/3 and therefore

$$880 = \frac{v}{L}, \ 1320 = \frac{3}{2}\frac{v}{L}.$$

Either of these gives

$$v = 880L = (880)(0.316) = 278 \text{ m/s}.$$

But

$$v = (T/\mu)^{\frac{1}{2}}$$

and μ = 0.00065 kg/m so that

$$T = \mu v^2 = (6.5 \times 10^{-4})(278)^2 = 50 \text{ N} \quad \underline{\text{Ans}}.$$

20-33

In the fundamental mode $\lambda = 2L = v/\nu$ and therefore

$$\nu = \frac{1}{2L}(T/\mu)^{\frac{1}{2}}.$$

To obtain six beats change the frequency of one wire by 6 Hz. Take differentials of the equation above and then divide the result by the equation itself to obtain

$$\frac{\Delta\nu}{\nu} = \frac{1}{2}\frac{\Delta T}{T}.$$

Since $\Delta\nu/\nu = 0.01$, $\Delta T/T = 0.02 = 2\%$.

20-38

(a) Let u be the speed of the jet. If θ is the half-angle of the cone,

$$\sin\theta = v/u = v/1.5v = 2/3,$$

$$\theta = 42^\circ \quad \underline{\text{Ans}}.$$

(b) The shock travels with the plane at a speed (1.5)(331 m/s) = 496.5 m/s. It must cover a distance $L = h\cot\theta$ = (5000 m)·(1.118) = 5590 m to reach the ground observer. Clearly it will take a time t = (5590 m)/(496.5 m/s) = 11 s Ans.

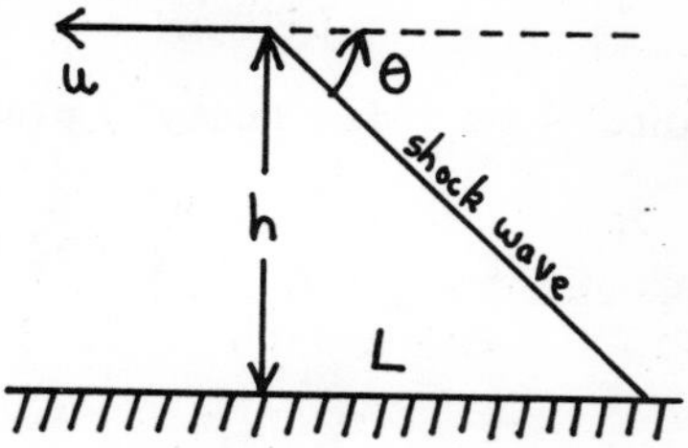

20-41

The frequency of each bleep as it strikes the cave wall is

$$\nu' = \nu\left(\frac{v}{v - v_s}\right),$$

where ν is the frequency of the bleep as produced by the bat. Also, ν' is the frequency of the reflected wave. The bat now "sees" sound of frequency ν' emitted from a source at which she is approaching at speed v_0. Hence, the frequency of this reflected wave as perceived by the approaching bat is

$$\nu'' = \nu'\left(\frac{v + v_0}{v}\right).$$

Eliminating ν' and noting that $v_0 = v_s$,

$$\nu'' = \nu\left(\frac{v + v_s}{v - v_s}\right) = \nu\left(\frac{1 + v_s/v}{1 - v_s/v}\right).$$

The ratio $v_s/v = 1/40$; ν = 39,000 Hz so that

$$\nu'' = (39{,}000)\frac{41/40}{39/40} = 41{,}000 \text{ Hz} \quad \text{Ans.}$$

20-44

Since the plane sees an approaching source the frequency ν' at which it receives the signal is

$$\nu' = \nu\left(1 + \frac{u}{c}\right)$$

where ν is the rest frequency of the source. The receiver also sees an approaching source (the plane) and it receives the approaching reflected microwaves as at a frequency ν'' given by

$$\nu'' = \nu'(1 + \frac{u}{c}) = \nu(1 + \frac{u}{c})^2.$$

Since $\frac{u}{c} \ll 1$ for today's planes,

$$\nu'' \approx \nu(1 + 2\,\frac{u}{c}),$$

$$\frac{\nu'' - \nu}{\nu} = 2\,\frac{u}{c}.$$

But $\nu = c/\lambda$ and therefore

$$u = \frac{\lambda}{2}(\nu'' - \nu) = \tfrac{1}{2}(0.1\text{ m})(990\text{ s}^{-1}) = 49.5\text{ m/s} \quad \underline{\text{Ans}}.$$

20-47

(a) The frequency the uncle hears is

$$\nu' = \nu(\frac{v}{v + v_s}) = (500)\frac{331}{331 + 10} = 485\text{ Hz} \quad \underline{\text{Ans}}.$$

(b) Since the relative velocity between the girl and the train is zero, she hears the rest frequency of the whistle; to wit, 500 Hz.

(c) The velocities relative to the air enter into the Doppler shift equations. The train moves at 20 m/s and the uncle 10 m/s relative to the air, both to the east (same sign). Since they are moving apart,

$$\nu' = (500)\frac{331 + 10}{331 + 20} = 486\text{ Hz} \quad \underline{\text{Ans}}.$$

(d) The girl is still at rest relative to the train and hears 500 Hz, as before.

CHAPTER 21

21-2

(a) The cooling of an object can occur by radiation, conduction, convection, evaporation etc., and these will be affected by the nature of the material, its surface area and temperature, the orientation of the object, the temperature of the surroundings, the presence or absence of air currents, etc. The units in the equation

$$\frac{d\Delta T}{dt} = -K(\Delta T)$$

are

$$\text{kelvins/seconds} = K(\text{kelvins});$$

therefore, the units of K must be $(\text{seconds})^{-1}$, or dimensions of 1/time.

(b) Rearrange Newton's law of cooling to read

$$\frac{d\Delta T}{\Delta T} = -K\,dt,$$

and integrate to get

$$\Delta T = Ce^{-Kt}$$

where C is a constant (i.e., independent of T and t). To evaluate C, set $\Delta T(t = 0) = \Delta T_0$. This gives $\Delta T_0 = C$. Hence,

$$\Delta T = \Delta T_0 e^{-Kt}.$$

21-14

The area of the plate is $A = ab$ before heating. After the rise in temperature, the area will be

$$A + \Delta A = (a + \Delta a)(b + \Delta b).$$

This may be written

$$A + \Delta A = a(1 + \frac{\Delta a}{a})\cdot b(1 + \frac{\Delta b}{b}) = ab(1 + \frac{\Delta a}{a} + \frac{\Delta b}{b} + \frac{\Delta a\Delta b}{ab}).$$

If $\Delta a/a$, $\Delta b/b \ll 1$, then the last term above is an order of magnitude smaller than either $\Delta a/a$ or $\Delta b/b$, and two orders of magnitude smaller than 1. Under these conditions, the last term can be dropped to give

$$A + \Delta A \approx ab(1 + \frac{\Delta a}{a} + \frac{\Delta b}{b}).$$

But $\Delta a = \alpha a\Delta T$ and $\Delta b = \alpha b\Delta T$ so that

$$A + \Delta A = A(1 + 2\alpha\Delta T),$$

$$\Delta A = 2\alpha A\Delta T \quad \underline{\text{Ans}}.$$

21-22

The volume of mercury increases by $\beta V_0 t$ and the volume of the bulb increases by $3\alpha V_0 t$ approximately. Hence, a volume

$$(\beta - 3\alpha)V_0 t$$

of mercury must leave the bulb. This enters the capillary, and if it reaches a height ℓ it will occupy a volume $A_0\ell$. Therefore

$$A_0\ell = (\beta - 3\alpha)V_0 t,$$

$$\ell = \frac{V_0}{A_0}(\beta - 3\alpha)t \quad \underline{\text{Ans}}.$$

21-26

(a) The angular momentum L is

$$L = I\omega = \tfrac{1}{2}MR^2\omega = \tfrac{1}{2}(0.5)(0.03)^2(60) = 0.0135\ \text{kg}\cdot\text{m}^2/\text{s} \quad \underline{\text{Ans}}.$$

From the work-energy theorem,

$$W = \Delta K = \tfrac{1}{2}I\omega^2 = \tfrac{1}{2}(\tfrac{1}{2}MR^2)\omega^2 = 0.405\ \text{J} \quad \underline{\text{Ans}}.$$

Assume that no torques act on the cylinder during heating. Then,

$$\Delta L = 0 \approx I\Delta\omega + \omega\Delta I$$

so that

$$\frac{\Delta\omega}{\omega} = -\frac{\Delta I}{I} = -2\alpha\Delta T$$

by Problem 21-25. Hence,

(b) $$\frac{\Delta\omega}{\omega} = -2(2 \times 10^{-5}\ /C°)(80\ C°) = -0.32\%\quad \underline{\text{Ans}};$$

(c) $$\frac{\Delta L}{L} = 0.00\%\quad \underline{\text{Ans}}.$$

(d) The fractional change in kinetic energy is given by

$$K = \tfrac{1}{2}I\omega^2,$$

$$\Delta K \approx \frac{\partial K}{\partial I}\Delta I + \frac{\partial K}{\partial \omega}\Delta\omega = \tfrac{1}{2}\omega^2\Delta I + I\omega\Delta\omega,$$

$$\frac{\Delta K}{K} = \frac{\Delta I}{I} + 2\,\frac{\Delta\omega}{\omega} = \frac{\Delta\omega}{\omega} = -0.32\%\quad \underline{\text{Ans}}.$$

<u>21-28</u>

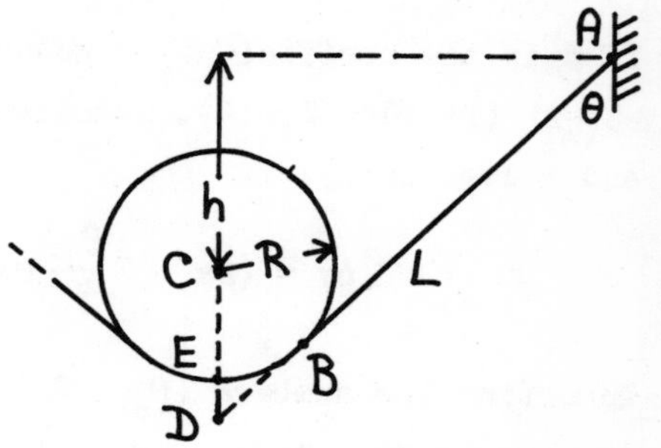

If h be the distance of C below the points of support, it is required that $\Delta h = 0$. Now 2L is the length of the belt. Since $L = ABE \approx ABD$, it follows that

$$h \approx L\cos\theta - R,$$

$$\Delta h = (\Delta L)\cos\theta - \Delta R = 0,$$

$$(\alpha_s L\Delta T)\cos\theta = \alpha_{al}R\Delta T,$$

$$R = (\alpha_s/\alpha_{al})L\cos\theta = \frac{11 \times 10^{-6}\ /C°}{23 \times 10^{-6}\ /C°}(2.5\ m)(\cos 50°),$$

$$R = 0.77\ m = 77\ cm\quad \underline{\text{Ans}}.$$

21-30

Let x be the length of each side of the cube, and let the cube be submerged to a depth y. The weight of the cube is $x^3\rho_a g$. Before heating, the bouyant force is $x^2 y\rho_m g$ and therefore

$$x^2 y\rho_m g = x^3\rho_a g,$$

$$y\rho_m = x\rho_a.$$

All quantities in the last equation change as the temperature rises, so that

$$y\Delta\rho_m + \rho_m\Delta y = x\Delta\rho_a + \rho_a\Delta x.$$

But the mass $M = x^3\rho_a$ of the cube remains constant:

$$\Delta M = 0 = 3x^2\rho_a\Delta x + x^3\Delta\rho_a,$$

$$x\Delta\rho_a = -3\rho_a\Delta x.$$

Therefore

$$y\Delta\rho_m + \rho_m\Delta y = -2\rho_a\Delta x.$$

If β is the coefficient of volume expansion of mercury, then $\Delta\rho_m = -\beta\rho_m\Delta T$ (Problem 21-19). Substitute this into the equation above and solve for Δy:

$$\Delta y = (\beta y - 2\frac{\rho_a}{\rho_m}\alpha x)\Delta T = (\beta - 2\alpha)x\frac{\rho_a}{\rho_m}\Delta T.$$

Inserting the numbers ($\rho_a = 2.7$ g/cm^3, $\rho_m = 13.6$ g/cm^3) gives $\Delta y = 0.266$ mm Ans.

21-31

(a) Let 2y be the horizontal span and x the sag of the cable. For a parabola, the length L of the cable is

$$L = (4x^2 + y^2)^{\frac{1}{2}} + \frac{y^2}{2x}\ln[\frac{2x + (4x^2 + y^2)^{\frac{1}{2}}}{y}].$$

The half-span $y = 4200/2 = 2100$ ft and is assumed to remain unchanged. At 50°F, $x = 470$ ft, so that the length of the cable at 50°F is given by the equation above with $y = 2100$, $x = 470$. This turns out to be $L = 4336$ ft. Hence,

$$\Delta L = \alpha L \Delta T = (6.5 \times 10^{-6})(4336)(130),$$

$$\Delta L = 3.66 \text{ ft} \quad \underline{\text{Ans.}}$$

(b) The change in the sag x cannot be computed from $\Delta x = \alpha x \Delta T$ because of the constraint on y. Instead, write

$$\Delta x = \frac{\Delta x}{\Delta L}(\Delta L) \approx \frac{\Delta L}{dL/dx}.$$

Use the first equation to compute dL/dx:

$$\frac{dL}{dx} = \frac{4x}{(4x^2 + y^2)^{\frac{1}{2}}} - \frac{y^2}{2x^2}\cdot\ln[\ ] + \frac{y^2}{2x}\,\frac{2 + 4x(4x^2 + y^2)^{-\frac{1}{2}}}{2x + (4x^2 + y^2)^{\frac{1}{2}}},$$

where

$$[\ ] = \frac{2x + (4x^2 + y^2)^{\frac{1}{2}}}{y}.$$

Numerically $dL/dx = 0.56436$ so that

$$\Delta x = 3.66/0.56436 = 6.5 \text{ ft} \quad \underline{\text{Ans.}}$$

<u>21-32</u>

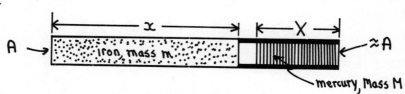

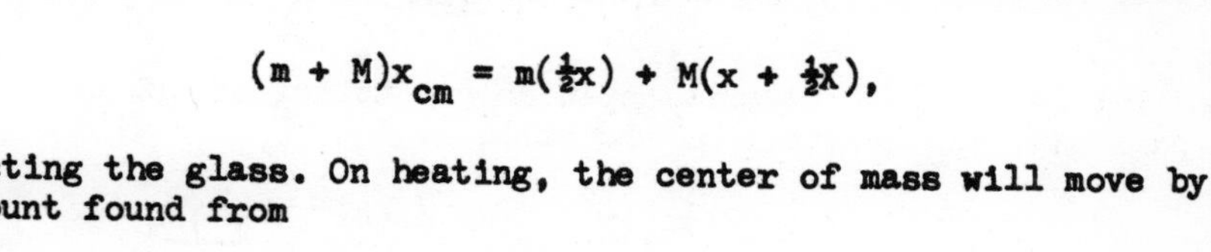

Relative to the left end of the tube, the center of mass is located by

$$(m + M)x_{cm} = m(\tfrac{1}{2}x) + M(x + \tfrac{1}{2}X),$$

neglecting the glass. On heating, the center of mass will move by an amount found from

$$(m + M)\Delta x_{cm} = m(\tfrac{1}{2}\Delta x) + M(\Delta x - \tfrac{1}{2}\Delta X),$$

$$(m + 2M)\Delta x - M\Delta X = 0,$$

if $\Delta x_{cm} = 0$. The change in length of the iron is

$$\Delta x = \alpha x \Delta T.$$

The mercury is contained in glass which also expands on heating. However, $\beta \gg 3\alpha_g$, so that the expansion of the glass can be ignored. This leaves a change in the length of the glass tube that did not originally contain mercury; this also will be neglected. The change in volume of the mercury is

$$\Delta V = \beta V \Delta T.$$

But $V = AX$ so that

$$\Delta(AX) = \beta(AX)\Delta T,$$

$$\Delta X = \beta X \Delta T.$$

Substituting ΔX and Δx gives

$$(m + 2M)(\alpha x) - M(\beta X) = 0,$$

the ΔT cancelling. But $M = XA\rho$, $m = xA\rho^*$ so that

$$(\beta\rho)X^2 - (2\rho\alpha x)X - (x^2\rho^*\alpha) = 0.$$

Putting in $\rho = 13.6 \times 10^3$ kg/m^3, etc., gives

$$18X^2 - 2.4X - 0.6944 = 0,$$

$$X = 0.274 \text{ m} \quad \underline{\text{Ans}}.$$

CHAPTER 22

22-4

Heat exchanges with the glass itself are ignored. The heat that is required to raise the temperature of the ice from -15°C to 0°C is

$$Q_1 = mc\Delta T = (100\ g)(\tfrac{1}{2}\ cal/g\cdot C°)(15\ C°) = 750\ cal.$$

Melting the ice at 0°C requires

$$Q_2 = (100\ g)(80\ cal/g) = 8000\ cal,$$

of heat. Thus, 8750 cal are needed to convert 100 g of ice at -15°C to liquid water at 0°C. However, if the 200 g of water originally in the glass at 25°C were cooled to 0°C, they would liberate

$$Q_3 = (200\ g)(1\ cal/g\cdot C°)(25\ C°) = 5000\ cal$$

of heat. Now this is greater than Q_1 but less than $Q_1 + Q_2$. Hence, the equilibrium temperature is 0°C, and not all of the ice melts.

22-6

The sphere, mass M, will pass through the ring, mass m, when the sphere's diameter D equals the inner diameter d of the ring. Let T be the equilibrium temperature; T may be found from

$$mc_c(T) = Mc_a(100 - T),$$

$$\frac{M}{m} = \frac{c_cT}{c_a(100 - T)},$$

$$\frac{M}{m} = \frac{0.0923T}{21.5 - 0.215T},$$

after substituting numerical values for the specific heat c_c of copper and c_a of aluminum. The diameters of the ring and sphere at temperature T will be, with d, D the initial diameters,

$$d + \Delta d = d + \alpha_c d(T) = d(1 + \alpha_c T),$$

$$d + \Delta d = (1.0)(1 + 17 \times 10^{-6}T);$$

$$D + \Delta D = D[1 - \alpha_a(100 - T)],$$

$$D + \Delta D = (1.002)[1 - 23 \times 10^{-6}(100 - T)].$$

Setting $d + \Delta d = D + \Delta D$, so that the sphere just passes through the ring, gives

$$1 + 17 \times 10^{-6}T = 1.002 - 23.046 \times 10^{-4} + 23.046 \times 10^{-6}T,$$

$$T = 50.38^\circ C.$$

Substituting this into the third equation, for M/m, yields

$$\frac{M}{m} = 0.436 \quad \underline{\text{Ans}}.$$

22-15

Consider a section of ice with a cross-sectional area of 1 cm^2. Since its thickness is 5 cm, the rate dQ/dt at which heat flows through it to the outside is

$$\frac{dQ}{dt} = kA\frac{dT}{dx} = (0.004)(1)\,\frac{0 - (-10)}{5} = 0.008 \text{ cal/s}.$$

Therefore it takes (80 cal/g)/(0.008 cal/s) = 10,000 seconds to freeze 1 g of water. In 10,000 s, then, the layer of ice will grow by x cm, where x is found from

$$(0.92 \text{ g/cm}^3)(1 \text{ cm}^2)(x \text{ cm}) = 1 \text{ g},$$

$$x = 1.087 \text{ cm},$$

each 10,000 s (since 1 cm^3 of water has a mass of 1 g). Hence, the hourly growth rate dx/dt is

$$\frac{dx}{dt} = \frac{3600 \text{ s/h}}{10{,}000 \text{ s}}(1.087 \text{ cm}) = 0.39 \text{ cm/h} \quad \underline{\text{Ans}}.$$

22-16

Let $T_1 > T_2$ with $r_1 < r_2$. In a steady state

$$\frac{dQ}{dt} = -kA\frac{dT}{dr}$$

will be a constant, H say, for all r. With $A = 4\pi r^2$,

$$H = -k(4\pi r^2)\frac{dT}{dr},$$

$$H\int_{r_1}^{r_2} dr/r^2 = -4\pi k\int_{T_1}^{T_2} dT,$$

$$H = \frac{dQ}{dt} = 4\pi k(T_1 - T_2)\frac{r_1 r_2}{r_2 - r_1}.$$

22-25

The work W done by friction is

$$W = \Delta K = 0 - \tfrac{1}{2}m_i v_i^2 = -\tfrac{1}{2}(50)(5.38)^2 = -723.61\ J.$$

This implies that there is available 723.61 J = 172.86 cal of energy to melt ice. Since it takes 80 cal to melt 1 g of ice, the mass of ice melted will be (172.86 cal)/(80 cal/g) = 2.16 g Ans.

22-30

In one-half liter there are 500 g = 0.5 kg of water. To begin boiling, this water must be brought to 100°C from 59°F = 15°C, that is, heated by 85 C°. The heat required is

$$Q = mc\Delta T = (500\ g)(1\ cal/g\cdot C^\circ)(85\ C^\circ) = 42{,}500\ cal.$$

If any additional heat is supplied, the water will begin to boil. Each shake produces

$$E = mg\Delta h = (0.5)(9.8)(0.3048) = 1.49352\ J$$

of energy, or 0.3568 cal of heat. Therefore 42,500/0.3568 = 119,114 shakes are needed. At the rate of 30 shakes/min = 0.5 shakes/s, this will take 238,228 s or 238,228/86,400 = 2.76 days.

22-33

The first law of thermodynamics is

$$Q = \Delta U + W.$$

Over a complete cycle $\Delta U = 0$ and therefore

$$W = Q = -(100\ \text{g})(80\ \text{cal/g}) = -8000\ \text{cal},$$

negative since heat left the gas to melt ice. W is the work done by the gas and therefore the work done on the gas is +8000 cal.

22-34

(a) The change in internal energy $\Delta U_{fi} = U_f - U_i$ between points i and f is the same regardless of path. Along path iaf, Q = 50 cal and W = 20 cal; hence $\Delta U_{fi} = Q - W = 50 - 20 = 30$ cal. Along path ibf,

$$\Delta U_{fi} = Q - W,$$

$$30 = 36 - W,$$

$$W = 6\ \text{cal}\quad \underline{\text{Ans}}.$$

(b) From (a), $\Delta U_{if} = U_i - U_f = -30$ cal. Hence, $\Delta U_{if} = -30 = Q - (-13)$, giving Q = -43 cal Ans.

(c) If $U_i = 10$ cal, then since from (a), $U_f - U_i = 30$, $U_f = 40$ cal.

(d) If $U_b = 22$ cal, $\Delta U_{bi} = U_b - U_i = 22 - 10 = 12$ cal. For the process ib, then, 12 = Q - W. Now, W = 6 cal for ibf, but W = 0 along bf since bf occurs at constant volume; hence W = 6 cal for ib. This gives 12 = Q - 6, or Q = 18 cal Ans. Along bf, $U_b = 22$ and $U_f = 40$ so that $U_f - U_b = 18$ cal. But W = 0 along bf as noted above, so that Q = 18 cal here also.

22-35

(a) The actual time that the ball is in contact with the floor is very short so that unless the floor is extraordinarily hot no heat will flow to the ball: $Q = 0$.

(b) By assumption, $\Delta U > 0$ so that, by the first law $Q = \Delta U + W$, $Q = 0$ gives $W < 0$ indicating that work has been done on the ball.

(c) The change in internal energy ΔU equals, in magnitude, the loss of mechanical energy which, if M is the mass of the ball, is

$$\Delta U = Mg\Delta h = M(9.8\ m/s^2)(9.5\ m) = 93M\ J$$

or 93 J/kg.

22-36

(a) Let m, ρ be the mass and density of the steam and V' its volume, V the volume of the chamber and M, A, v the mass, area and speed of the piston. Although $V' \neq V$, $dV'/dt \approx dV/dt$ as the water level, due to steam condensation, rises slowly compared to the fall of the piston. Then, disregarding a minus sign,

$$\frac{dV'}{dt} = \frac{dV}{dt} = Av = \frac{d}{dt}\left(\frac{m}{\rho}\right) = \frac{1}{\rho}\frac{dm}{dt},$$

$$\frac{dm}{dt} = \rho Av = (6 \times 10^{-4})(2)(0.3) = 3.6 \times 10^{-4}\ g/s \quad \underline{Ans.}$$

(c) Since 540 cal are liberated for each gram of steam that condenses,

$$\frac{dQ}{dt} = -\left(\frac{dm}{dt}\right)(540) = -(3.6 \times 10^{-4})(540),$$

$$\frac{dQ}{dt} = -0.1944\ cal/s = -0.8138\ J/s \quad \underline{Ans.}$$

(b) Differentiate the first law to obtain

$$\frac{dQ}{dt} = \frac{dU}{dt} + \frac{dW}{dt}.$$

To evaluate dW/dt, note that

$$\frac{dW}{dt} = p\,\frac{dV}{dt} = p(-Av).$$

The pressure p is given by

$$p = p_{atm} + \frac{Mg}{A},$$

so that

$$\frac{dW}{dt} = -(p_{atm}A + Mg)v.$$

Numerically, $p_{atm}A = (1.013 \times 10^5 \text{ Pa})(2 \times 10^{-4} \text{ m}^2) = 20.26$ N and $Mg = 19.6$ N; since $v = 0.003$ m/s,

$$\frac{dW}{dt} = -0.1196 \text{ J/s}.$$

Combining this with (c) gives

$$\frac{dU}{dt} = -0.8138 - (-0.1196),$$

$$\frac{dU}{dt} = -0.6942 \text{ J/s} \quad \underline{\text{Ans}}.$$

CHAPTER 23

23-4

For a fixed amount of gas undergoing a change from p_1, V_1, T_1 to p_2, V_2, T_2, the ideal gas law $pV = nRT$ applied to each state gives

$$p_1V_1 = nRT_1, \quad p_2V_2 = nRT_2.$$

Dividing these equations yields

$$\frac{p_1V_1}{p_2V_2} = \frac{T_1}{T_2}.$$

Let point 1 apply at the surface and 2 indicate the bottom of the lake. Then $p_1 = p_{atm} = 1.013 \times 10^5$ Pa, and

$$p_2 = p_{atm} + \rho gh = p_{atm} + (1000\ kg/m^3)(9.8\ m/s^2)(40\ m),$$

$$p_2 = (1.013 + 3.92) \times 10^5 = 4.933 \times 10^5\ Pa.$$

The temperatures are $T_1 = 20 + 273 = 293$ K, $T_2 = 4 + 273 = 277$ K. Hence,

$$\frac{1.013 \times 10^5}{4.933 \times 10^5}\frac{V_1}{20} = \frac{293}{277},$$

$$V_1 = 103\ cm^3 \quad \underline{Ans};$$

the units of V_1 are cm^3 since the units of V_2 are cm^3.

23-11

Let l refer to the left arm and r to the right. Since the temperature does not change,

$$p_{lf}V_{lf} = p_0V_{li},$$

$$p_{lf} = p_0(V_{li}/V_{lf}) = \frac{50}{44} p_0.$$

Similarly,

$$p_{rf} = \frac{30}{26} p_0.$$

The mercury in the left column is 2 cm higher than in the right; hydrostatic equilibrium requires that

$$p_{lf} + \rho gh = p_{rf}.$$

With $\rho = 13.6 \times 10^3$ kg/m^3, $g = 9.8$ m/s^2 and $h = 0.02$ m, the quantity $\rho gh = 2666$ Pa. The preceding equation then yields

$$\frac{50}{44} p_0 + 2666 = \frac{30}{26} p_0,$$

$$p_0 = 1.52 \times 10^5 \text{ Pa} \quad \underline{\text{Ans}}.$$

23-12

The initial pressure is $p_i = 14.7 + 15 = 29.7$ lb/in^2. Since the expansion is isothermal (T constant),

$$p_iV_i = p_fV_f,$$

$$(29.7)(5) = (14.7)V_f,$$

$$V_f = 10.10 \text{ ft}^3.$$

The work W_e done during this expansion is

$$W_e = \int_i^f p\, dV = nRT\cdot\ln(V_f/V_i) = p_iV_i\ln(V_f/V_i) = (29.7)(5)\ln(\frac{10.1}{5}),$$

$$W_e = 104.44 \text{ ft}^3\cdot\text{lb/in}^2.$$

For the compression at constant pressure,

$$W_c = p\Delta V = p_f(V_i - V_f) = (14.7)(5 - 10.1) = -74.97 \text{ ft}^3\cdot\text{lb/in}^2.$$

Therefore the total work $W = W_e + W_c$ is

$$W = 29.47\ \text{ft}^3\cdot\text{lb}/(\tfrac{1}{144}\ \text{ft}^2),$$

$$W = 4244\ \text{ft}\cdot\text{lb}\quad \underline{\text{Ans}}.$$

23-13

The momentum imparted to the wall on each collision is

$$\Delta p = m\Delta v_p = 2mv_p = 2mv\cos 45°,$$

where v_p is the component of the velocity perpendicular to the wall. Let N be the number of collisions per unit time; then the pressure P on the wall is

$$P = \frac{F}{A} = \frac{N\Delta p}{A} = \frac{2Nmv\cos 45°}{A},$$

$$P = [(2)(10^{23})(3.3 \times 10^{-24})(10^5)\cos 45°]/(2),$$

$$P = 2.33 \times 10^4\ \text{dyne/cm}^2 = 2330\ \text{Pa}\quad \underline{\text{Ans}}.$$

23-21

(a) The escape speed and root-mean-square speed are

$$v_e = \left(\frac{2GM}{R}\right)^{\frac{1}{2}} = (2gR)^{\frac{1}{2}},$$

$$v_{rms} = \left(\frac{3kT}{m}\right)^{\frac{1}{2}}.$$

If $v_e = v_{rms}$, then

$$T = \frac{2}{3}\frac{Rgm}{k}.$$

For the earth, $R = 6.4 \times 10^6$ m, $g = 9.8\ \text{m/s}^2$. H_2 has a molecular mass of 2 so that $m = 2(1.66 \times 10^{-27}\ \text{kg})$; the equation above then gives T = 10,000 K. O_2 has a molecular mass of 32, or 16 times greater than H_2, giving T = 160,000 K for O_2.

(b) On the moon $g = 0.16(9.8\ \text{m/s}^2)$ and $R = 1.74 \times 10^6$ m. The equation for T then gives T = 440 K for H_2 and 7000 K for O_2.

(c) Although 1000 K $\ll$ 10,000 K, there are enough molecules in the "tail" of the Maxwell-Boltzmann speed distribution to ensure depletion of molecular hydrogen over the two or three billion years since the earth's present atmosphere was formed. Most of the O_2, fortunately, is retained.

23-24

(a) The molecular mass M of water (H_2O) is 2(1) + 16 = 18. The number n of molecules per gram is

$$n = N_0/M = 6.02 \times 10^{23}/18 = 3.3444 \times 10^{22}.$$

Thus,

$$\epsilon = \frac{540}{n} = (540 \text{ cal/g})/(3.3444 \times 10^{22} \text{ g}^{-1}),$$

$$\epsilon = 161 \times 10^{-22} \text{ cal} = 6.76 \times 10^{-20} \text{ J} \quad \underline{\text{Ans}}.$$

(b) The average translational kinetic energy of a molecule is

$$K = \frac{3}{2}kT = \frac{3}{2}(1.38 \times 10^{-23})(300) = 6.21 \times 10^{-21} \text{ J}.$$

Evidently, $\epsilon/K = 10.9$ Ans.

23-29

The work W done by the gas is

$$W = \int_i^f p \cdot dV = \int_i^f \frac{nRT}{V}\, dV = nRT\int_i^f \frac{dV}{V},$$

$$W = nRT \cdot \ln(V_f/V_i),$$

since, for an isothermal process, T is constant and can be drawn out from the integral. The change in internal energy is

$$\Delta U = \frac{3}{2} nR\, \Delta T = 0$$

for an isothermal process. Then, by the first law $\Delta U = Q - W$,

$$0 = Q - W,$$

$$Q = W = RT\cdot\ln(V_f/V_i),$$

for n = 1, i.e., one mole.

23-38

(a) Use the adiabatic equation

$$p_1V_1^{\gamma} = p_2V_2^{\gamma},$$

with p in atmospheres and V in liters:

$$(1)(4)^{3/2} = p_2(1)^{3/2},$$

$$p_2 = 4^{3/2} = 8\ \text{atm}\ \ \underline{\text{Ans}}.$$

(b) For an ideal gas pV = nRT. Combine this with the adiabatic equation to eliminate p:

$$\left(\frac{nRT_1}{V_1}\right)V_1^{\gamma} = \left(\frac{nRT_2}{V_2}\right)V_2^{\gamma},$$

$$T_1V_1^{\gamma-1} = T_2V_2^{\gamma-1},$$

$$(300)(4)^{\frac{1}{2}} = T_2(1)^{\frac{1}{2}},$$

$$T_2 = 600\ \text{K}\ \ \underline{\text{Ans}}.$$

23-40

(a) Process 1-2: since the volume does not change, $W = \int pdV = 0$; $\Delta U = \frac{3}{2}nR\Delta T = \frac{3}{2}(1)(8.314)(600 - 300) = 3741.3$ J. From the first law, $Q = \Delta U + W = 3741.3 + 0 = 3741.3$ J.

Process 2-3: this is adiabatic, meaning that Q = 0. $\Delta U = \frac{3}{2}R\Delta T = \frac{3}{2}(8.314)(455 - 600) = -1808.295$ J. The first law gives $W = -\Delta U = +1808.295$ J.

Process 3-1: the pressure is constant and therefore $W = p\Delta V = nR\Delta T$ (by the ideal gas law pV = nRT); hence W = (1)(8.314)(300 -

455) = -1288.67 J. $\Delta U = \frac{3}{2}(8.314)(300 - 455) = -1933.005$ J. The first law then gives $Q = \Delta U + W = -3221.675$ J.

Whole cycle: adding the results above gives $\Delta U = 3741.3 - 1808.295 - 1933.005 = 0$ (as expected); $Q = 3741.3 + 0 - 3221.675 = 519.625$ J; $W = 0 + 1808.295 - 1288.67 = 519.625$ J $= Q$, as expected.

(b) The volume at point 1 is given from the ideal gas law $pV = nRT$:

$$(1.013 \times 10^5)(V_1) = (1)(8.314)(300),$$
$$V_1 = 0.02462 \text{ m}^3.$$

At point 2, $V_2 = V_1 = 0.02462$ m^3. Also,

$$\frac{p_1V_1}{p_2V_2} = \frac{T_1}{T_2},$$
$$\frac{(1)}{p_2} = \frac{300}{600},$$
$$p_2 = 2 \text{ atm}.$$

At point 3, $p_3 = p_1 = 1$ atm, so that

$$\frac{p_1V_1}{p_3V_3} = \frac{T_1}{T_3},$$
$$\frac{0.02462}{V_3} = \frac{300}{455},$$
$$V_3 = 0.03734 \text{ m}^3.$$

As a check, since 2-3 is an adiabat, the values above should, within roundoff error, satisfy

$$p_2V_2^{\gamma} = p_3V_3^{\gamma},$$
$$(2)(0.02462)^{5/3} = (1)(0.03734)^{5/3},$$
$$(2)(0.00208) = 0.00417.$$

23-41

The constant pressure expansion is plotted as a straight line parallel to the V-axis. For the isothermal process

$$p_0V_0 = nRT = p_1V_1,$$

so that

$$p_1 = p_0(V_0/V_1),$$

the graph of the isotherm being a hyperbola. For the adiabatic process (Q = 0)

$$p_1' = p_0(V_0/V_1)^{\gamma},$$

since pV^{γ} = constant. But $\gamma > 1$ and $V_1 > V_0$ so that $p_1 > p_1'$. Hence the curves are positioned as shown. Since the area under each is the work done, these works can be ordered immediately: $W_a > W_b > W_c > 0$. For an ideal gas ΔU is proportional to ΔT. Hence, $\Delta U_b = 0$. Since $pV = nRT$ the expansion at constant pressure is accompanied by an increase in temperature giving $\Delta U_a > 0$. For the adiabat the first law $Q = \Delta U + W$ shows that as $Q_c = 0$ and $W_c > 0$, then $\Delta U_c < 0$. Therefore, for the constant pressure process the change in internal energy and the work done must be greater than for the other processes and for the adiabatic process they must be least. The first law then implies that the same statement may be made also for Q.

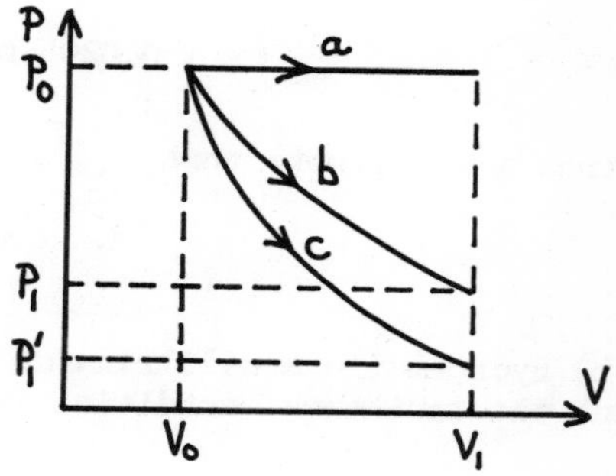

23-42

The numerical data needed are: $p_0 = 1.013 \times 10^5$ Pa, a = 0.45 m, $\rho = 13.6 \times 10^3$ kg/m^3, g = 9.8 m/s^2, h = 0.1 m, $\gamma = 7/5$. Also, let A be the cross-sectional area of the tube and x the distance the mercury drops.

(a) Hydrostatic equilibrium in the rotated state requires that

$$p_b = p_t + \rho gh.$$

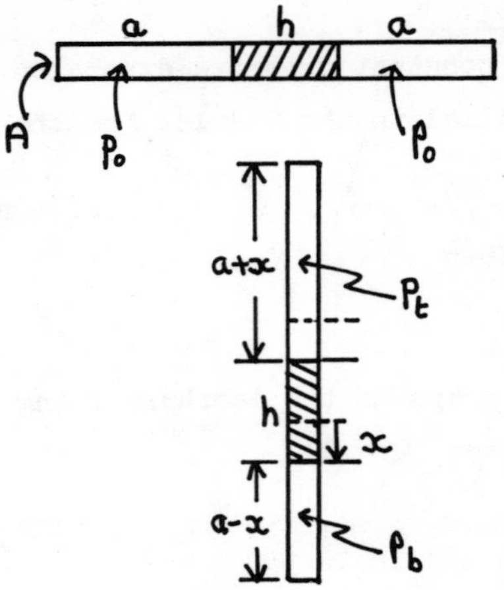

The isothermal condition yields

$$pV = \text{constant},$$

$$p_b(a - x)A = p_0 aA,$$

$$p_t(a + x)A = p_0 aA.$$

Solve these for p_b, p_t and substitute into the first equation to obtain

$$\frac{1}{1 - x/a} = \frac{1}{1 + x/a} + \rho gh/p_0,$$

$$x^2 + (2p_0 a/\rho gh)x - a^2 = 0,$$

$$x^2 + 6.8405x - 0.2025 = 0,$$

$$x = 0.0295 \text{ m} = 2.95 \text{ cm} \quad \underline{\text{Ans}}.$$

Since $x \ll 1$, note that

$$x \approx \frac{\rho gha}{2p_0}.$$

(b) Hydrostatic equilibrium still holds for the adiabatic case, but the isothermal condition is replaced with

$$pV^\gamma = \text{constant},$$

$$p_b[(a - x)A]^\gamma = p_0(aA)^\gamma$$

$$p_b = p_0\left(\frac{a}{a - x}\right)^\gamma.$$

Similarly,

$$p_t = p_0\left(\frac{a}{a + x}\right)^\gamma.$$

Substitute these into the hydrostatic equation to obtain

$$\left(\frac{1}{1 - x/a}\right)^{\gamma} = \left(\frac{1}{1 + x/a}\right)^{\gamma} + \rho gh/p_0.$$

Like the isothermal situation, it is anticipated that $x/a \ll 1$, so that an expansion of the first two terms is appropriate. Therefore,

$$1 + \gamma\cdot\frac{x}{a} \approx 1 - \gamma\cdot\frac{x}{a} + \rho gh/p_0,$$

$$x = \frac{\rho gha}{2\gamma p_0} = 0.0211 \text{ m} = 2.11 \text{ cm} \quad \underline{\text{Ans}}.$$

Compare this with the isothermal expression above. The more rapidly the tube is turned the better the adiabatic assumption.

23-47

Equation 23-3 is $p = \frac{1}{3}\rho\overline{v^2}$. From equipartition, $\frac{1}{2}m\overline{v^2} = \frac{3}{2}kT$. Hence,

$$p = \frac{1}{3}\rho\left(\frac{3kT}{m}\right) = \frac{\rho}{m}\cdot kT.$$

Since ρ = mass of gas/volume of gas and m is the mass of each particle of the gas, then ρ/m = number of particles/volume of gas = N/V, so that

$$p = \frac{N}{V}kT.$$

Therefore, if two gasses a and b have the same temperature and pressure, this last equation gives

$$\frac{N_a}{N_b} = \frac{V_a}{V_b}.$$

Finally, if $V_a = V_b$, then $N_a = N_b$.

23-48

Let V be the volume of the room and n the number of molecules per unit volume of the air inside. Since the pressure remains unchanged

$$p_0 = n_1kT_1 = n_2kT_2.$$

The internal energy $U = \frac{5}{2}NkT = \frac{5}{2}(nV)kT$ for diatomic molecules. Hence,

$$U_2 = \frac{5}{2}(n_2V)kT_2 = \frac{5}{2}V(\frac{n_1T_1}{T_2})kT_2 = \frac{5}{2}(n_1V)kT_1 = U_1$$

as asserted. Of course, although the internal energy does not change, the temperature of the remaining air has increased, so the room feels warmer.

23-52

(b) Let m be the mass of each atom. Momentum conservation for a head-on and therefore straight-line collision, requires that

$$mv + m(0) = mu + mV.$$

If ΔE is the energy difference between the ground and first excited states, then for an inelastic collision,

$$\tfrac{1}{2}mv^2 = \tfrac{1}{2}mu^2 + \tfrac{1}{2}mV^2 + \Delta E.$$

Eliminating V yields

$$\tfrac{1}{2}mv^2 = \tfrac{1}{2}mu^2 + \tfrac{1}{2}m(v - u)^2 + \Delta E,$$

$$v = u + \frac{\Delta E}{mu}.$$

Hence, the initial kinetic energy is

$$K = \tfrac{1}{2}mv^2 = \tfrac{1}{2}mu^2 + \frac{1}{2m}(\Delta E)^2u^{-2} + \Delta E.$$

For the minimum K, set $\partial K/\partial u = 0$; this gives $u^2 = \Delta E/m$. Putting this into the equation above for K yields $K_{min} = 2\Delta E$.

(a) Since $K = 13$ eV $< 2\Delta E$, the collision must be elastic.

CHAPTER 24

24-7

(a) The mass of a hydrogen molecule is $m = 2(1.66 \times 10^{-27}\ kg) = 3.32 \times 10^{-27}$ kg. Since T = 4000 K and $k = 1.38 \times 10^{-23}$ J/K, the root-mean-square speed will be

$$v_{rms} = (3kT/m)^{\frac{1}{2}} = 7.06 \times 10^3 \text{ m/s} \quad \underline{\text{Ans}}.$$

(b) When the hydrogen molecule and argon atoms touch, their distance of closest approach, center to center, will be

$$d = r_H + r_A = \tfrac{1}{2}(1 + 3) \times 10^{-8} = 2 \times 10^{-8} \text{ cm} \quad \underline{\text{Ans}},$$

assuming they can be pictured as rigid spheres.

(c) By definition of mean free path, the collision frequency f is

$$f = v/\ell = \sqrt{2}\pi n_A d^2 v_{rms} = 5.02 \times 10^{10} \text{ s}^{-1} \quad \underline{\text{Ans}},$$

using $v_{rms} = 7.06 \times 10^5$ cm/s from (a), $d = 2 \times 10^{-8}$ cm from (b) and $n_A = 4 \times 10^{19}$ cm^{-3}.

24-8

Consider a group n* of particles at some instant and let any particle that suffers a collision be removed from the group. Let n be the number of particles still in the group after each has traveled a distance x. Assume that the number dn of particles lost from the group as each travels dx is

$$dn = - Pn(x)dx,$$

where P is a constant. Then,

$$\frac{dn}{n} = - Pdx,$$

$$n = n^* e^{-Px}.$$

Thus the probability that any particle travels x without a collision is

$$\text{Prob.} = n/n^* = e^{-Px}.$$

The mean free path is

$$\bar{\ell} = \frac{1}{n^*}\int x\,dn,$$

since dn is the number of particles that have survived a distance x but do not survive the distance x + dx: i.e., dn is the number of particles with mean free path x, so that

$$\bar{\ell} = \int_0^\infty x(-Pn^* e^{-Px})dx/n^* = \frac{1}{P},$$

and therefore

$$\text{Prob.} = e^{-x/\bar{\ell}}.$$

24-9

In the laboratory frame, the velocities of N particles are distributed uniformly on the surface of a sphere in a v_x, v_y, v_z coordinate system. The center of the sphere is at the origin and the radius of the sphere is v. There are

$$N(\theta) = \left(\frac{N}{4\pi}\cdot 2\pi\sin\theta\right)\Delta\theta = \frac{N}{2}\sin\theta\Delta\theta$$

particles with velocities between θ and θ + Δθ, as shown. As seen from one of the particles (one with $v_y = -v$, $v_x = v_z = 0$) the distribution is exactly the same except that the origin is moved to 0', a "distance" v from 0

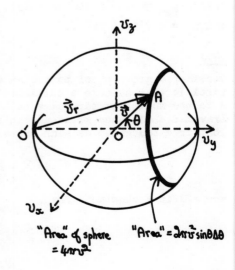

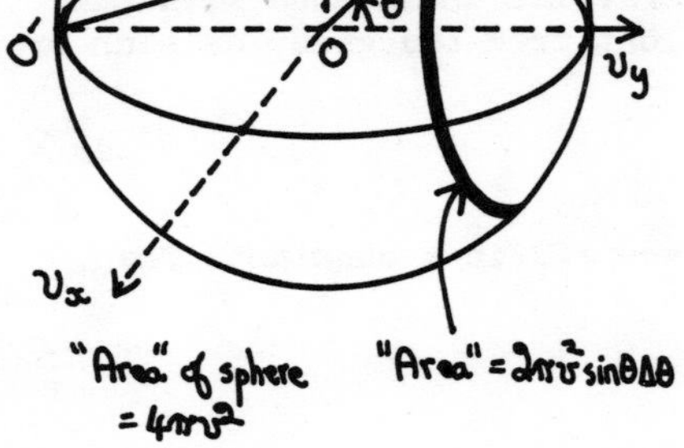

along the negative v_y-axis. The relative speed v_r of a particle with a laboratory velocity v_A may be obtained by applying the law of cosines to triangle OAO':

$$v_r^2 = v^2 + v^2 + 2v\cdot v\cos\theta = 2v^2(1 + \cos\theta),$$

$$v_r = \sqrt{2}v(1 + \cos\theta)^{\frac{1}{2}}.$$

The average value of v_r is

$$\overline{v_r} = \frac{1}{N}\int_0^{\pi} v_r(\theta)N(\theta) = \frac{1}{N}\int_0^{\pi}\sqrt{2}v(1 + \cos\theta)^{\frac{1}{2}}\cdot\frac{N}{2}\cdot\sin\theta\, d\theta,$$

$$\overline{v_r} = \frac{1}{\sqrt{2}}\, v\, \frac{(1 + \cos\theta)^{3/2}}{3/2}\Big]_0^{\pi} = \frac{4}{3}\, v.$$

<u>24-13</u>

(a) By definition,

$$N\bar{v} = v_1 + v_2 + \ldots + v_N;\; Nv_{rms}^2 = v_1^2 + v_2^2 + \ldots + v_N^2.$$

For any particle,

$$v_i = \bar{v} + (v_i - \bar{v}) = \bar{v} + \epsilon_i.$$

Thus,

$$Nv_{rms}^2 = (\bar{v} + \epsilon_1)^2 + (\bar{v} + \epsilon_2)^2 + \ldots + (\bar{v} + \epsilon_N)^2.$$

$$Nv_{rms}^2 = N\bar{v}^2 + 2\bar{v}\Sigma\epsilon_i + \Sigma\epsilon_i^2.$$

But

$$\Sigma\epsilon_i = \Sigma(v_i - \bar{v}) = \Sigma v_i - N\bar{v} = N\bar{v} - N\bar{v} = 0,$$

and therefore

$$v_{rms}^2 = \bar{v}^2 + \frac{1}{N}\Sigma\epsilon_i^2.$$

Since $\epsilon_i^2 \geq 0$, then $v_{rms} \geq \bar{v}$.

(b) If all the $\epsilon_i = 0$, $v_{rms} = \bar{v}$: all speeds are the same.

23-14

(a) The area under the curve equals the total number N of particles:

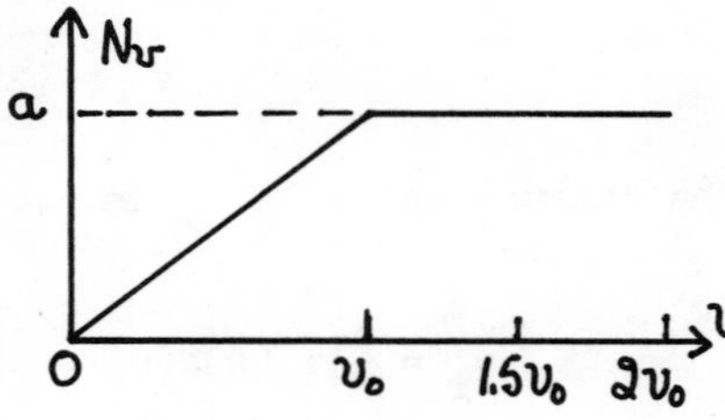

$$N = \tfrac{1}{2}av_0 + av_0 = \tfrac{3}{2}av_0,$$

$$a = 2N/3v_0 \quad \underline{\text{Ans.}}$$

(b) The area under the curve between $1.5v_0$ and $2v_0$ equals the number of particles with speeds between these limits; hence, this number is

$$\tfrac{1}{2}av_0 = \tfrac{1}{2}(2N/3v_0)v_0 = \frac{N}{3} \quad \underline{\text{Ans.}}$$

(c) The distribution function is $N_v = av/v_0$ for $0 \leq v \leq v_0$ and $N_v = a$ for $v_0 \leq v \leq 2v_0$. Therefore, by definition,

$$\bar{v} = \frac{1}{N}\int vN_v\,dv = \frac{1}{N}\int_0^{v_0} v\left(\frac{av}{v_0}\right)dv + \frac{1}{N}\int_{v_0}^{2v_0} va\,dv,$$

$$\bar{v} = \frac{1}{N}(av_0^2/3) + \frac{1}{N}(3av_0^2/2) = \frac{1}{N}(11av_0^2/6),$$

$$\bar{v} = \frac{11}{9}v_0 \quad \underline{\text{Ans.}}$$

24-15

(a) From the ideal gas law

$$pV = nRT,$$

$$(3 \times 10^5 \text{ Pa})(10^{-3} \text{ m}^3) = n(8.314 \text{ J/K})(300 \text{ K}),$$

$$n = 0.12028 \text{ moles.}$$

Hence, the number N of atoms is

$$N = nN_0 = 7.24 \times 10^{22} \quad \underline{\text{Ans}},$$

where N_0 is Avogadro's number.

(b) The average speed is

$$\bar{v} = \left(\frac{8}{\pi}\frac{kT}{m}\right)^{\frac{1}{2}}.$$

But $m = (40)(1.66 \times 10^{-27}$ kg) and $k = 1.38 \times 10^{-23}$ J/K. These give $\bar{v} = 398$ m/s.

(c) Consider a small sphere, surface area A, suspended in a gas. The collision cross-section of this sphere for any atom of the gas is $\frac{1}{4}A$. Hence, the number of particles that strike the sphere per unit time with speeds between v and v + dv is

$$dn = \tfrac{1}{4}AvN(v)dv.$$

Integrating over v gives

$$n = \tfrac{1}{4}AN\bar{v}/V.$$

Now let A be the area of the hole. Substituting the other numerical data gives

$$n = 7.20 \times 10^{20} \quad \underline{\text{Ans}}.$$

(d) Since all atoms striking the hole leave the gas

$$-\frac{dn}{dt} = \frac{dN}{dt} = -\left(\tfrac{1}{4}A\bar{v}/V\right)N = -(9.95 \times 10^{-3})N \approx -0.01N,$$

$$N = N_0e^{-0.01t},$$

where now N_0 is the original number of argon atoms. If $N = N_0e^{-1}$,

$$0.01t = 1,$$

$$t = 100 \text{ s} \quad \underline{\text{Ans}}.$$

24-26

Van der Waal's equation is

$$(P + \frac{n^2 a}{V^2})(\frac{V}{n} - b) = RT.$$

Now let $V = v_{cr}v = 3bv$, $P = p_{cr}p = (a/27b^2)p$, $T = T_{cr}t = (8a/27bR)t$ and substitute into the above to obtain

$$(\frac{a}{27b^2}\,p + \frac{n^2 a}{9b^2 v^2})(\frac{3bv}{n} - b) = R\,\frac{8a}{27bR}\,t,$$

$$(\frac{p}{3} + \frac{n^2}{v^2})(3\frac{v}{n} - 1) = \frac{8}{3}t,$$

independent of a and b.

CHAPTER 25

25-13

(a)

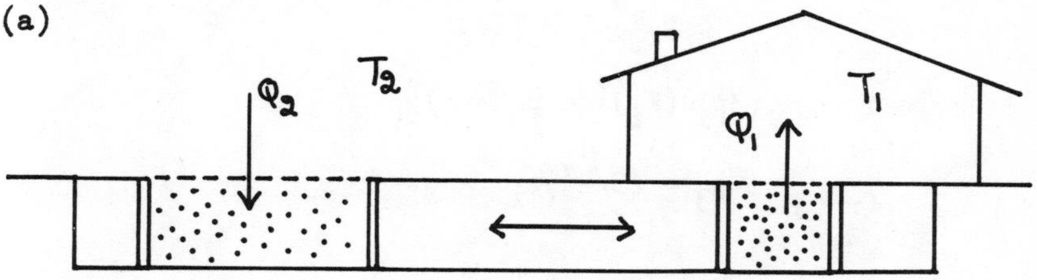

(b) First let the gas in the pump expand so that the gas temperature is decreased and lower than the outside temperature T_2. Therefore, a quantity of heat Q_2 is extracted from the outside air at T_2. Then move the piston into the house and compress the gas until its temperature is higher than inside the house, which is at T_1. A larger quantity of heat Q_1 is delivered to the inside of the house. Thus, in principle there is no difference of significance between the heat pump and the refrigerator. In practical use, in a refrigerator the quantity of heat Q_2 is extracted at T_2 through the vaporization of a liquid instead of the expansion of a gas as in the heat pump.

(c) By the first law, $Q_1 = Q_2 + W$, the change in internal energy being zero over one cycle.

(d) The heat pump can be reversed by reversing the expansion and compression stages.

(e) The advantages of the pump are that the heat Q_1 delivered is greater than W, the energy paid for (to the power company), and that the pump can be operated to heat and cool.

25-15

(a) Apply the ideal gas law to points 1 and 2. Since $p_2 = 3p_1$ and $V_2 = V_1$, this gives

$$p_1V_1/p_2V_2 = T_1/T_2,$$

$$p_1V_1/(3p_1)V_1 = T_1/T_2,$$

$$T_2 = 3T_1 \quad \underline{\text{Ans}}.$$

Points 2 and 3 are connected by the adiabatic relation

$$p_2V_2^\gamma = p_3V_3^\gamma,$$

$$(3p_1)(V_1)^\gamma = p_3(4V_1)^\gamma,$$

$$p_3 = (3/4^\gamma)p_1 \quad \underline{\text{Ans}}.$$

Also,

$$p_1V_1/p_3V_3 = T_1/T_3,$$

$$p_1V_1/(3/4^\gamma)p_1(4V_1) = T_1/T_3,$$

$$T_3 = (3/4^{\gamma-1})T_1 \quad \underline{\text{Ans}}.$$

For point 4,

$$p_1V_1^\gamma = p_4V_4^\gamma,$$

$$p_1V_1^\gamma = p_4(4V_1)^\gamma,$$

$$p_4 = p_1/4^\gamma \quad \underline{\text{Ans}}.$$

Finally,

$$p_1V_1/p_4V_4 = T_1/T_4,$$

$$p_1V_1/(4^{-\gamma}p_1)(4V_1) = T_1/T_4,$$

$$T_4 = T_1/4^{\gamma-1} \quad \underline{\text{Ans}}.$$

(b) From (a), $T_1 < T_2$ and $T_4 < T_1$. Hence, heat Q_{12} enters along path 1-2 and heat $Q_{34} > 0$ leaves along path 3-4. No heat enters or leaves along adiabats. Although the paths along which heat is transferred are not isotherms, the efficiency e still can be defined as

$$e = 1 - Q_{34}/Q_{12}.$$

The paths 1-2 and 3-4 are constant volume processes; hence,

$$Q_{12} = nC_v(T_2 - T_1).$$

But $C_p - C_v = R$ and $C_p/C_v = \gamma$ so that $C_v = R(\gamma - 1)$, giving

$$Q_{12} = nR(\gamma - 1)\cdot 2T_1.$$

Similarly,

$$Q_{34} = nR(\gamma - 1)(2T_1/4^{\gamma-1}),$$

since $Q_{34} = nC_v(T_3 - T_4)$ and using the results from (a) for T_4 in terms of T_1. The efficiency now becomes

$$e = 1 - \frac{nR(\gamma - 1)(2T_1/4^{\gamma-1})}{nR(\gamma - 1)\cdot 2T_1} = 1 - 4^{1-\gamma} \quad \underline{\text{Ans}}.$$

<u>25-20</u>

If the heating of the water and the cooling of the aluminum were carried out reversibly, the entropy change of each could be found as follows:

$$\Delta S = \int \frac{dQ}{T} = \int_{T_i}^{T_f} mc_p dT/T = mc_p \int_{T_i}^{T_f} dT/T = mc_p \cdot \ln(T_f/T_i).$$

To find the equilibrium temperature T, set the heat lost by the aluminum equal to the heat gained by the water. Each of these can be written $mc_p\Delta T$, so that

$$(200)(0.215)(100 - T) = (50)(1)(T - 20),$$

$$T = 57°C = 330\ K.$$

The total entropy change is

$$\Delta S = \Delta S_{water} + \Delta S_{Al},$$

$$\Delta S = (50)(1)\ln\frac{330}{293} + (200)(0.215)\ln\frac{330}{373} = 5.946 - 5.267,$$

$$\Delta S = 0.68 \text{ cal/K} \quad \underline{\text{Ans}}.$$

25-22

The equilibrium temperature can safely be taken as the temperature of the lake, $15^\circ C$ = 288 K. By Problem 25-20,

$$\Delta S_{ice} = (10)(0.5)\ln\frac{273}{263} = 0.1866 \text{ cal/K},$$

$$\Delta S_{water} = (10)(1)\ln\frac{288}{273} = 0.5349 \text{ cal/K},$$

the water being the liquid formed by the melted ice. During the melting,

$$\Delta S_{melting} = \frac{Q}{T} = \frac{(10)(80)}{273} = 2.9304 \text{ cal/K}$$

the melting taking place at the constant temperature of 273 K. Turning now to the lake, although its temperature does not change it loses heat ΔQ required to warm the ice to $0^\circ C$, melt the ice and then bring the water formed to the lake temperature. This heat is

$$\Delta Q = (10)(0.5)(10) + (10)(80) + (10)(1)(15) = 1000 \text{ cal}.$$

Hence,

$$\Delta S_{lake} = \frac{-\Delta Q}{T} = -\frac{1000}{288} = -3.4722 \text{ cal/K}.$$

The total entropy change is the sum of the four listed above, i.e.,

$$\Delta S = 0.18 \text{ cal/K} \quad \underline{\text{Ans}}.$$

25-24

(a) If T is constant, then since $pV = nRT$, the work W will be

$$W = \int p\,dV = \int\frac{nRT}{V}dV = nRT\int\frac{dV}{V} = nRT\cdot\ln(V_2/V_1),$$

$$W = (4)(8.314)(400)\ln 2 = 9221 \text{ J} \quad \underline{\text{Ans}}.$$

(b) With T constant,

$$\Delta S = \frac{Q}{T} = \frac{W}{T} = \frac{9221}{400} = 23\ \text{J/K} \quad \underline{\text{Ans}},$$

since at constant T, $\Delta U = 0$ and $Q = W$ by the first law.

(c) For an adiabatic process $Q = 0$ so that $\Delta S = 0$ also.

25-25

(a) By definition of heat reservoirs, their temperatures do not change although they may gain or lose heat. The temperatures are $T_H = 400$ K and $T_C = 300$ K. The heat leaves the hot reservoir; thus,

$$\Delta S_H = \frac{Q}{T} = \frac{-1200}{400} = -3\ \text{cal/K},$$

$$\Delta S_C = \frac{Q}{T} = \frac{+1200}{300} = +4\ \text{cal/K}.$$

Evidently the entropy change of the system is $4 - 3 = 1$ cal/K.

(b) If the rod simply transmits the heat, absorbing none itself, then its entropy change is zero, as assumed above.

25-28

Infinitesimal cycles implies reversible processes. Also, for the limiting works, use Carnot engines.

(a) If Q_r is the heat delivered to the reservoir at T_1,

$$W = Q - Q_r,$$

so that the entropy change of the reservoir is

$$\Delta S = +\frac{Q_r}{T_1}.$$

But if ΔS_b is the entropy change of the body, $\Delta S_r + \Delta S_b = 0$, and then

$$W = Q - T_1\Delta S_r = Q + T_1\Delta S_b = Q + T_1(S_1 - S_2),$$

$$W_{max} = Q - T_1(S_2 - S_1).$$

(b) Similarly, if Q_d is the heat delivered to the reservoir at T_1,

$$W = Q_d - Q = T_1 \Delta S_r - Q = T_1(-\Delta S_b) - Q = -(S_0 - S_1)T_1 - Q,$$

$$W_{min} = T_1(S_1 - S_0) - Q.$$

25-30

(a) The probability P of dealing any one hand is

$$P = 1/(\text{number of hands that can be dealt}).$$

The number of hands that can be dealt is

$$\frac{52!}{13!(52 - 13)!} = 6.35 \times 10^{11}.$$

Thus, $w_i = 1.57 \times 10^{-12}$. The initial entropy is

$$S_i = k \ln w_i = (1.38 \times 10^{-23}) \cdot \ln(1.57 \times 10^{-12})/(4.186),$$

$$S_i = 9.0 \times 10^{-23} \text{ cal/K}.$$

The final entropy is

$$S_f = k \cdot \ln(1) = 0.$$

Thus the entropy change is -9.0×10^{-23} cal/K.

(b) From previous problems and examples in the text, it is seen that this is very much smaller than typical thermodynamic entropy changes.

CHAPTER 26

26-8

The third charge must lie between the two charges +q, +4q (i.e., on the line joining them), for if it did not it would feel a force of repulsion if it was positive and attraction if negative. In fact, the charge must be negative: each of the given charges +q, +4q, could not be in equilibrium otherwise, since each would feel repulsive forces from the other two charges. So let the third charge be -Q, Q > 0, located a distance x from the +q charge; then, for this charge to be in equilibrium,

$$\frac{1}{4\pi\epsilon_0}\frac{qQ}{x^2} = \frac{1}{4\pi\epsilon_0}\frac{Q(4q)}{(\ell - x)^2},$$

$$x = \frac{\ell}{3} \quad \underline{\text{Ans}}.$$

To evaluate Q, require that the +q charge, say, also be in equilibrium:

$$\frac{1}{4\pi\epsilon_0}\frac{qQ}{x^2} = \frac{1}{4\pi\epsilon_0}\frac{q(4q)}{\ell^2},$$

$$Q = 4q\left(\frac{x}{\ell}\right)^2 = \frac{4}{9}\,q \quad \underline{\text{Ans}}.$$

The same result is obtained when equilibrium of the +4q charge is examined.

26-9

(a) Draw a free-body diagram of one of the balls; the electrical force F is of repulsion. Assuming equilibrium,

$$T\cos\theta - mg = 0,$$

$$T\sin\theta - F = 0.$$

Hence,

$$\tan\theta = F/mg.$$

Now set $\tan\theta \approx \sin\theta = \frac{1}{2}x/\ell$, and

$$F = (1/4\pi\epsilon_0)q^2/x^2$$

to obtain

$$\frac{\frac{1}{2}x}{\ell} \quad \frac{1}{4\pi\epsilon_0}\frac{q^2}{x^2},$$

$$x = [\frac{1}{2\pi\epsilon_0}\frac{q^2\ell}{mg}]^{1/3}.$$

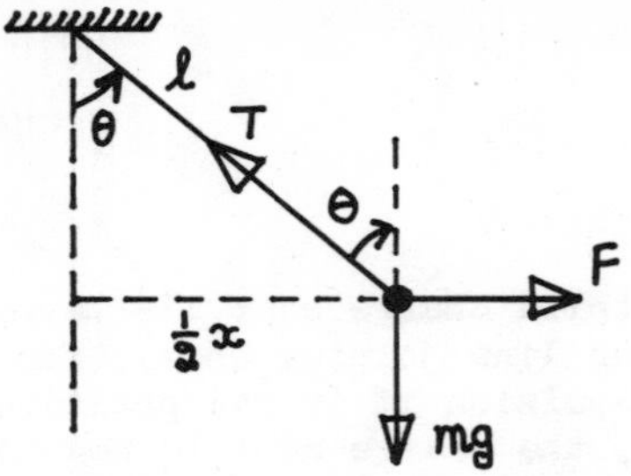

(b) Setting $x = 5$ cm $= 0.05$ m, $\ell = 120$ cm $= 1.2$ m, $m = 10$ g $= 0.01$ kg, $g = 9.8\ m/s^2$, $1/4\pi\epsilon_0 = 9 \times 10^9\ N\cdot m^2/C^2$ gives $q = 2.4 \times 10^{-8}$ C. Since repulsive forces (F) are involved, the charges on the balls must either be both positive or both negative.

26-13

Assume that the spheres are small enough so that the charges are always uniformly distributed over their surfaces, allowing them to be treated as point charges (radii << 0.5 m). Let the initial charges be Q_1, $-Q_2$, with $Q_1, Q_2 > 0$. Then, before being connected by the wire,

$$\frac{1}{4\pi\epsilon_0}\frac{Q_1Q_2}{r^2} = F,$$

where $r = 0.5$ m. Thus,

$$Q_1Q_2 = (0.108)(0.5)^2/(9 \times 10^9) = 3 \times 10^{-12}.$$

When the spheres are connected with the wire they and the wire form a single conductor carrying a charge $Q_1 - Q_2$. When equilibrium is reached, the wire will carry virtually no charge since its surface area is very small compared with the areas of the spheres; these spheres, being identical, will each carry half the net charge: $\frac{1}{2}(Q_1 - Q_2)$. Then,

$$\frac{1}{4\pi\epsilon_0} \cdot \frac{[\frac{1}{2}(Q_1 - Q_2)]^2}{r^2} = 0.0360.$$

$$Q_1 - Q_2 = \pm 2 \times 10^{-6}\ C.$$

But it was determined above that $Q_1 = 3 \times 10^{-12}/Q_2$. Combining this with the previous equation,

$$Q_2^2 \pm (2 \times 10^{-6})Q_2 - 3 \times 10^{-12} = 0,$$

$$Q_2 = -3 \times 10^{-6}\ \text{C},\ Q_1 = 1 \times 10^{-6}\ \text{C, upper sign;}$$

$$Q_2 = 3 \times 10^{-6}\ \text{C},\ Q_1 = -1 \times 10^{-6}\ \text{C, lower sign.}$$

Hence the initial charges on the spheres are 3×10^{-6} C, 1×10^{-6} C but of opposite sign.

26-17

(a) It is desired that

$$GM_eM_m/r^2 = (1/4\pi\epsilon_0)q^2/r^2,$$

$$q = (G4\pi\epsilon_0 M_e M_m)^{\frac{1}{2}}.$$

Since the mass of the moon $M_m \approx \frac{1}{81} M_e$, this gives

$$q = \left(\frac{6.67 \times 10^{-11}}{9 \times 10^9}\right)^{\frac{1}{2}}(5.98 \times 10^{24})/9,$$

$$q = 5.72 \times 10^{13}\ \text{C} \quad \underline{\text{Ans}}.$$

(b) The distance between earth and moon does not enter numerically because the gravitational and electrostatic forces are both inverse square in the distance, so that it drops out.

(c) Each hydrogen atom contributes 1.6×10^{-19} C of positive charge. Hence $(5.72 \times 10^{13})/(1.6 \times 10^{-19}) = 3.575 \times 10^{32}$ atoms are needed. In mass this is $(3.575 \times 10^{32})(1.66 \times 10^{-27}\ \text{kg})) = 5.935 \times 10^5$ kg. On earth, this is equivalent to 13.09×10^5 lb = 650 tons.

26-25

Compute the force F on the charge located at the origin of the coordinate system shown. All charges are equal:

$$q_i = q,\ i = 1, 2, \ldots 7,$$

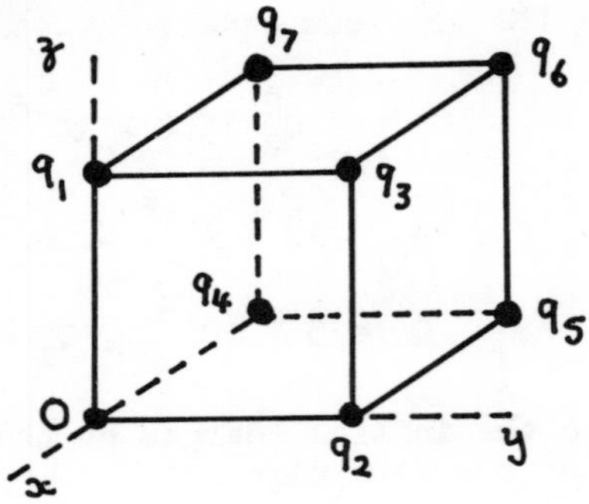

and all the forces are repulsive. The force $\vec{F}_i$ due to charge q_i is

$$\vec{F}_i = -(q^2/4\pi\epsilon_0)\vec{r}_i/r_i^3,$$

where $\vec{r}_i$ is the position vector of the ith charge with respect to the origin O. From the sketch,

$$\vec{r}_1 = a\vec{k},\ r_1 = a;$$
$$\vec{r}_2 = a\vec{j},\ r_2 = a;$$
$$\vec{r}_3 = a\vec{j} + a\vec{k},\ r_3 = a\sqrt{2};$$
$$\vec{r}_4 = -a\vec{i},\ r_4 = a;$$
$$\vec{r}_5 = -a\vec{i} + a\vec{j},\ r_5 = a\sqrt{2};$$
$$\vec{r}_6 = -a\vec{i} + a\vec{j} + a\vec{k},\ r_6 = a\sqrt{3};$$
$$\vec{r}_7 = -a\vec{i} + a\vec{k},\ r_7 = a\sqrt{2}.$$

Let $C = (1/4\pi\epsilon_0)q^2/a^3$; the force on the charge becomes

$$\vec{F} = \Sigma\vec{F}_i = -C[a\vec{k} + a\vec{j} + \frac{1}{2\sqrt{2}}(a\vec{j} + a\vec{k}) - a\vec{i} + \frac{1}{2\sqrt{2}}(-a\vec{i} + a\vec{j})$$
$$+ \frac{1}{3\sqrt{3}}(-a\vec{i} + a\vec{j} + a\vec{k}) + \frac{1}{2\sqrt{2}}(-a\vec{i} + a\vec{k})],$$
$$\vec{F} = -Ca(1 + \frac{1}{\sqrt{2}} + \frac{1}{3\sqrt{3}})(-\vec{i} + \vec{j} + \vec{k}),$$
$$\vec{F} = (0.15116)\frac{q^2}{\epsilon_0 a^2}(\vec{i} - \vec{j} - \vec{k}).$$

(a) Since $|\vec{i} - \vec{j} - \vec{k}| = \sqrt{3}$, $F = 0.262q^2/\epsilon_0 a^2$.

(b) $\vec{i} - \vec{j} - \vec{k} = -\vec{r}_6/a$, and therefore the direction of $\vec{F}$ is along the body diagonal, but directed away from the cube.

26-26

The forces on the rod are the weight W and the two electrostatic forces

$$\frac{1}{4\pi\epsilon_0}\frac{qQ}{h^2},\quad \frac{1}{4\pi\epsilon_0}\frac{2qQ}{h^2}$$

on the left and right ends, respectively. It is assumed that no force acts at the bearing. For equilibrium the forces must sum to zero, so that, since the electrostatic forces act upward and the weight W downward,

$$W = 3\,\frac{1}{4\pi\epsilon_0}\frac{qQ}{h^2}.$$

Taking moments about the left end,

$$Wx = \frac{1}{4\pi\epsilon_0}\frac{2qQ}{h^2}\,\ell.$$

Dividing the last two equations gives $x = 2\ell/3$ Ans. Also, from the first equation,

$$h = \left(\frac{3}{4\pi\epsilon_0}\frac{qQ}{W}\right)^{\frac{1}{2}} \quad \underline{\text{Ans}}.$$

26-27

For the mass m of the electron substitute the reduced mass μ

$$\mu = \frac{mM}{m + M},$$

where M is the mass of the proton. With this done the proton can be considered at rest throughout the motion. By the work-energy theorem,

$$W = \Delta K = \tfrac{1}{2}\mu(2v_0)^2 - \tfrac{1}{2}\mu v_0^2 = \tfrac{3}{2}\mu v_0^2.$$

By analogy with a corresponding gravitational problem, for which $W = GMm/r$, set

$$W = \frac{1}{4\pi\epsilon_0}\frac{e^2}{r},$$

r the desired distance. Hence

$$\tfrac{3}{2}\mu v_0^2 = \frac{1}{4\pi\epsilon_0}\frac{e^2}{r}.$$

Since $M \approx 2000m$, $\mu \approx m$ and therefore

$$r = \frac{2}{3}\frac{1}{4\pi\epsilon_0}(e^2/mv_0^2).$$

Substitution of numerical values gives $r = 1.6 \times 10^{-9}$ m <u>Ans</u>.

CHAPTER 27

27-5

(a) Since $\vec{F} = q\vec{E}$, the electric field E is

$$E = \frac{F}{q} = \frac{3 \times 10^{-6}}{2 \times 10^{-9}} = 1500 \text{ N/C} \quad \underline{\text{Ans}}.$$

(b) In part (a) the charge of the particle was negative (the sign was disregarded since only the magnitude of the field was required) and therefore $\vec{F}$ is directed opposite to $\vec{E}$, that is, upwards. If a proton is placed in this field, the force it feels will be directed parallel to $\vec{E}$ (i.e., upward) since the charge (e) of the proton is positive. The magnitude of the force will be

$$F_e = eE = (1.6 \times 10^{-19} \text{ C})(1500 \text{ N/C}) = 2.4 \times 10^{-16} \text{ N} \quad \underline{\text{Ans}}.$$

(c) The gravitational force on the proton is

$$F_g = mg = (1.67 \times 10^{-27} \text{ kg})(9.8 \text{ m/s}^2) = 1.6366 \times 10^{-26} \text{ N} \quad \underline{\text{Ans}}.$$

(d) The ratio of these forces is

$$F_e/F_g = \frac{2.40 \times 10^{-16}}{1.6366 \times 10^{-26}} = 1.47 \times 10^{10} \quad \underline{\text{Ans}}.$$

27-6

The period of a simple pendulum (small conducting sphere with a radius much less than the length of the string) is

$$T = 2\pi(\ell/g_e)^{\frac{1}{2}},$$

where g_e is the acceleration due to the "effective" gravitational field. This will be

$$g_e = g \pm a,$$

g being the acceleration due to gravity proper and 'a' the acceleration imparted by the electric force. This latter is qE/m since $F_e = qE = ma$. Therefore,

$$g_e = g \pm qE/m.$$

If a is directed down (like g) then $g_e = g + qE/m$; this will be the case with the lower plate charged negatively (the electric field is then down and the sphere, being charged positively, experiences an electrical force in the same direction as E). With the lower plate charged positively, a is directed upward, which is opposite to g. Hence gravity is reduced effectively and $g_e = g - qE/m$. If E is large enough ($qE/m > g$), "gravity is reversed" and the pendulum oscillates about the topmost point of its circular trajectory. Of course, with $qE/m = g$, "gravity is neutralised" and the period is infinite (no tendency to oscillate). To summarize:

(a) lower plate positive:

$$T = 2\pi\left(\frac{\ell}{g - qE/m}\right)^{\frac{1}{2}}, \quad qE/m < g;$$

$$T = \infty, \quad qE/m = g;$$

$$T = 2\pi\left(\frac{\ell}{qE/m - g}\right)^{\frac{1}{2}}, \quad g < qE/m.$$

(b) Lower plate negative:

$$T = 2\pi\left(\frac{\ell}{g + qE/m}\right)^{\frac{1}{2}}.$$

27-24

The equilibrium position of the electron is at the center of the ring. The force on the electron a distance x away from this point along the ring's axis is

$$F = eE = \frac{1}{4\pi\epsilon_0}\frac{eqx}{(a^2 + x^2)^{3/2}},$$

and is attractive, being directed back toward the ring's center. If $x \ll a$, the x^2 in the denominator can be deleted to give

$$F = \frac{1}{4\pi\epsilon_0}\frac{eq}{a^3}\,x.$$

By Newton's second law,

$$\frac{1}{4\pi\epsilon_0}\frac{eq}{a^3}x = -m\frac{d^2x}{dt^2}.$$

This has the same form as the equation for simple harmonic motion,

$$\frac{d^2x}{dt^2} + \omega^2 x = 0.$$

Comparing the last two equations gives

$$\omega = [\frac{eq}{4\pi\epsilon_0 ma^3}]^{\frac{1}{2}}.$$

27-25

(a)

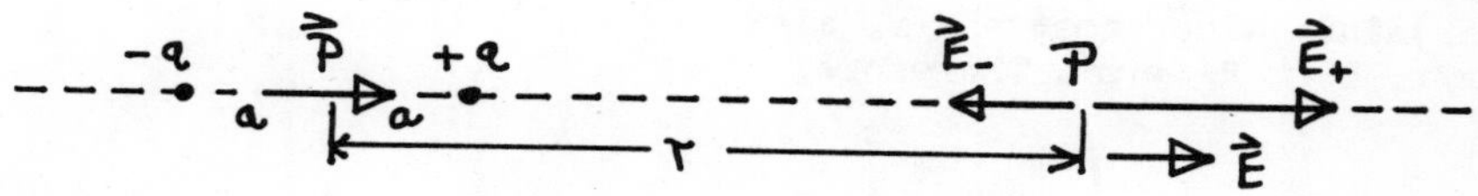

Let the point P be a distance r from the center of the dipole, along its axis, so that P is a distance r-a from the positive charge and r+a from the negative. The fields from these charges are oppositely directed, so that if $\vec{E} = \vec{E}_+ + \vec{E}_-$, then

$$E = E_+ - E_-,$$

$$E = \frac{1}{4\pi\epsilon_0}\frac{q}{(r-a)^2} - \frac{1}{4\pi\epsilon_0}\frac{q}{(r+a)^2},$$

$$E = \frac{q}{4\pi\epsilon_0}[\frac{1}{(r-a)^2} - \frac{1}{(r+a)^2}] = \frac{q}{4\pi\epsilon_0}[\frac{1}{r^2(1-a/r)^2} - \frac{1}{r^2(1+a/r)^2}],$$

$$E = \frac{q}{4\pi\epsilon_0 r^2}[\frac{1}{(1-a/r)^2} - \frac{1}{(1+a/r)^2}],$$

$$E \approx \frac{q}{4\pi\epsilon_0 r^2}[(1+2a/r) - (1-2a/r)] = \frac{q}{4\pi\epsilon_0 r^2}(4a/r),$$

$$E = \frac{2aq}{2\pi\epsilon_0 r^3} = \frac{p}{2\pi\epsilon_0}\frac{1}{r^3},$$

if $a/r \ll 1$ and $p = 2aq$.

(b) Since $\vec{p}$ points from -q to +q, then $\vec{E}$ is parallel to $\vec{p}$.

27-29

Due to the element of charge λdz located at Q as shown, the contribution to the electric field at P is

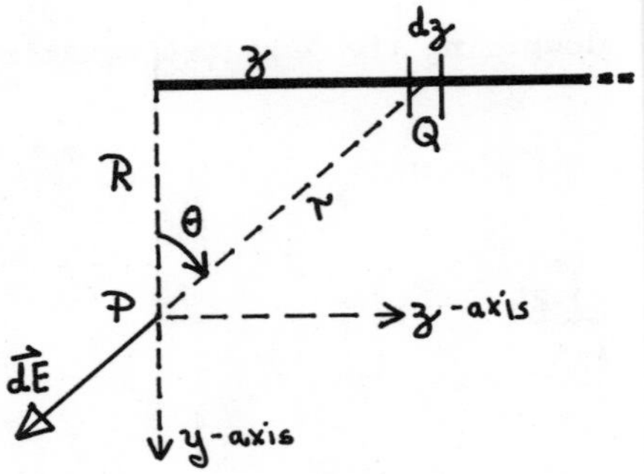

$$dE = \frac{1}{4\pi\epsilon_0}\frac{\lambda dz}{r^2},$$

along QP extended. Taking the components, $dE_y = (dE)\cos\theta$, $dE_z = -(dE)\sin\theta$; also, $\cos\theta = R/r$, $\sin\theta = z/r$, $r^2 = R^2 + z^2$. Therefore,

$$E_y = \int_0^\infty \frac{\lambda}{4\pi\epsilon_0}\frac{1}{R^2 + z^2}\frac{R}{(R^2 + z^2)^{\frac{1}{2}}}\,dz = \frac{\lambda}{4\pi\epsilon_0 R};$$

$$E_z = -\int_0^\infty \frac{\lambda}{4\pi\epsilon_0}\frac{1}{R^2 + z^2}\frac{z}{(R^2 + z^2)^{\frac{1}{2}}}\,dz = -\frac{\lambda}{4\pi\epsilon_0 R}.$$

In magnitude, then, $E_y = E_z$ and therefore $\theta = 45°$, where θ is the angle made by $\vec{E} = \vec{E}_y + \vec{E}_z$ with the rod.

27-30

Consider the contribution to the field from a narrow circular strip, radius r, of charge that is aligned perpendicular to the axis of the hemisphere, as shown. The area of the strip is

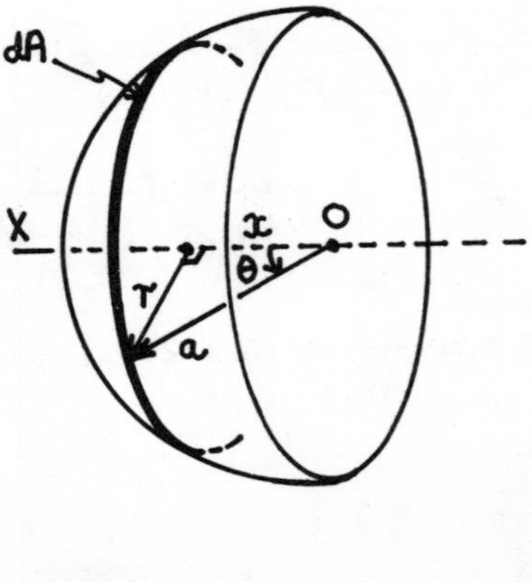

$$dA = 2\pi r(a\,d\theta) = 2\pi(a\sin\theta)(a\,d\theta),$$

$$dA = 2\pi a^2\sin\theta\,d\theta,$$

and it carries a charge

$$dq = \frac{q}{2\pi a^2}dA = q\sin\theta d\theta,$$

since q is the charge and $2\pi a^2$ the area of the hemisphere. By Example 5, the field at O due to this strip is

$$dE = \frac{1}{4\pi\epsilon_0}\frac{(dq)\,x}{(r^2 + x^2)^{3/2}}$$

along the negative X-direction. But $x = a\cos\theta$, $r^2 = a^2 - x^2$ so that

$$dE = \frac{1}{4\pi\epsilon_0}\frac{(q\sin\theta d\theta)(a\cos\theta)}{a^3}.$$

Integrating over similar strips to cover the hemisphere,

$$E = \frac{q}{4\pi\epsilon_0 a^2}\int_0^{\frac{1}{2}\pi}\sin\theta\cos\theta d\theta = \frac{q}{8\pi\epsilon_0 a^2} \quad \underline{\text{Ans}}.$$

27-32

This problem is similar to number 27-30 above; here the circular strips of radius x have area

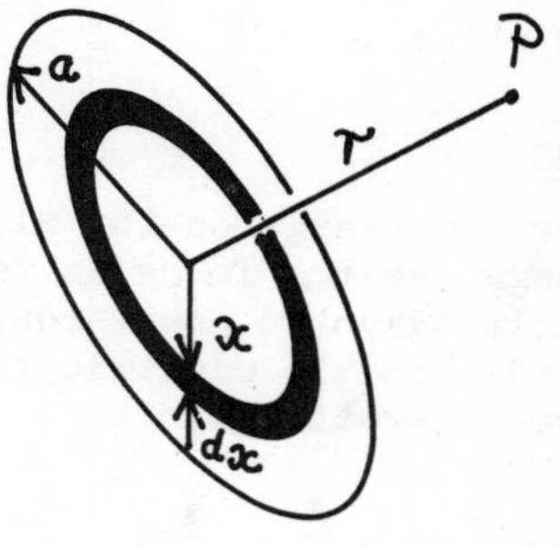

$$dA = 2\pi x dx$$

and carry a charge

$$dq = \sigma dA = \sigma 2\pi x dx.$$

By Example 5, the field at P due to the strip shown is

$$dE = \frac{1}{4\pi\epsilon_0}\frac{(\sigma 2\pi x dx)\,r}{(r^2 + x^2)^{3/2}}$$

along OP extended, if the charge is positive. Hence, the field due to the entire disc is

$$E = \frac{2\pi\sigma r}{4\pi\epsilon_0}\int_0^a \frac{x\,dx}{(r^2 + x^2)^{3/2}} = \frac{\sigma}{2\epsilon_0}\left[1 - \frac{r}{(a^2 + r^2)^{\frac{1}{2}}}\right] \quad \underline{\text{Ans.}}$$

<u>27-33</u>

The distances of P to the two positive charges are r-a and r+a, and the distance to the two negative charges is r. Taking account of the directions of the three contributions to the field at P, and carrying out the approximation, for large distances, used in Problem 27-25 to the third term in the series yields

$$E = \frac{1}{4\pi\epsilon_0}\left[\frac{q}{(r - a)^2} + \frac{q}{(r + a)^2} - \frac{2q}{r^2}\right],$$

$$E = \frac{q}{4\pi\epsilon_0 r^2}\left[\frac{1}{(1 - a/r)^2} + \frac{1}{(1 + a/r)^2} - 2\right],$$

$$E \approx \frac{q}{4\pi\epsilon_0 r^2}[(1 + 2a/r + 3a^2/r^2) + (1 - 2a/r + 3a^2/r^2) - 2],$$

$$E = \frac{q}{4\pi\epsilon_0 r^2}(6a^2/r^2) = \frac{3Q}{4\pi\epsilon_0 r^4},$$

if $Q = 2qa^2$.

<u>27-38</u>

Since the charge on the electron is negative the force it feels will be directed downwards, opposite to the electric field. The acceleration is

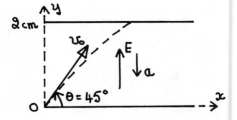

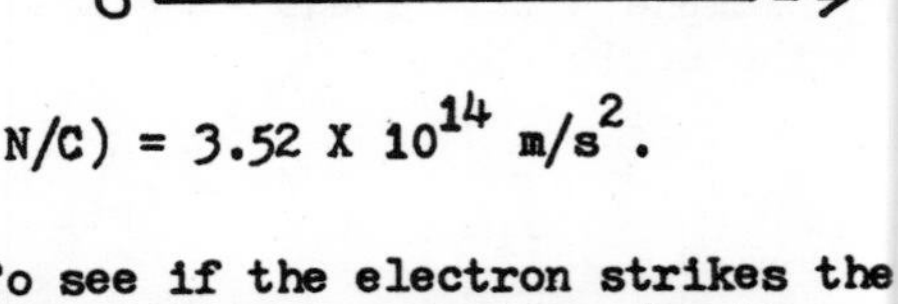

$$a = \frac{F}{m} = \left(\frac{e}{m}\right)E,$$

$$a = (1.76 \times 10^{11}\ \text{C/kg})(2000\ \text{N/C}) = 3.52 \times 10^{14}\ \text{m/s}^2.$$

Clearly, gravity can be ignored. To see if the electron strikes the top plate solve for t, the time required from projection for the electron to reach a height y = 2 cm = 0.02 m. Since the acceleration is constant,

$$y = \tfrac{1}{2}at^2 + v_{0y}t.$$

But $v_{0y} = v_0\sin45° = 3\sqrt{2} \times 10^6$ m/s; set $y = 0.02$ m to obtain

$$0.02 = \tfrac{1}{2}(-3.52 \times 10^{14})t^2 + (3\sqrt{2} \times 10^6)t,$$

$$t = 6.428 \times 10^{-9}\ \text{s},$$

choosing the smaller solution of the quadratic equation (the other applies to the descending part of the trajectory if the motion is not interrupted by the upper plate). The horizontal distance x traveled in this time is

$$x = v_{0x}t = (3\sqrt{2} \times 10^6\ \text{m/s})(6.428 \times 10^{-9}\ \text{s}) = 0.0273\ \text{m},$$

$$x = 2.73\ \text{cm}.$$

Since 2.7 cm < 10 cm, (a) the electron will strike the upper plate (b) at a distance 2.73 cm from the left edge.

<u>27-41</u>

(a) At the point P shown in the sketch the electric field is

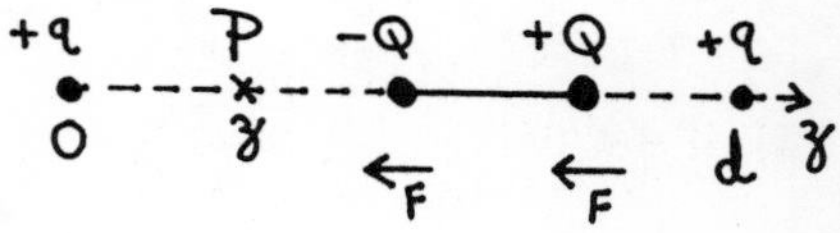

$$E = \frac{1}{4\pi\epsilon_0}\frac{q}{z^2} - \frac{1}{4\pi\epsilon_0}\frac{q}{(d - z)^2},$$

and therefore

$$\frac{dE}{dz} = -\frac{q}{2\pi\epsilon_0}\left[\frac{1}{z^3} + \frac{1}{(d - z)^3}\right].$$

If P is midway between the charges, $z = \tfrac{1}{2}d$ and dE/dz becomes

$$\frac{dE}{dz} = -\frac{8q}{\pi\epsilon_0 d^3}\quad \underline{\text{Ans.}}$$

(b) A dipole has extent and therefore the force it feels cannot be determined from the value of the electric field at just one point. For example, when oriented as shown above, the force on the dipole is to the left.

27-42

The torque on a dipole is $\tau = pE\sin\theta$ where θ is the displacement from equilibrium. But $\tau = I\alpha = -pE\sin\theta$, the minus sign added now since the torque tends to return the dipole to its equilibrium $\theta = 0$ orientation. For small displacements $\sin\theta \approx \theta$, and hence

$$\tau = I\alpha$$

$$-pE\theta = I\frac{d^2\theta}{dt^2},$$

$$\frac{d^2\theta}{dt^2} + \left(\frac{pE}{I}\right)\theta = 0.$$

By comparison with the equation for simple harmonic motion

$$\frac{d^2x}{dt^2} + \omega^2 x = 0,$$

it is evident that

$$\nu = \frac{\omega}{2\pi} = \frac{1}{2\pi}\left(\frac{pE}{I}\right)^{\frac{1}{2}} \quad \underline{\text{Ans}}.$$

28-8

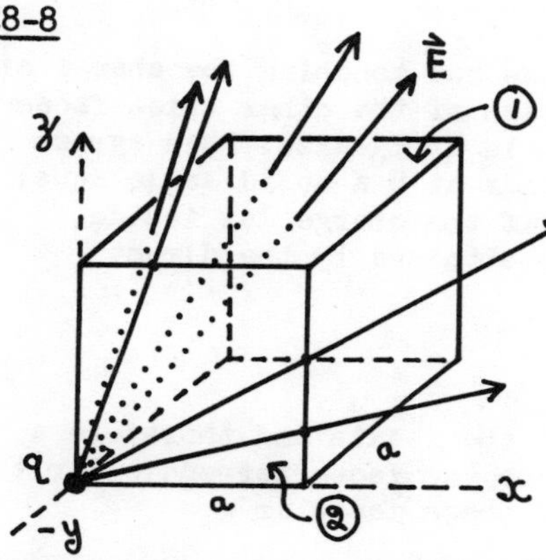

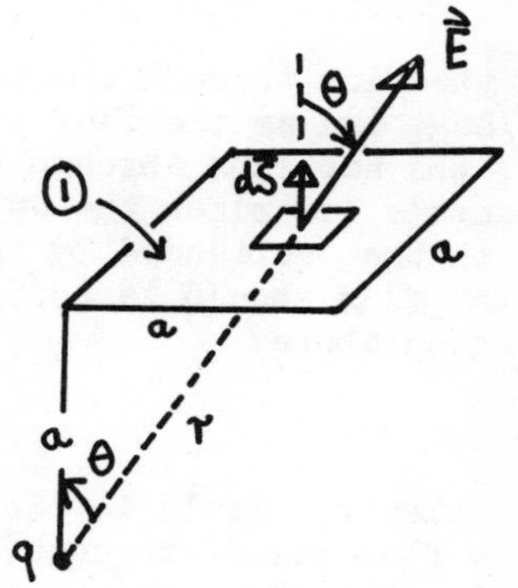

Clearly the flux through each of the three faces (such as number 2) that meet at the charge is zero since $\vec{E}$ is perpendicular to $d\vec{S}$ on these surfaces. The flux through each of the other surfaces (represented by surface 1) are identical and can be computed as follows. The electric field strength is

$$E = \frac{1}{4\pi\epsilon_0}\frac{q}{r^2} = \frac{1}{4\pi\epsilon_0}\frac{q}{a^2 + x^2 + y^2};$$

also,

$$\cos\theta = \frac{a}{r}, \quad dS = dxdy.$$

Therefore $\Phi_E = \int\vec{E}\cdot d\vec{S}$ is simply

$$\Phi_E = \frac{aq}{4\pi\epsilon_0}\int_0^a\int_0^a \frac{dx\ dy}{(a^2 + x^2 + y^2)^{3/2}},$$

$$\Phi_E = \frac{aq}{4\pi\epsilon_0}\int_0^a \frac{x\ dy}{(a^2 + y^2)(a^2 + x^2 + y^2)^{\frac{1}{2}}}\Bigg|_0^a,$$

$$\Phi_E = \frac{a^2 q}{4\pi\epsilon_0}\int_0^a \frac{dy}{(a^2 + y^2)(2a^2 + y^2)^{\frac{1}{2}}} = \frac{q}{4\pi\epsilon_0}\tan^{-1}\left[\frac{y}{(2a^2 + y^2)^{\frac{1}{2}}}\right]_0^a = \frac{q}{24\epsilon_0}.$$

Hence, the flux through the three faces not touching the charge at 0 is $q/8\epsilon_0$, and as the flux through each of the other three faces is nil, the net flux through the cube is $q/8\epsilon_0$ also. This agrees with Gauss's law since the cube subtends at 0 a solid angle equal to 1/8 th that subtended by the cube if the charge lay inside. Hence the flux should be $(q/\epsilon_0)/8$, as obtained by the direct calculation above.

28-10

(a) The electric field is parallel to the x-axis and therefore a non-zero flux passes through only those two faces perpendicular to this axis. The flux through either of these faces is

$$\Phi = \pm ES = \pm bx^{\frac{1}{2}}a^2.$$

For the left-hand face x = a and the negative sign is appropriate since the lines of force enter the cube through this face. The lines leave via the right-hand face where x = 2a; the net flux, then, is

$$\Phi_E = b(2a)^{\frac{1}{2}}a^2 - b(a)^{\frac{1}{2}}a^2 = ba^{5/2}(\sqrt{2} - 1).$$

Numerically this is

$$\Phi_E = (800\ \mathrm{N/C\cdot m^{\frac{1}{2}}})\ (0.1\ \mathrm{m})^{5/2}\ (\sqrt{2} - 1) = 1.048\ \mathrm{N\cdot m^2/C} \quad \underline{\text{Ans}}.$$

(b) By Gauss's law the charge q inside the cube must be

$$q = \epsilon_0\Phi_E = (8.854 \times 10^{-12}\ \mathrm{C^2/N\cdot m^2})(1.048\ \mathrm{N\cdot m^2/C}),$$

$$q = 9.3 \times 10^{-12}\ \mathrm{C} \quad \underline{\text{Ans}}.$$

28-20

The field due to q is

$$E_q = \frac{1}{4\pi\epsilon_0}\frac{q}{r^2}.$$

The field due to the sphere, for a < r < b, is found from Gauss's law by constructing a spherical Gaussian surface centered at r = 0. Then,

$$\epsilon_0 E \cdot 4\pi r^2 = q_{enc} = \int \rho dV = \int_a^r \frac{A}{r} 4\pi r^2 dr = 2\pi A(r^2 - a^2),$$

so that

$$E = \frac{A}{2\epsilon_0}(1 - \frac{a^2}{r^2})$$

radially in or out. In order that $\vec{E}_q + \vec{E}$ be independent of r, it must be true that

$$\frac{Aa^2}{2\epsilon_0} = \frac{q}{4\pi\epsilon_0},$$

$$A = \frac{q}{2\pi a^2} \quad \underline{\text{Ans}},$$

A and q having the same sign.

28-21

(a) From Equation 28-9,

$$E = \frac{1}{4\pi\epsilon_0}\frac{qr}{R^3}.$$

But the charge density is $\rho = q/(\frac{4}{3}\pi R^3)$ so that $E = \rho r/3\epsilon_0$. If the charge is positive, $\vec{E}$ is radially out. Since $\vec{r}$ is radially out also

$$\vec{E} = \frac{\rho}{3\epsilon_0}\vec{r} \quad \underline{\text{Ans}}.$$

(b) Assume the charge is positive. If A be any point in the cavity, then from superposition it is expected that, at A,

(Field $\vec{E}'$ due to sphere of charge density ρ centered at O' and of radius R')

+ (Field $\vec{E}_0$ due to the given distribution of charge)

= (Field $\vec{E}$ due to sphere of charge density ρ centered at O and of radius R);

here O' is the center of the cavity of radius R', O and R being the center and radius of the sphere in which the cavity is located.

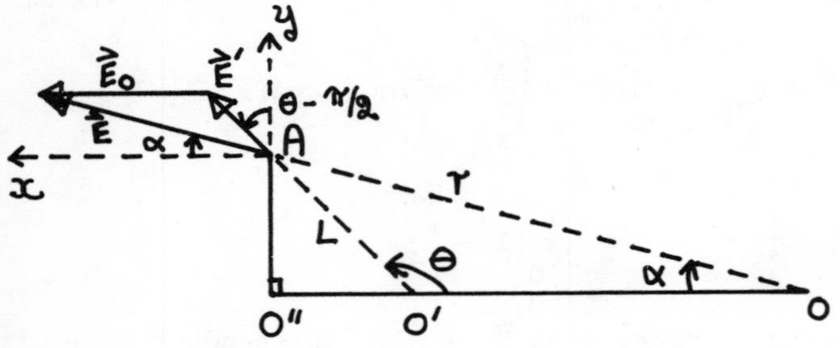

Therefore

$$\vec{E}_0 = \vec{E} - \vec{E}'.$$

Let OO' = a, OA = r, O'A = L; also, locate a point O" such that AO" is perpendicular to OO'. Choose the x-axis parallel to OO' and the y-axis perpendicular to it in the plane OO'A. Then

$$E_x = E\cos\alpha, \quad E_y = E\sin\alpha,$$

$$E'_x = E'\sin(\theta - \tfrac{\pi}{2}), \quad E'_y = E'\cos(\theta - \tfrac{\pi}{2}).$$

Then, from (a),

$$E_{0y} = E_y - E'_y = \frac{\rho r}{3\epsilon_0}\sin\alpha - \frac{\rho L}{3\epsilon_0}\sin\theta = \frac{\rho}{3\epsilon_0}(r\cdot\sin\alpha - L\cdot\sin\theta) = 0$$

by the sine rule applied to triangle AO'O. Turning to the x-component,

$$E_{0x} = E_x - E'_x = E\cos\alpha - E'\sin(\theta - \tfrac{\pi}{2}) = \frac{\rho}{3\epsilon_0}(r\cdot\cos\alpha + L\cdot\cos\theta),$$

$$E_{0x} = \frac{\rho}{3\epsilon_0}[r\cdot\cos\alpha - L\cdot\cos(\pi - \theta)] = \frac{\rho}{3\epsilon_0}(OO" - O'O") = \frac{\rho}{3\epsilon_0}\cdot OO',$$

$$E_{0x} = \frac{\rho}{3\epsilon_0}\, a.$$

Thus, vectorially, with $\vec{a}$ a vector from O to O', charge positive,

$$\vec{E}_0 = \frac{\rho}{3\epsilon_0}\vec{a}.$$

28-27

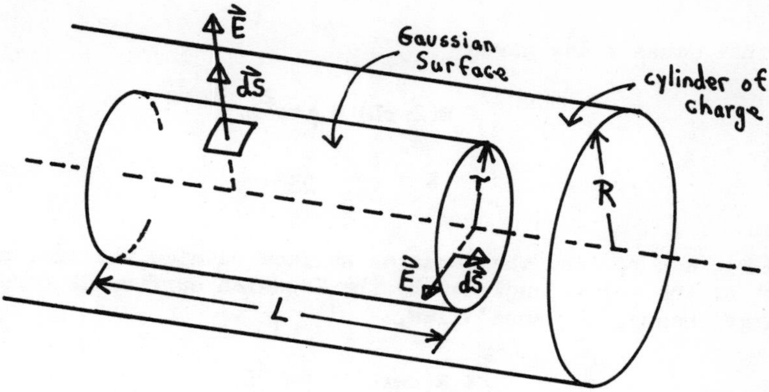

(a) By symmetry it is expected that the field will be azimuthally symmetric around the cylinder axis, the lines being radially inward if the charge is negative and outward if the charge is positive. Positive charge is assumed here. Construct a coaxial, cylindrical Gaussian surface (ends are included) of radius r and length L. Then

$$\epsilon_0 \oint \vec{E}\cdot\vec{dS} = q_{enc},$$

where q_{enc} is the charge inside the Gaussian surface over which the integral is taken. For convenience, the integral is divided into two parts, over the curved portion of the cylindrical surface and over the two end caps:

$$\epsilon_0 \int_{\substack{\text{curved}\\ \text{surface}}} \vec{E}\cdot\vec{dS} + \epsilon_0 \int_{\text{ends}} \vec{E}\cdot\vec{dS} = q_{enc}.$$

Now $\vec{E}\cdot\vec{dS} = E(dS)\cos\sphericalangle(\vec{E},\vec{dS})$. On the flat ends the angle between $\vec{E}$ and $\vec{dS}$ is 90°, the cosine of which is zero. Hence, the integral over the ends (indeed, over each end) is nil. On the curved surface the angle is zero (or 180° if the charge is negative), the cosine of which is +1 (-1). Also, since all points on the curved surface are at the same distance from the axis of the rod, E is independent of location on the curved surface. Applying these considerations,

$$\int \vec{E}\cdot\vec{dS} = \int E dS = E\int dS = E 2\pi r L,$$

over the curved part. Since the entire Gaussian surface is inside the rod, the charge enclosed, which equals the product of charge density and volume of Gaussian surface occupied by charge, becomes

$$q_{enc} = (\rho)(\pi r^2 L),$$

so that Gauss's law gives

$$\epsilon_0 E(2\pi rL) = \rho\pi r^2 L,$$

$$E = \frac{\rho r}{2\epsilon_0} \quad \underline{\text{Ans}}.$$

(b) For $R < r$, draw the Gaussian surface outside the rod; now only part of the volume enclosed by the Gaussian surface is occupied by charge. Hence, by Gauss's law,

$$\epsilon_0 E(2\pi rL) = \rho\pi R^2 L,$$

$$E = \frac{\rho R^2}{2\epsilon_0 r} \quad \underline{\text{Ans}},$$

since $\pi R^2 L$ is the volume of the Gaussian surface that contains charge.

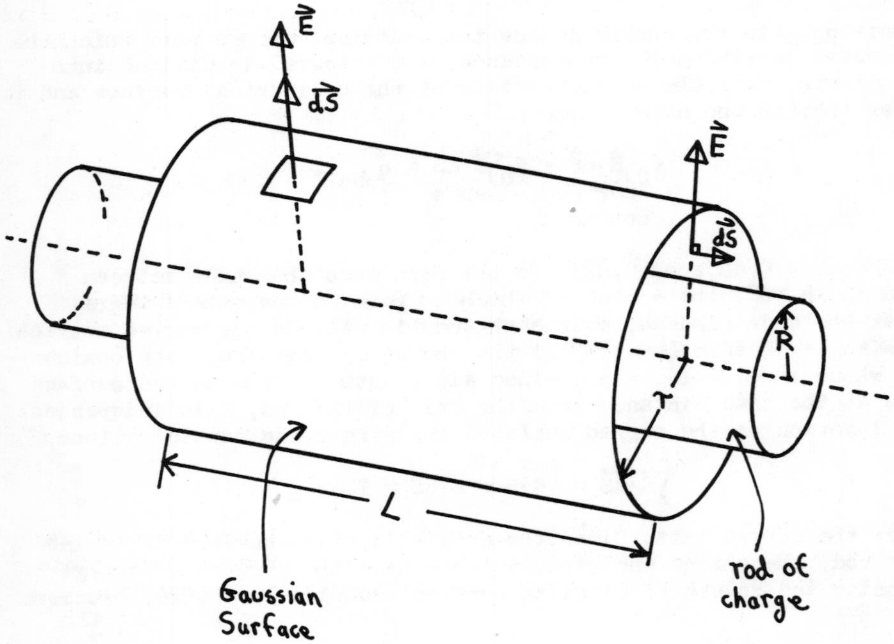

28-29

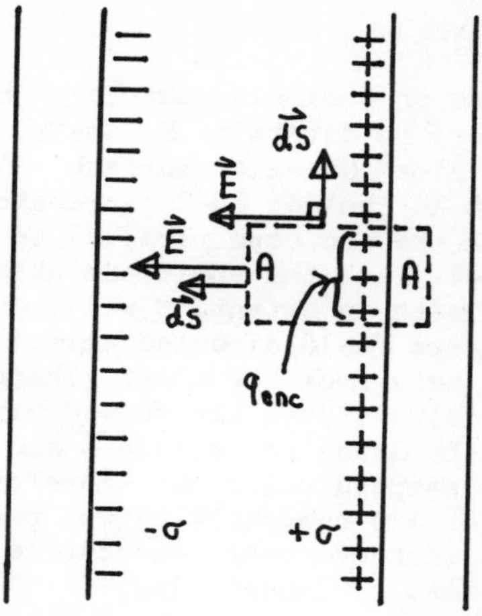

The restriction on points to be considered is tantamount to regarding the plates as infinite in two dimensions (thickness unaffected). The field lines, by symmetry, must pass, perpendicular to the sheets, from the positive to the negative charges. Under electrostatic conditions no lines can penetrate into the conducting plates, and since there is no fringing of the lines around the edges, there being no edges in effect, the field must be zero at all points not between the plates: i.e., (a) E = 0 and (c) E = 0 Ans.

(b) To calculate E at points between the sheets, construct a right rectangular cylindrical "pillbox" Gaussian surface (any shape cross sectional area A), with end caps parallel to the plates, as shown. The flux, $\int \vec{E}\cdot\vec{dS}$, over that part of the Gaussian surface lying inside the plate is zero as E = 0 there. The flux is also zero over the curved part of the surface found outside the plates: this is because here $\vec{E}$ is at right angles to $\vec{dS}$. Over the left end cap, however, $\vec{E}$ and $\vec{dS}$ are parallel, and as all points on the cap are at the same distance from the plate, Gauss's law yields

$$\epsilon_0\oint \vec{E}\cdot\vec{dS} = \epsilon_0\int_{\substack{\text{left}\\\text{cap}}} \vec{E}\cdot\vec{dS} = \epsilon_0\int E\,dS = \epsilon_0 E\int dS = \epsilon_0 EA = q_{enc}.$$

The charge enclosed by the Gaussian surface resides on that part of the plate lying inside the surface: this is q_{enc} = (charge per area) X (area) = σA. Therefore,

$$\epsilon_0 EA = \sigma A,$$

$$E = \sigma/\epsilon_0 \quad \text{Ans},$$

directed to the left. It may appear as though the presence of the negative plate was ignored, but this is not so for, if that plate was absent, the positive charge would then be found equally distributed on both sides of the plate, making $E = \frac{1}{2}\sigma/\epsilon_0$ at all points not inside the plate.

28-31

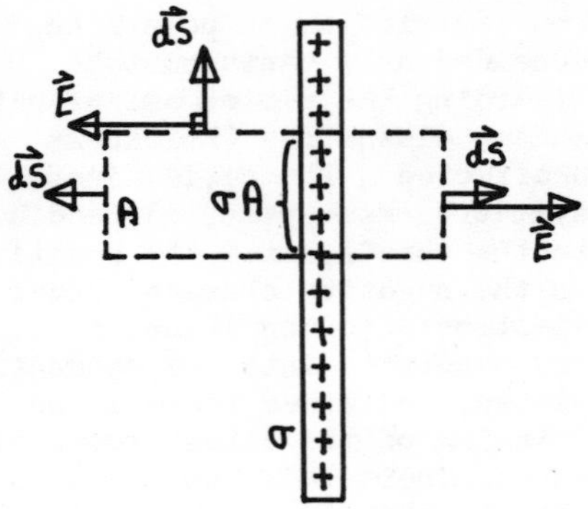

Focus on a single nonconducting sheet. To determine E construct a "pillbox" Gaussian surface, like that in Problem 28-29, oriented with the end caps parallel to the sheet. With the sheet essentially infinite in extent, E will be a uniform field directed normal to and away from the sheet (charge positive). Over the curved part of the Gaussian surface $\vec{E}$ and $\vec{dS}$ are perpendicular and therefore $\int \vec{E}\cdot d\vec{S} = 0$ over this curved part. The flux over each end cap is EA, so that by Gauss's law,

$$\epsilon_0(2EA) = q_{enc} = \sigma A,$$

$$E = \sigma/2\epsilon_0.$$

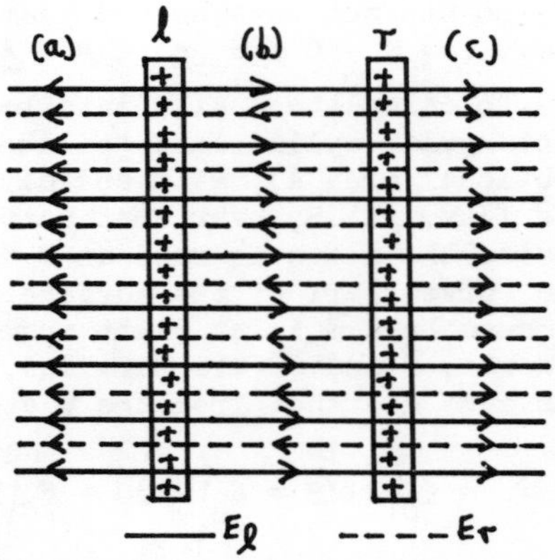

With both sheets present the total electric field $\vec{E} = \vec{E}_r + \vec{E}_\ell$, E_r and E_ℓ due to the right and left sheets. These last are equal in magnitude (both being $\sigma/2\epsilon_0$). However, it is clear (b) that between the sheets their directions are opposite giving E = 0 there. At all points not between the sheets the fields are parallel so that at these points $E = \sigma/2\epsilon_0 + \sigma/2\epsilon_0 = \sigma/\epsilon_0$, to the left in region (a) and to the right in region (c).

28-32

(a) Consider the slab to be of very large area and the charge positive. By symmetry, the field inside the slab will be directed at right angles to the slab faces, but it will vary in magnitude with distance x from the median plane of the slab. Construct a "pillbox" Gaussian surface oriented as shown. The flux over the curved part of the surface is zero, and since each end cap is at the same distance from the median plane,

$$\Phi_E = \oint \vec{E}\cdot d\vec{S} = 2EA,$$

where E = E(x) is the electric field strength at a distance x from the median plane and A is the area of each end cap. The charge enclosed by the surface is the volume charge density ρ of the slab multiplied by the volume enclosed by the Gaussian surface, since this entire volume contains charge; hence,

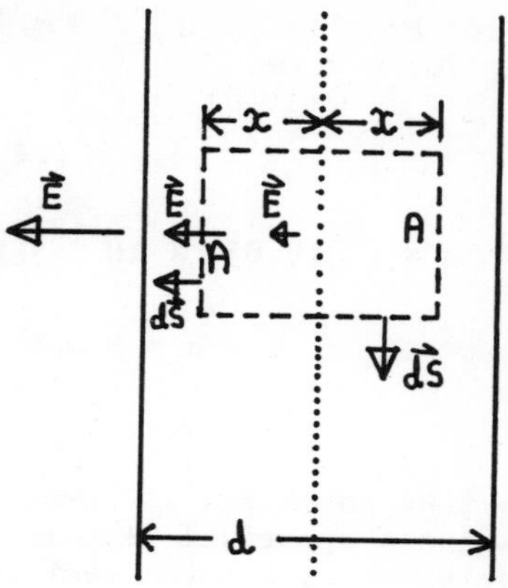

$$q_{enc} = \rho(2x\ A),$$

Thus, by Gauss's law,

$$\epsilon_0(2EA) = \rho(2xA),$$

$$E = \rho x/\epsilon_0 \quad \underline{\text{Ans}}.$$

(b) Since the field outside the slab is directed at right angles to it, the analysis for the electric field outside a single, non-conducting sheet, given in Problem 28-31, is valid here, so that $E = \sigma/2\epsilon_0$. Now σ is the charge per unit area of the slab, that is, the charge contained in a cylinder extending across the full width d of the slab and having a cross-sectional area of 1 m^2. The volume of the cylinder is $d(1)$ m^3. Hence, equating expressions for the charge contained in this cylinder expressed in terms of σ on the one hand and ρ on the other gives

$$\sigma(1) = \rho[d(1)],$$

so that $\sigma = \rho d$. Therefore $E = \rho d/2\epsilon_0$.

28-34

Draw a free-body diagram for the sphere. For equilibrium,

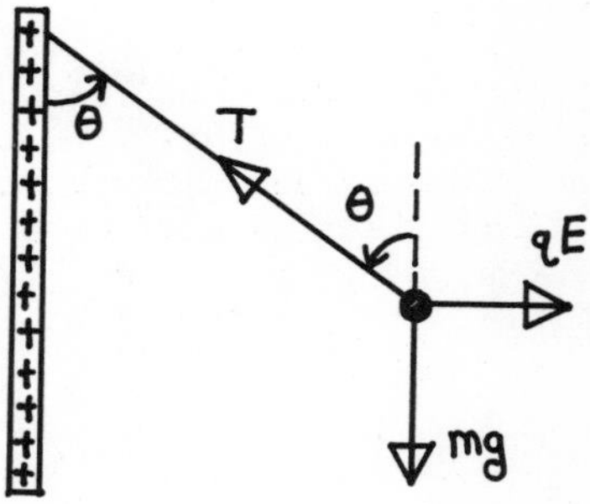

$$T\cos\theta = mg,$$

$$T\sin\theta = qE.$$

Dividing the equations gives

$$\tan\theta = \frac{qE}{mg}.$$

But, from Problem 28-31, $E = \sigma/2\epsilon_0$. Putting this into the above and solving for σ gives

$$\sigma = (2\epsilon_0 mg\tan\theta)/q,$$

$$\sigma = (2)(8.854 \times 10^{-12})(10^{-6})(9.8)(1/\sqrt{3})/(2 \times 10^{-8}),$$

$$\sigma = 5.0 \times 10^{-9}\ C/m^2 \quad \underline{Ans}.$$

28-35

Assume that no charge is located in the immediate neighbourhood of P. Imagine a spherical Gaussian surface of very small radius with its center at P. Let the test charge +q be displaced from P to any point on the Gaussian surface. For stable equilibrium the charge must experience a force that tends to push it back to P; i.e., the electrostatic field E must have an inward directed normal component everywhere on the Gaussian surface. But, by Gauss's law, this means that net negative charge, responsible for setting up E, must reside inside the Gaussian surface. This is contrary to assumption. A similar argument may be made for negative test charges.

CHAPTER 29

29-2

(a) Let the point P at which the potential is desired be at a distance r = a from the center of the sphere of charge, r < R. Then,

$$V_P = -\int_{\infty}^{P} \vec{E}\cdot\vec{d\ell} + V_{\infty} = -\int_{\infty}^{P} \vec{E}\cdot\vec{d\ell},$$

choosing the potential at infinity $V_{\infty} = 0$. For convenience, integrate along a straight line from infinity to P. Now $\vec{d\ell}$ points in the direction of integration, opposite to $\vec{dr}$ which points outward from the sphere since the center of the sphere is at r = 0. Hence,

$$\vec{E}\cdot\vec{d\ell} = E\,d\ell\cos\pi = -E\,d\ell = -E(-dr) = E\,dr,$$

assuming the charge is positive so that the electric field points outward. Substituting the last result into the integral for V_P gives

$$V_P = -\int_{\infty}^{a} E\,dr.$$

The electric field E(r) has different analytical forms for r < R, r > R:

$$E(r) = \frac{1}{4\pi\epsilon_0}\frac{q}{r^2},\ r \geq R;\ E(r) = \frac{1}{4\pi\epsilon_0}\frac{q}{R^3}r,\ r \leq R.$$

For this reason, the integral must be divided into two parts, one for points exterior and one for points interior to the sphere:

$$V_P = -\int_{\infty}^{R} \frac{q}{4\pi\epsilon_0}\frac{dr}{r^2} - \int_{R}^{a} \frac{q/R^3}{4\pi\epsilon_0}r\,dr,$$

$$V_P = -\frac{q}{4\pi\epsilon_0}\left[\left(-\frac{1}{R} + 0\right) + \frac{1}{R^3}\left(\tfrac{1}{2}a^2 - \tfrac{1}{2}R^2\right)\right],$$

$$V_P = \frac{q(3R^2 - a^2)}{8\pi\epsilon_0 R^3}.$$

Substitution of r for a gives the desired result.

(b) The nonzero value for a = 0 is reasonable since work is required to move a test charge from infinity to the center of the sphere.

29-4

(a) The work done on the electron is equal to the change in its kinetic energy; if the electron started from rest, W = eV = K, where e is the magnitude of the charge on the electron and K is its final kinetic energy. Since $K = \frac{1}{2}mv^2$ classically, if v = c then

$$\tfrac{1}{2}mc^2 = eV,$$

$$V = \tfrac{1}{2}\left(\frac{e}{m}\right)^{-1}c^2 = \tfrac{1}{2}(1.76 \times 10^{11})^{-1}(3 \times 10^8)^2,$$

$$V = 2.557 \times 10^5 \text{ V} \quad \underline{\text{Ans}}.$$

(b) The rest energy mc^2 of the electron is 0.511 MeV, as can be verified by direct calculation. Putting e = 1, the kinetic energy of the electron after being accelerated through the potential difference of (a) is 0.2557 MeV. Since W = K is relativistically correct,

$$0.2557 = (0.511)[(1 - v^2/c^2)^{-\frac{1}{2}} - 1],$$

$$1.500 = (1 - v^2/c^2)^{-\frac{1}{2}},$$

$$v = \frac{\sqrt{5}}{3}c = 0.745c \quad \underline{\text{Ans}}.$$

29-13

Put the origin at the middle charge. Then,

$$V = \frac{1}{4\pi\epsilon_0}\frac{q}{r - a} + \frac{1}{4\pi\epsilon_0}\frac{q}{r} + \frac{1}{4\pi\epsilon_0}\frac{-q}{r + a} = \frac{q}{4\pi\epsilon_0}\frac{r^2 + 2ar - a^2}{(r^2 - a^2)\,r}.$$

If a << r, this becomes, upon ignoring the a^2 terms,

$$V = \frac{q}{4\pi\epsilon_0}\left(\frac{r + 2a}{r^2}\right) = \frac{1}{4\pi\epsilon_0}\left(\frac{q}{r} + \frac{2aq}{r^2}\right) \quad \underline{\text{Ans.}}$$

The first term is the potential due to a point charge (the middle charge) and the second is the potential due to a dipole (the outer charges) at large distances compared to the dipole charge separation, along the dipole axis.

29-14

Each hydrogen atom has a charge +e in the nucleus and -e on the electron; the oxygen atom has a charge +8e in the nucleus and -8e due to its electrons. Thus, transferring the two electrons from the hydrogen atoms to the oxygen atom leaves each hydrogen atom with a charge +e, and the oxygen atom with a net charge -2e. This arrangement can be considered as two dipoles, each of dipole moment ed (d the OH separation), at an angle of 104°. The vector sum of these is

$$\vec{p}_t = \vec{p}_1 + \vec{p}_2,$$

$$p_t = 2p\cos\theta = 2(ed)\cos 52^\circ = 1.9 \times 10^{-29}\ \text{C}\cdot\text{m} \quad \underline{\text{Ans.}}$$

The direction is along the symmetry axis of the molecule, away from the oxygen atom. In reality, the electrons are "shared" between the atoms, so that the model used above is approximate only.

29-25

Let W_e = work to be done against the electrostatic force; this is

$$W_e = (-q)(V_2 - V_1) = -q\left(\frac{1}{4\pi\epsilon_0}\frac{Q}{r_2} - \frac{1}{4\pi\epsilon_0}\frac{Q}{r_1}\right) = \frac{Qq}{4\pi\epsilon_0}\left(\frac{1}{r_1} - \frac{1}{r_2}\right).$$

The initial kinetic energy of the electron can supply some of this work, however. For uniform circular motion,

$$\frac{1}{4\pi\epsilon_0}\frac{Qq}{r^2} = m\frac{v^2}{r}$$

so that

$$K = \tfrac{1}{2}mv^2 = \tfrac{1}{2}\frac{1}{4\pi\epsilon_0}\frac{Qq}{r}$$

Therefore, the kinetic energy available for work is

$$K_1 - K_2 = \frac{1}{2}\frac{Qq}{4\pi\epsilon_0}\left(\frac{1}{r_1} - \frac{1}{r_2}\right).$$

This implies that the work W that the external agent must do is

$$W = W_e - (K_1 - K_2) = \frac{Qq}{8\pi\epsilon_0}\left(\frac{1}{r_1} - \frac{1}{r_2}\right) \quad \underline{\text{Ans}}.$$

29-26

If r_{ij} is the distance between q_i and q_j it is required that

$$U = \frac{1}{4\pi\epsilon_0}\left[\frac{q_1q_2}{r_{12}} + \frac{q_1q_3}{r_{13}} + \frac{q_2q_3}{r_{23}}\right] = 0.$$

There are an infinite number of possible arrangements. For example, suppose that the charges are placed at the vertices of an equilateral triangle, so that $r_{12} = r_{13} = r_{23}$. The above condition then reduces to

$$q_1q_2 + q_1q_3 + q_2q_3 = 0.$$

Again, there are an infinite number of choices of charges that can satisfy this equation. Suppose $q_1 = q_2 = q$, say; then

$$q^2 + 2qq_3 = 0,$$

$$q_3 = -\tfrac{1}{2}q,$$

is one possibility.

29-29

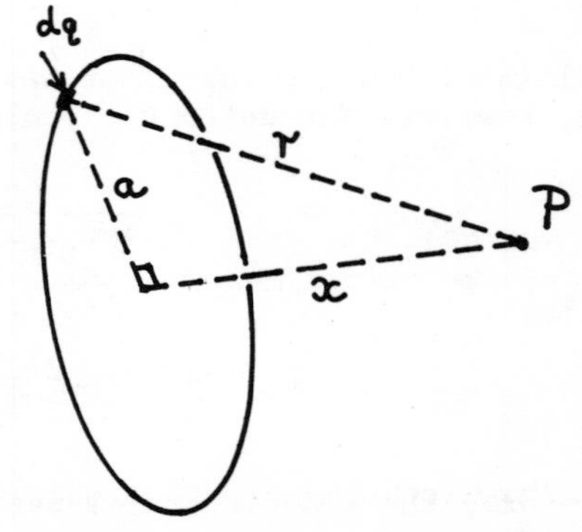

(a) Let $d\ell$ be an element of length on the ring; due to this element, carrying a charge dq, the potential at P is, since $q/2\pi a$ is the charge per unit length on the ring,

$$dV = \frac{1}{4\pi\epsilon_0}\frac{dq}{r} = \frac{1}{4\pi\epsilon_0}\frac{(q/2\pi a)\,d\ell}{(a^2 + x^2)^{\frac{1}{2}}}.$$

The total potential at P is

$$V = \int_{\text{ring}} dV = \frac{1}{4\pi\epsilon_0}\frac{(q/2\pi a)}{(a^2 + x^2)^{\frac{1}{2}}}\int d\ell,$$

noting that a and x are independent of the element's position on the ring. Now, $\int d\ell = 2\pi a$, the circumference of the ring, and hence

$$V = \frac{1}{4\pi\epsilon_0}\frac{q}{(a^2 + x^2)^{\frac{1}{2}}} \quad \underline{\text{Ans}}.$$

(b) By symmetry, the component of **E** at the axis is zero in any direction perpendicular to the axis. Thus,

$$E = E_x = -\frac{dV}{dx} = \frac{1}{4\pi\epsilon_0}\frac{xq}{(a^2 + x^2)^{3/2}} \quad \underline{\text{Ans}},$$

as found in the quoted example.

29-33

(a) With $\lambda dx = dq$ = charge on an element of length dx, the potential V at P is found from

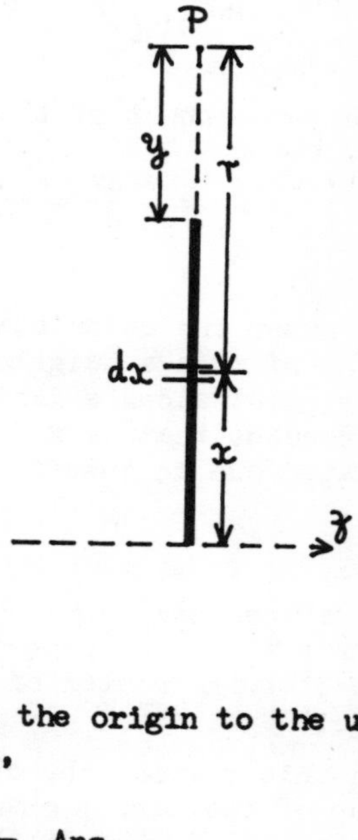

$$4\pi\epsilon_0 V = \int\frac{\lambda dx}{r} = \int_0^L \frac{\lambda dx}{(L + y) - x},$$

$$4\pi\epsilon_0 V = -\lambda\cdot\ln(L + y - x)\Big|_0^L,$$

$$4\pi\epsilon_0 V = -\lambda[\ln y - \ln(L + y)],$$

$$V = \frac{\lambda}{4\pi\epsilon_0}\ln\left(\frac{L + y}{y}\right) \quad \underline{\text{Ans}}.$$

(b) Now call the axis the y-axis and move the origin to the upper end of the charge segment, as shown. Then,

$$E_y = -\frac{dV}{dy} = \frac{\lambda}{4\pi\epsilon_0}\frac{L}{y(L + y)} \quad \underline{\text{Ans}}.$$

(c) The direction of the electric field at P due to any element of the segment of charge is directed along the y (previously the x) axis; hence $E_z = 0$ Ans. (This result cannot be derived from the potential obtained in (a) for that expression is valid only at one specific value of z, to wit z = 0, and therefore dV/dz cannot be calculated from it.)

29-34

(a) The potential dV at P due to the element of length dx shown that carries a charge $dq = \lambda dx = (kx)dx$ will be

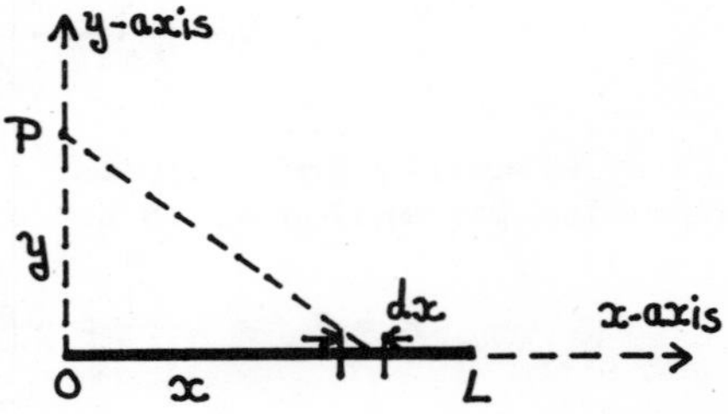

$$dV = \frac{1}{4\pi\epsilon_0}\frac{(kx)dx}{(x^2 + y^2)^{\frac{1}{2}}},$$

and therefore

$$V = \frac{k}{4\pi\epsilon_0}\int_0^L \frac{x\ dx}{(x^2 + y^2)^{\frac{1}{2}}} = \frac{k}{4\pi\epsilon_0}[(L^2 + y^2)^{\frac{1}{2}} - y] \quad \text{Ans.}$$

(b) The y-component of the electric field at P is

$$E_y = -\frac{dV}{dy} = \frac{k}{4\pi\epsilon_0}[1 - \frac{y}{(L^2 + y^2)^{\frac{1}{2}}}] \quad \text{Ans.}$$

(c) In order the calculate $E_x = -dV/dx$, V must be known as a function of x in a neighbourhood about P, but in (a) V was computed for any point along a line, each point of which has the same single x coordinate, that is x = 0. Hence, the derivative cannot be calculated due to insufficient "x information".

29-45

Let Q be the total charge on the shell.

(a) For $r > r_2$, the potential is identical with that of a point charge Q at the center of the actual distribution: that is, $V = Q/4\pi\epsilon_0 r$.

(b) In this region, the charge distribution can be considered as made up of two parts: charge Q_1 interior to r and Q_2 exterior to r. The potential V at r is the sum $V_1 + V_2$ of the potentials due to each of these distributions. The point considered is on the surface of Q_1 so that $V_1 = Q_1/4\pi\epsilon_0 r$. To find V_2 it is necessary to compute the potential V* in the cavity of a thick shell of charge. Consider

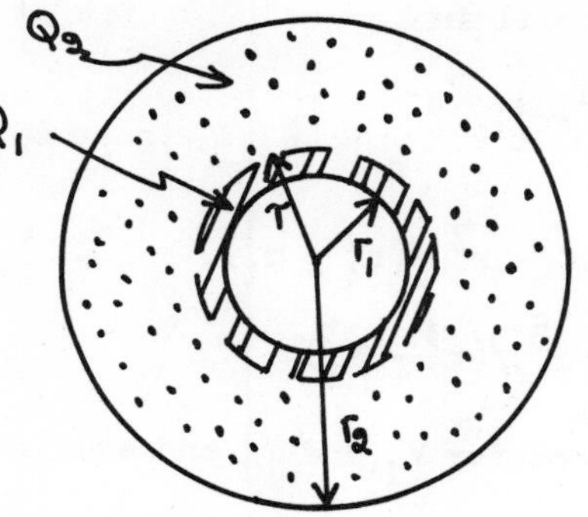

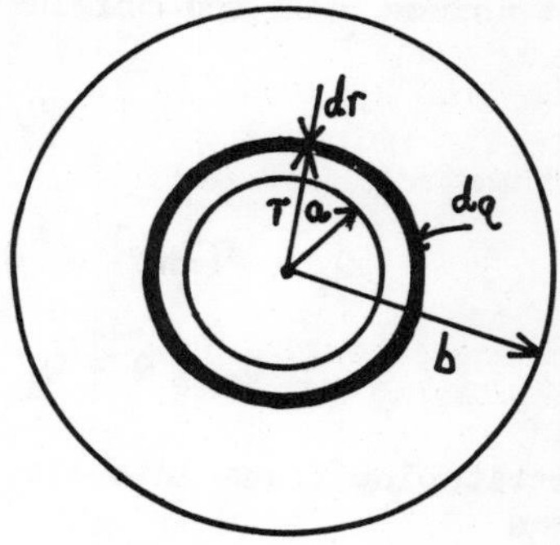

such a shell carrying charge q and of inner radius a and outer radius b. The shell can be divided into many very thin shells, such as the one shown of radius r and thickness dr. If the charge on this shell is dq, the potential dV* due to the shell is

$$dV^* = dq/4\pi\epsilon_0 r.$$

But,

$$dq = [q/(\tfrac{4}{3}\pi b^3 - \tfrac{4}{3}\pi a^3)](4\pi r^2 dr)$$

the term in square brackets being the charge per unit volume on the thick shell and $4\pi r^2 dr$ being the volume of the thin shell of radius r. Hence,

$$dV^* = \frac{q\ rdr}{\frac{4}{3}\pi\epsilon_0(b^3 - a^3)}$$

and

$$V^* = \int_a^b dV^* = \frac{3q}{8\pi\epsilon_0}\frac{b^2 - a^2}{b^3 - a^3}.$$

Applying this result to V_2, put $b = r_2$, $a = r$, $q = Q_2$ to get

$$V_2 = \frac{3Q_2}{8\pi\epsilon_0}\frac{r_2^2 - r^2}{r_2^3 - r^3}.$$

It remains to calculate Q_1 and Q_2 in terms of Q. The charge per unit volume ρ on the original thick shell is

$$\rho = Q/(\tfrac{4}{3}\pi r_2^3 - \tfrac{4}{3}\pi r_1^3),$$

so that

$$Q_1 = \rho(\tfrac{4}{3}\pi r^3 - \tfrac{4}{3}\pi r_1^3) = Q(r^3 - r_1^3)/(r_2^3 - r_1^3),$$

$$Q_2 = Q - Q_1 = Q(r_2^3 - r^3)/(r_2^3 - r_1^3).$$

Substituting these into the expressions for V_1 and V_2 and adding gives

$$V = \frac{\rho}{3\epsilon_0}(\tfrac{3}{2}r_2^2 - \tfrac{1}{2}r^2 - \frac{r_1^3}{r}) \quad \underline{\text{Ans}}.$$

(c) For $r < r_1$, i.e., inside the cavity, use the result from (b), but now set $r = r_1$ (the inner boundary of the original shell), so that

$$V = \frac{\rho}{2\epsilon_0}(r_2^2 - r_1^2) \quad \underline{\text{Ans}}.$$

(d) The appropriate potentials must agree at the boundaries of the regions in which they are valid: that is, $V_a = V_b$ at $r = r_2$ and $V_b = V_c$ at $r = r_1$.

29-48

(a) Immediately upon being connected with the wire charge is transferred until the potentials of the spheres are the same since, when connected, the spheres and wire form a single conductor. Let r, R be the radii of the smaller and larger sphere, R = 2r, and q, Q their final charges. Then,

$$V_r = \frac{q}{4\pi\epsilon_0 r} = V_R = \frac{Q}{4\pi\epsilon_0 R},$$

$$q = \tfrac{1}{2}Q.$$

Also,

$$q + Q = 2(3 \times 10^{-8}\ \text{C}) = 6 \times 10^{-8}\ \text{C}.$$

Solving the last two equations yields

$$q = 2 \times 10^{-8}\ \text{C},\ Q = 4 \times 10^{-8}\ \text{C}.$$

As each sphere started with 3×10^{-8} C, it is evident that 10^{-8} C has been transported from the smaller to the larger sphere Ans.

(b) The final charges are, from (a), 2×10^{-8} C on the smaller sphere and twice that on the larger Ans.
The potentials are the same:

$$V_R = V_r = (9 \times 10^9\ \mathrm{N \cdot m^2/C^2}) \frac{2 \times 10^{-8}\ \mathrm{C}}{6 \times 10^{-2}\ \mathrm{m}} = 3000\ \mathrm{V} \quad \text{Ans.}$$

29-50

(a) For a spherical charge distribution the potential at the surface will be

$$V = \frac{q}{4\pi\epsilon_0 R} = (9 \times 10^9) \frac{3 \times 10^{-11}}{R} = 500,$$

$$R = 5.4 \times 10^{-4}\ \mathrm{m} = 0.54\ \mathrm{mm} \quad \text{Ans.}$$

(b) The radius of the combined drops is found from

$$\tfrac{4}{3}\pi r^3 = 2(\tfrac{4}{3}\pi R^3),$$

$$r = 2^{1/3} R.$$

The total charge is $2q = 6 \times 10^{-11}$ C. If the drops act as conductors (due to impurities, say) so that the charge becomes uniformly distributed throughout the new drop, the surface potential will become

$$V^* = \frac{2q}{4\pi\epsilon_0 2^{1/3} R} = 794\ \mathrm{V} \quad \text{Ans.}$$

29-52

(a) The charge q_α of an α-particle is +2e and its mass $m_\alpha \approx 4m_p$, $e = 1.6 \times 10^{-19}$ C being the charge on the protom and m_p is the proton mass. Let V be the accelerating potential and K kinetic energy; then,

$$K_\alpha = q_\alpha V = (2e)V = 2(1.6 \times 10^{-19}\ \mathrm{C})(10^6\ \mathrm{V}) = 3.2 \times 10^{-13}\ \mathrm{J} \quad \text{Ans.}$$

(b) Similarly, $K_p = eV = 1.6 \times 10^{-13}$ J Ans.

(c) Evidently $K_\alpha = 2K_p$ and therefore

$$K_p = \tfrac{1}{2}m_p v_p^2 = \tfrac{1}{2}K_\alpha = \tfrac{1}{2}(\tfrac{1}{2}m_\alpha v_\alpha^2) = \tfrac{1}{2}[\tfrac{1}{2}(4m_p)v_\alpha^2],$$

$$v_p^2 = 2v_\alpha^2.$$

Thus the proton, due to its greater charge to mass ratio, has the greater final speed.

<u>29-54</u>

(a) The potential of the shell and the electric field near the shell's surface are

$$V = \frac{1}{4\pi\epsilon_0}\frac{q}{r}, \quad E = \frac{1}{4\pi\epsilon_0}\frac{q}{r^2}.$$

Hence,

$$E = V/r.$$

For $E < 10^8$ V/m, it is necessary that $r > V/E = (9 \times 10^6\ \text{V})/(10^8\ \text{V/m}) = 0.09$ m = 9 cm.

(b) The work done in bringing up to the machine a charge Q is QV. Therefore the power supplied must be

$$P = \frac{dW}{dt} = V\cdot\frac{dQ}{dt} = (9 \times 10^6\ \text{V})(3 \times 10^{-4}\ \text{C/s}) = 2700\ \text{W} \quad \underline{\text{Ans}}.$$

(c) If the surface charge density is σ and x denotes a length of the belt, then, since $Q = \sigma A = \sigma wx$,

$$\frac{dQ}{dt} = \sigma\frac{dA}{dt} = \sigma\frac{d(wx)}{dt} = \sigma wv,$$

$$\sigma = (dQ/dt)/wv = 2 \times 10^{-5}\ \text{C/m}^2 \quad \underline{\text{Ans}}.$$

30-5

The capacitance of a parallel flat-plate capacitor is $\epsilon_0 A/d$, where d is the plate separation. If the plates of the upper capacitor are a distance d apart, the plates of the lower capacitor must be separated by a - (b + d). Hence, the equivalent capacitance C_e is found from

$$\frac{1}{C_e} = \frac{1}{C_1} + \frac{1}{C_2},$$

$$\frac{1}{C_e} = \frac{1}{\epsilon_0 A/d} + \frac{1}{\epsilon_0 A/(a - b - d)} = \frac{1}{\epsilon_0 A}(d + a - b - d) = \frac{a - b}{\epsilon_0 A},$$

$$C_e = \frac{\epsilon_0 A}{a - b},$$

independent of d and therefore of the position of the center piece.

30-14

The initial charges are given to the left and final charges to the right of each capacitor in the figure. Clearly,

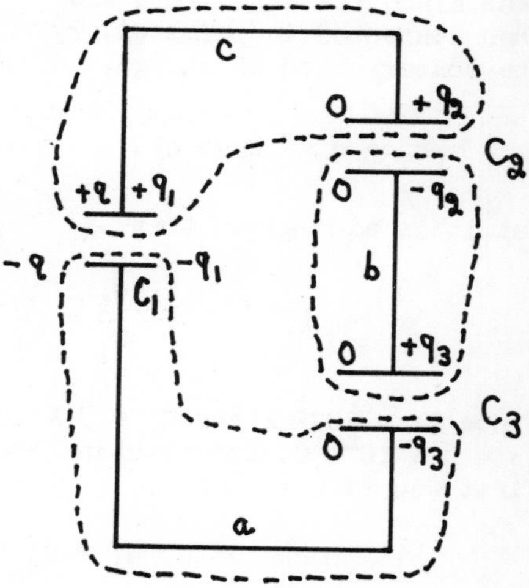

$$q = C_1 V_0.$$

By conservation of charge applied to conductor a (if no charge jumps the gap),

$$-q = -q_1 - q_3;$$

from conductor b,

$$0 = -q_2 + q_3;$$

for conductor c,

$$+q = +q_1 + q_2.$$

Also, since electrostatic fields are conservative, the total drop in potential around the circuit must be zero. As the q's are taken to be positive, this indicates that

$$0 = \frac{q_2}{C_2} + \frac{q_3}{C_3} - \frac{q_1}{C_1}.$$

Solving the equations above simultaneously gives (notice that only two of the three charge equations are independent),

$$q_1 = \frac{C_1C_2 + C_1C_3}{C_1C_2 + C_1C_3 + C_2C_3} C_1V_0 \quad \underline{\text{Ans}},$$

$$q_2 = q_3 = \frac{C_2C_3}{C_1C_2 + C_1C_3 + C_2C_3} C_1V_0 \quad \underline{\text{Ans}}.$$

<u>30-17</u>

Let starred quantities represent final values (switches closed). The final polarities must be the same since the capacitors are then connected in parallel. By the conservation of charge,

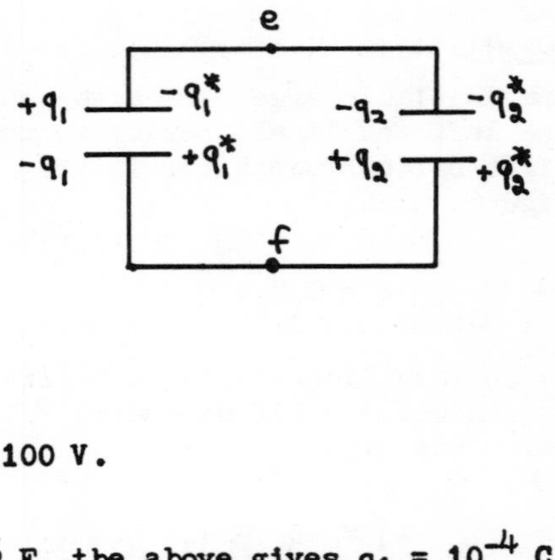

$$-q_1 + q_2 = q_1^* + q_2^*.$$

But it is also given that

$$\frac{q_1}{C_1} = \frac{q_2}{C_2} = 100 \text{ V}.$$

Since $C_1 = 10^{-6}$ F and $C_2 = 3 \times 10^{-6}$ F, the above gives $q_1 = 10^{-4}$ C, $q_2 = 3 \times 10^{-4}$ C. Substituting these numerical values into the very first equation yields

$$q_1^* + q_2^* = 2 \times 10^{-4}.$$

However, in the final arrangement also,

$$V^* = \frac{q_1^*}{C_1} + \frac{q_2^*}{C_2}$$

so that

$$q_2^* = 3q_1^*,$$

just as $q_2 = 3q_1$. Solving the two equations for q_1^* and q_2^* gives

(b) $$q_1^* = 0.5 \times 10^{-4}\ \text{C} \quad \underline{\text{Ans}},$$

(c) $$q_2^* = 1.5 \times 10^{-4}\ \text{C} \quad \underline{\text{Ans}}.$$

(a) Finally,

$$V_{ef} = V^* = (1.5 \times 10^{-4}\ \text{C})/(3 \times 10^{-6}\ \text{F}) = 50\ \text{V} \quad \underline{\text{Ans}}.$$

<u>30-18</u>

Apply a potential difference V between points x and y (say, by connecting a battery) thereby charging the capacitors. By the conservation of charge applied to conductors <u>a</u> and <u>b</u>

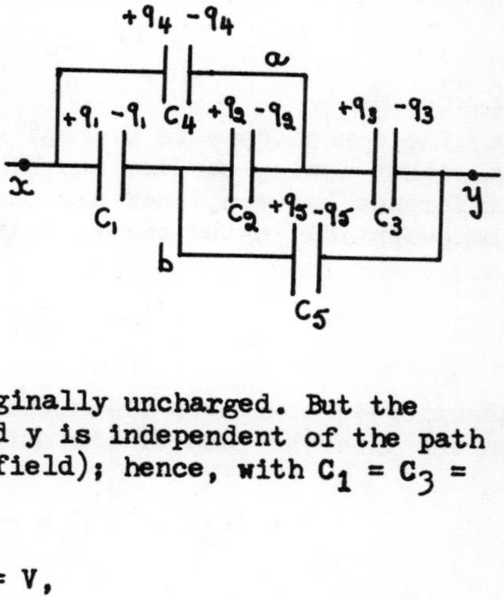

$$-q_4 - q_2 + q_3 = 0,$$

$$-q_1 + q_2 + q_5 = 0,$$

the capacitors being considered originally uncharged. But the potential difference V between x and y is independent of the path taken to evaluate it (conservative field); hence, with $C_1 = C_3 = C_4 = C_5 = C$,

$$\frac{q_4}{C} + \frac{q_3}{C} = V,$$

$$\frac{q_1}{C} + \frac{q_2}{C_2} + \frac{q_3}{C} = V,$$

$$\frac{q_1}{C} + \frac{q_5}{C} = V.$$

From these five equations the charges q_1, q_2, q_3, q_4, q_5 can be found in terms of C, C_2, V. However, it is not necessary to solve for all of them since the equivalent capacitance C_e is

$$C_e = \frac{q_1 + q_4}{V} = \frac{q_3 + q_5}{V};$$

that is, the magnitude of the charge on either terminal x or y (they will be equal, again by charge conservation) divided by the potential difference between the terminals. Solving the equations it is found that

$$q_1 = q_3 = q_4 = q_5 = CV/2, \quad q_2 = 0.$$

Thus,

$$C_e = \frac{CV}{V} = C,$$

independent of C_2. Numerically this is $C_e = 4\mu F$.

30-22

(a) The capacitance is $C = q/V$ where q = magnitude of the charge on either facing surface and V = absolute value of the potential difference between inner and outer surfaces. But, from Chapter 29, the potential in the region between the shells is

$$V(r) = \frac{1}{4\pi\epsilon_0}\frac{q}{r} + \frac{1}{4\pi\epsilon_0}\frac{(-q)}{b},$$

assuming +q charge on the outer surface of the inner shell and -q on the inner surface of the outer shell. Hence,

$$V = V(a) - V(b) = \frac{q}{4\pi\epsilon_0}\left(\frac{1}{a} - \frac{1}{b}\right) = \frac{q}{4\pi\epsilon_0}\left(\frac{b - a}{ab}\right).$$

V may also be found by evaluating $-\int \vec{E}\cdot\vec{dr}$, $E = q/4\pi\epsilon_0 r^2$ from one shell to the other. The capacitance is

$$C = \frac{q}{V} = 4\pi\epsilon_0 \frac{ab}{b - a}.$$

(b) If a = R and b "=" ∞, the a = R in the denominator can be deleted (since $R \ll \infty$) giving

$$C = 4\pi\epsilon_0 R,$$

as in Example 3 for the capacitance of an isolated sphere.

30-25

Consider two narrow parallel strips, one on each plate, located the same distance x from the left end of the capacitor, as shown. They form a small parallel plate capacitor having capacitance

$$dC = \epsilon_0 \frac{dA}{d^*} = \epsilon_0 \frac{a\,dx}{d + x\theta}$$

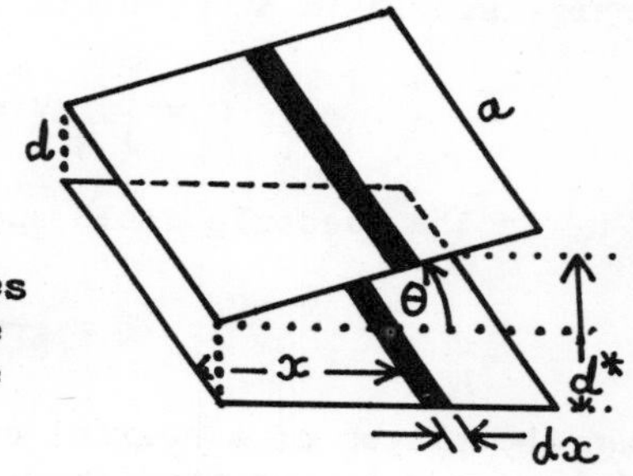

provided $\theta \ll 1$ rad. The real condenser is constructed of many such minute capacitors, all connected in parallel since plates carrying the same sign charge are contiguous. Hence the capacitance C of the condenser is

$$C = \int dC = \epsilon_0 a \int_0^a \frac{dx}{d + x\theta} = \frac{\epsilon_0 a}{\theta} \ln(1 + \frac{a\theta}{d}) = \frac{\epsilon_0 a}{\theta}[\frac{a\theta}{d} - \frac{1}{2}(\frac{a\theta}{d})^2 + \ldots],$$

$$C \approx \frac{\epsilon_0 a^2}{d}(1 - \frac{1}{2}\frac{a\theta}{d}) \quad \text{Ans.}$$

30-33

With the plates disconnected, no charge can be transferred from one plate to the other. If the plate separation remains small compared with the dimensions of the plates, the electric field will remain uniform and independent of the distance between the plates.

(a)

$$V_f = E_f d_f = E(2d) = 2(Ed) = 2V \quad \text{Ans.}$$

(b) Initially,

$$U_i = \frac{1}{2}C_i V_i^2 = \frac{1}{2}(\epsilon_0 A/d)V^2 \quad \text{Ans.}$$

Both the capacitance and potential difference change as the plates are pulled apart; hence,

$$U_f = \tfrac{1}{2}C_fV_f^2 = \tfrac{1}{2}(\epsilon_0A/d_f)(2V)^2 = \tfrac{1}{2}(\epsilon_0A/2d)(2V)^2 = \epsilon_0AV^2/d \quad \underline{\text{Ans}}.$$

(c) The work W required to pull the plates apart is

$$W = U_f - U_i = U_i = \tfrac{1}{2}(\epsilon_0A/d)V^2 \quad \underline{\text{Ans}}.$$

30-37

The energy stored in an electric field occupying a volume V is

$$U = \int u_E dV = \int \tfrac{1}{2}\epsilon_0 E^2 dV$$

where u_E is the electric field energy density. From Example 2,

$$E = \frac{q}{2\pi\epsilon_0 L}\,\frac{1}{r},$$

between the plates of a coaxial cylindrical capacitor of length L and charge q; r is measured from the axis. The volume element dV is taken to be the volume contained in a very thin cylindrical shell of radius r, thickness dr and length L, so that

$$dV = 2\pi rL\, dr.$$

Hence, the energy stored within a region between the inner shell of radius $\underline{a}$ and a cylindrical surface of radius $\underline{s}$ is

$$U_s = \int_a^s \tfrac{1}{2}\epsilon_0\left(\frac{q}{2\pi\epsilon_0 Lr}\right)^2 2\pi rL dr = \frac{q^2}{4\pi\epsilon_0 L}\int_a^s \frac{dr}{r} = \frac{q^2}{4\pi\epsilon_0 L}\ln\left(\frac{s}{a}\right).$$

Using the expression for the capacitance C derived in Example 2, the energy stored between the plates is

$$U = \tfrac{1}{2}q^2/C = \tfrac{1}{2}\,\frac{q^2}{2\pi\epsilon_0 L}\ln\left(\frac{b}{a}\right).$$

Setting $U_s = \tfrac{1}{2}U$, and substituting the expressions found above gives

$$s = (ab)^{\frac{1}{2}}.$$

30-38

Move one plate with zero acceleration holding the other plate in place. As the charge q on the plates does not change, neither does the electric field $E = q/A\epsilon_0$. But $V = Ex$ will increase; hence,

$$dW = Fdx = U_f - U_i = \tfrac{1}{2}q(V_f - V_i),$$

$$Fdx = \tfrac{1}{2}q[(x + dx)E - xE] = \tfrac{1}{2}qEdx,$$

$$F = \tfrac{1}{2}qE = \frac{1}{2}\frac{q^2}{A\epsilon_0} \quad \underline{\text{Ans}}.$$

30-40

The work done by the bubble in expanding is

$$W = \int p_b dV,$$

p_b being the pressure of the air in the bubble. If surface tension is ignored, then initially (uncharged bubble) this equals the atmospheric pressure p. Assuming this pressure to remain constant,

$$W = p\Delta V = p\,\frac{4\pi}{3}(R^3 - R_0^3).$$

The electric field set up by a charge q distributed uniformly over a spherical surface exists throughout all space external to the surface. If the sphere has radius s, the stored electric energy is

$$U = \int \tfrac{1}{2}\epsilon_0 E^2 dV = \int_s^\infty \tfrac{1}{2}\epsilon_0\left(\frac{q}{4\pi\epsilon_0 r^2}\right)^2(4\pi r^2 dr) = \frac{q^2}{8\pi\epsilon_0}\int_s^\infty r^{-2}dr = \frac{q^2}{8\pi\epsilon_0}\,\frac{1}{s}.$$

Thus, if the surface expands from radius R_0 to radius R with q held fixed, the stored energy decreases by

$$\Delta U = \frac{q^2}{8\pi\epsilon_0}\left(\frac{1}{R_0} - \frac{1}{R}\right).$$

Setting $W = \Delta U$ gives

$$\frac{q^2}{8\pi\epsilon_0}\frac{R - R_0}{RR_0} = \frac{4}{3}\pi p(R^3 - R_0^3),$$

$$q^2 = \frac{32}{3}\pi^2\epsilon_0 pRR_0(R^2 + RR_0 + R_0^2).$$

This calculation is somewhat artificial since it is not likely that the pressure of the air inside the bubble remains constant; also, surface tension has been ignored. If it is assumed that the pressure in the bubble is inversely proportional to the volume of the bubble and surface tension is not ignored, then the result obtained will be

$$q^2 = 32\pi^2\epsilon_0 pR(R^3 - R_0^3).$$

Numerically, with $p = 1$ atm $= 1.013 \times 10^5$ Pa, $R_0 = 0.020$ m, $R = 0.021$ m, $\epsilon_0 = 8.854 \times 10^{-12}$ $C^2/N{\cdot}m^2$, the original formula gives

$$q = 7.1 \times 10^{-6}\text{ C} \quad \underline{\text{Ans}}.$$

30-43

Let q, $-q$ be the charges on the plates separated by a distance d. If V is the potential difference across the plates, then,

$$C = \frac{q}{V}.$$

The uniform electric fields in the dielectrics are $\sigma/k\epsilon_0$, so that

$$E_1 = \frac{q/A}{k_1\epsilon_0},\ E_2 = \frac{q/A}{k_2\epsilon_0}.$$

Therefore,

$$C = \frac{q}{E_1(\frac{1}{2}d) + E_2(\frac{1}{2}d)} = \frac{2q}{d}\left(\frac{q/A}{k_1\epsilon_0} + \frac{q/A}{k_2\epsilon_0}\right)^{-1} = \frac{2A\epsilon_0}{d}\frac{k_1k_2}{k_1 + k_2}.$$

For an air capacitor, $k_1 = k_2 = 1$, and the above reduces to $C = A\epsilon_0/d$, as expected. If $k_1 = k_2 = k$ then, also as anticipated, the expression becomes $C = k\epsilon_0 A/d$.

30-48

For the values before the slab is introduced, see Example 9.

(d) The capacitance with the slab in place does not depend on the conditions under which the slab is introduced, but only on the geometry of the capacitor and the dielectric between its plates. Hence, from Example 9, part (f), C = 16 pF Ans.

(a) Since the battery remains connected, V = 100 volts. Thus,

$$q = CV = (16 \times 10^{-12}\ \mathrm{F})(100\ \mathrm{V}) = 1.6 \times 10^{-9}\ \mathrm{C} \quad \underline{\text{Ans}}.$$

(b) The electric field in the gap is

$$E_0 = \frac{q}{A\epsilon_0} = \frac{1.6 \times 10^{-9}\ \mathrm{C}}{(10^{-2}\ \mathrm{m}^2)(8.85 \times 10^{-12}\ \mathrm{C^2/N\cdot m^2})} = 1.8 \times 10^4\ \mathrm{N/C} \quad \underline{\text{Ans}}.$$

(c) The electric field in the dielectric is

$$E_d = \frac{1}{k}E_0 = \frac{1}{7}(1.8 \times 10^4\ \mathrm{N/C}) = 2.6 \times 10^3\ \mathrm{N/C} \quad \underline{\text{Ans}}.$$

30-51

Let the electric field in the gaps be E and the field in the dielectric be E_d. Now $E_d = E/k$ and therefore the potential difference across the plates is

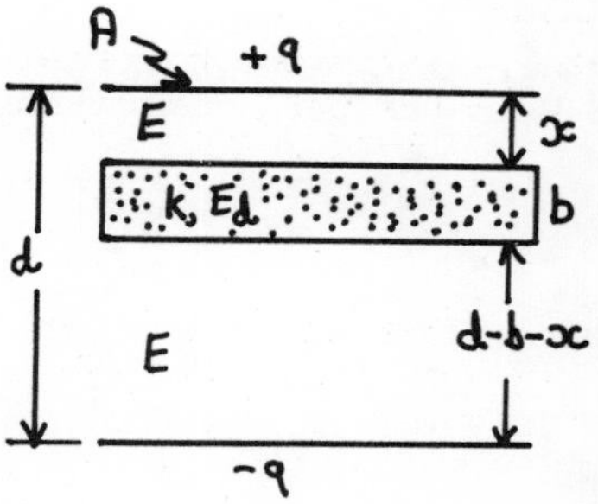

$$V = xE + bE_d + (d - b - x)E,$$

$$V = E(d - b) + E_d b,$$

$$V = E(d - b + \frac{b}{k}).$$

But, for a parallel-plate condenser,

$$E = \sigma/\epsilon_0 = q/A\epsilon_0,$$

A the area of each plate. Thus,

$$V = \frac{q}{A\epsilon_0}\left[d + b\left(\frac{1}{k} - 1\right)\right] = \frac{q}{A\epsilon_0 k}\left[kd - b(k - 1)\right],$$

and hence the capacitance is

$$C = \frac{q}{V} = \frac{k\epsilon_0 A}{kd - b(k - 1)}.$$

The cases $b = 0$ and $k = 1$ each correspond to an air capacitor for which $C = \epsilon_0 A/d$ by the above formula, as expected. If $b = d$ the dielectric fills entirely the space between the plates giving $C = k\epsilon_0 A/d$, again as expected.

CHAPTER 31

31-3

(a) Since each alpha-particle carries a charge $2(1.6 \times 10^{-19}\ C)$ and $1\ A = 1\ C/s$, the number of alpha-particles striking the surface each second is

$$\frac{0.25 \times 10^{-6}\ C/s}{3.2 \times 10^{-19}\ C} = 7.81 \times 10^{11}.$$

Hence, in three seconds, $3(7.81 \times 10^{11}) = 2.34 \times 10^{12}$ particles strike.

(b) The kinetic energy $K = (20\ MeV)(1.6 \times 10^{-13}\ J/MeV) = \frac{1}{2}mv^2$. The mass m of an alpha-particle is about four times a proton mass, so that $m = 4(1.66 \times 10^{-27}\ kg)$. Hence,

$$3.2 \times 10^{-12} = \tfrac{1}{2}(6.64 \times 10^{-27})v^2,$$

$$v = 3.1046 \times 10^7\ m/s.$$

The time t required to travel 20 cm = 0.20 m is

$$t = \frac{0.20\ m}{3.1046 \times 10^7\ m/s} = 6.442 \times 10^{-9}\ s.$$

From (a), 7.81×10^{11} particles pass any point each second; it follows that in time t, $(7.81 \times 10^{11})(6.442 \times 10^{-9}) = 5031$ particles pass, and therefore there are this many particles in a 20 cm length of the beam.

(c) If the accelerating potential is V, then $K = qV$, so that 20 MeV = (+2)V, giving $V = 10^7$ volts.

31-6

Let σ be the charge density sought. On a length L of the belt of width w there is present an amount of charge $Q = \sigma(wL)$. If the belt is moving, the current i due to the moving charges will be

$$i = \frac{dQ}{dt} = \frac{d(\sigma wL)}{dt} = \sigma w \frac{dL}{dt} = \sigma wv,$$

v the speed of the belt. Therefore,

$$10^{-4} = \sigma(0.5)(30),$$

$$\sigma = 6.67 \times 10^{-6} \text{ C/m}^2 \quad \underline{\text{Ans}}.$$

31-8

The potential of a conducting sphere carrying a charge q is

$$V = \frac{q}{4\pi\epsilon_0 R}.$$

If q is changing, V will also, and at the rate

$$\frac{dV}{dt} = \frac{1}{4\pi\epsilon_0 R}\frac{dq}{dt} = \frac{1}{4\pi\epsilon_0 R}\, i = \frac{1}{4\pi\epsilon_0 R}(i_{in} - i_{out})$$

i being the net current. If i does not change with time,

$$\frac{\Delta V}{\Delta t} = \frac{1}{4\pi\epsilon_0 R}\, i,$$

$$\frac{1000}{\Delta t} = (9 \times 10^9)(0.1)^{-1}(1.000002 - 1),$$

$$\Delta t = 5.56 \times 10^{-3} \text{ s} \quad \underline{\text{Ans}}.$$

31-15

Since the materials of the two conductors are the same, $\rho_A = \rho_B$. Also, it is given that $\ell_A = \ell_B$. Therefore, with $R = \rho\ell/A$ for each,

$$\frac{R_A}{R_B} = \frac{A_B}{A_A} = \frac{\pi(d_o^2 - d_i^2)/4}{\pi d^2/4} = \frac{d_o^2 - d_i^2}{d^2} = \frac{2^2 - 1^2}{1^2} = 3 \quad \underline{\text{Ans}}.$$

31-16

The length of the composite resistor is still ℓ, the length of any single resistor, but the area through which the current can flow is now nine times the area A of a single wire. The new resistance is

$$R = \rho\frac{\ell}{A} = \frac{\rho\ell}{9\pi d^2/4}.$$

The area of the equivalent single wire is $\frac{1}{4}\pi D^2$, but its length is still ℓ. For this wire to have the same resistance as the composite given above, the diameter D must be chosen so that

$$\frac{\rho\ell}{9\pi d^2/4} = \frac{\rho\ell}{\pi D^2/4},$$

$$D = 3d \quad \underline{\text{Ans}}.$$

<u>31-21</u>

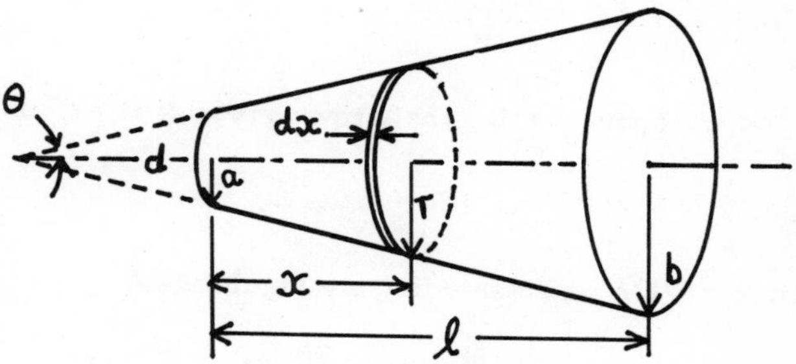

(a) Consider a thin slice, perpendicular to the axis of the cone, of thickness dx and at a distance x from the narrow end of the cone; the cross-sectional area of this slice is πr^2 and its resistance dR will be

$$dR = \rho\frac{d\ell}{A} = \rho\frac{dx}{\pi r^2}.$$

Let θ be the semi-angle of the cone from which the resistor was cut. Let the vertex of the cone be at a distance d from the narrow end of the resistor, as shown. Then,

$$\tan\theta = \frac{a}{d} = \frac{r}{x + d} = \frac{b}{\ell + d}.$$

From the first equation,

$$\frac{a}{d} = \frac{r}{x + d},$$

it is found that

$$d = \frac{ax}{r - a}.$$

Substituting this into the second equation

$$\frac{r}{x + d} = \frac{b}{\ell + d},$$

gives

$$r = \frac{b - a}{\ell} x + a.$$

Thus, the resistance R of the object is

$$R = \int dR = \frac{\rho}{\pi}\int_0^{\ell}\left(\frac{b - a}{\ell} x + a\right)^{-2} dx = \rho\frac{\ell}{\pi ab} \quad \underline{\text{Ans}}.$$

(b) For zero taper a = b, the above gives $R = \rho\ell/\pi a^2 = \rho\ell/A$, as expected.

31-24

(a) With $R = \rho\ell/A$ and assuming small changes,

$$\Delta R = \frac{\partial R}{\partial \rho}(\Delta\rho) + \frac{\partial R}{\partial \ell}(\Delta\ell) + \frac{\partial R}{\partial A}(\Delta A),$$

and therefore

$$\frac{\Delta R}{R} = \frac{\Delta\rho}{\rho} + \frac{\Delta\ell}{\ell} - \frac{\Delta A}{A}.$$

Let a be the coefficient of (linear) thermal expansion, so that

$$\Delta\ell = a\ell(\Delta T),\ \Delta A = 2aA(\Delta T),$$

where ΔT is the change in temperature. Therefore, the percent change in length is $a(\Delta T) = (1.7 \times 10^{-5})(1) = 0.0017\%$, and the percent change in area is twice that, or 0.0034%. The percent change in R is, from the above,

$$\frac{\Delta R}{R} = (\alpha + a - 2a)\Delta T = (\alpha - a)\Delta T = 0.3883\%,$$

since $\alpha = 3.9 \times 10^{-3}\ /\text{C}^\circ$.

(b) The relative change in resistivity is much larger than those in length and area.

31-36

The rate of thermal (Joule) heating is $P = V^2/R$. As the temperature drops, R changes but V does not; hence,

$$\Delta P \approx \frac{\partial P}{\partial R}(\Delta R) = -\frac{V^2}{R^2}(\Delta R) = -\frac{P}{R}(\Delta R).$$

But $R = \rho L/A$ so that $\Delta R = (\Delta\rho)L/A$; dividing these equations gives

$$\frac{\Delta R}{R} = \frac{\Delta\rho}{\rho} = \bar{\alpha}(\Delta T).$$

Therefore,

$$\Delta P = -P\bar{\alpha}(\Delta T) = -(500\ \text{W})(4 \times 10^{-4}\ /\text{C}^\circ)(-600\ \text{C}^\circ) = +120\ \text{W},$$

and it follows that at 200°C, P = 500 + 120 = 620 W Ans.

31-38

(a) The resistance of the conductor is $R = R_i + R_c$, where R_i, R_c are the resistances of the iron and carbon sections. If their cross-sectional areas are the same,

$$R = \rho_i \ell_i/A + \rho_c \ell_c/A,$$

$$\Delta R = \frac{1}{A}(\ell_i \Delta\rho_i + \ell_c \Delta\rho_c) = 0,$$

the last by supposition. But

$$\Delta\rho_i = \alpha_i \rho_i \Delta T,\ \Delta\rho_c = \alpha_c \rho_c \Delta T$$

so that the thicknesses must be chosen so that

$$\ell_i \alpha_i \rho_i + \ell_c \alpha_c \rho_c = 0,$$

$$\ell_c/\ell_i = -\alpha_i \rho_i/\alpha_c \rho_c = 1/35 \quad \text{Ans},$$

using the entries in Table 31-1.

(b) The desired ratio is

$$\frac{P_c}{P_i} = \frac{i^2R_c}{i^2R_i} = \frac{\rho_c \ell_c}{\rho_i \ell_i} = -\frac{\alpha_i}{\alpha_c} = 10 \quad \underline{\text{Ans}}.$$

31-39

(a) The charge q accelerated during each pulse is

$$q = iT = (0.5\ \text{A})(0.1 \times 10^{6}\ \text{s}) = 5 \times 10^{-8}\ \text{C}.$$

Each electron carries 1.6×10^{-19} C, so that the number n of electrons accelerated is

$$n = (5 \times 10^{-8})/(1.6 \times 10^{-19}) = 3.125 \times 10^{11} \quad \underline{\text{Ans}}.$$

(b) In one second a total charge

$$Q = (500\ \text{pulses})(5 \times 10^{-8}\ \text{C/pulse}) = 25 \times 10^{-6}\ \text{C}$$

is accelerated. The average current is

$$\bar{i} = \frac{Q}{t} = 25 \times 10^{-6}\ \text{A} \quad \underline{\text{Ans}}.$$

(c) The accelerating voltage V must be

$$V = K/e = (50\ \text{MeV})/(1) = 5 \times 10^{7}\ \text{volts}$$

since the charge on the electron is one quantum unit. The power output is $P = iV$; per pulse this is

$$P = (0.5\ \text{A})(5 \times 10^{7}\ \text{V}) = 2.5 \times 10^{7}\ \text{W} \quad \underline{\text{Ans}};$$

over one second the average power is

$$\bar{P} = \bar{i}V = (25 \times 10^{-6}\ \text{A})(5 \times 10^{7}\ \text{V}) = 1250\ \text{W} \quad \underline{\text{Ans}}.$$

32-7

(a) The current in the circuit is

$$i = \frac{\mathcal{E}}{r + R},$$

and therefore the rate of thermal (Joule) heating is

$$P = i^2R = \frac{\mathcal{E}^2R}{(r + R)^2}.$$

To find the value of R that maximizes P, set dP/dR = 0 and solve for R:

$$\frac{dP}{dR} = 0 = \mathcal{E}^2 \frac{r - R}{(r + R)^3},$$

which gives R = r Ans.

(b) The maximum power dissipation is

$$P(R = r) = i^2r = \mathcal{E}^2r/(r + r)^2 = \mathcal{E}^2/4r \quad \text{Ans.}$$

32-10

Let there be n^2 resistors. They cannot be connected either all in series or all in parallel since the equivalent resistance in the former case is n^2R and in the latter R/n^2 (R the resistance of each resistor), and these equal R only for n = 1. But a single resistor cannot tolerate a current greater than $(P/R)^{\frac{1}{2}} = (1\ \text{W}/10\ \Omega)^{\frac{1}{2}} = 10^{-\frac{1}{2}}$ A, and a current equal to $(5\ \text{W}/10\ \Omega)^{\frac{1}{2}} = 1/\sqrt{2}$ A is required. Suppose the n^2 resistors are arranged into n sets, each set being n resistors connected in parallel, the sets being connected in series. The equivalent resistance $R_e = n(R/n) = R = 10\ \Omega$. The current i entering the combination must be $1/\sqrt{2}$ A in order that

$i^2R_e = 5$ W as demanded. The current passing through each resistor is $i/n = 1/\sqrt{2}n$, so that each resistor dissipates

$$P = (1/\sqrt{2}n\ \text{A})^2(10\ \Omega) = 5/n^2\ \text{W}.$$

This must be 1 W at the most; hence, $5/n^2 \le 1$, so that the minimum $n = 3$. Thus, at least nine resistors are needed, connected as shown below.

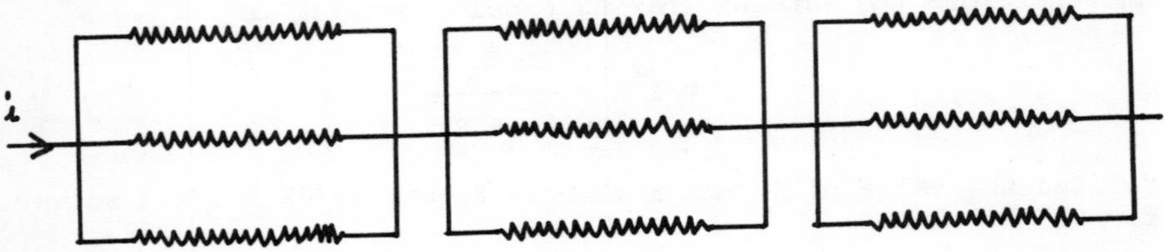

32-13

(a) Let the potential difference be $V = V_A - V_B > 0$ since energy is removed from the circuit. As $P = iV$, $50\ \text{W} = (1\ \text{A})V$, giving $V = 50$ volts Ans.

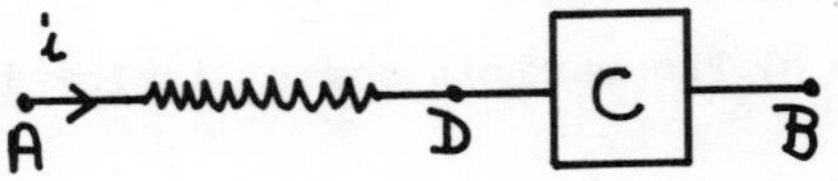

(b) Consider a point D between the resistor and element C; now

$$V_A - V_D = iR = 2\ \text{volts},$$

so that

$$V_A - V_B = (V_A - V_D) + (V_D - V_B),$$

$$50 = 2 + (V_D - V_B),$$

$$V_D - V_B = 48\ \text{volts}\quad \underline{\text{Ans}}.$$

(c) Since $V_D > V_B$, the negative terminal must be at B.

32-14

Let the points a and b represent the terminals of battery 1. The current i flows clockwise in the diagram and

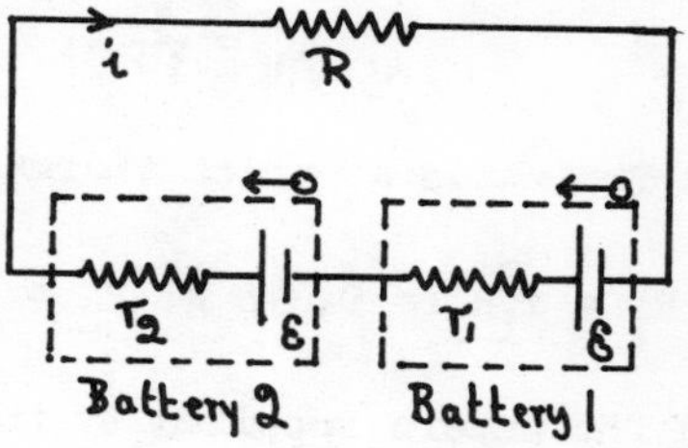

$$V_a + \mathcal{E} - ir_2 - iR = V_b,$$

$$V_a - V_b = i(r_2 + R) - \mathcal{E}.$$

If $V_a - V_b = 0$, the last equation gives

$$i(r_2 + R) = \mathcal{E}.$$

But, by the loop theorem, $i = 2\mathcal{E}/(r_1 + r_2 + R)$ so that the condition becomes

$$\frac{2\mathcal{E}}{r_1 + r_2 + R}(r_2 + R) = \mathcal{E},$$

$$R = r_1 - r_2 \quad \underline{\text{Ans}}.$$

<u>32-23</u>

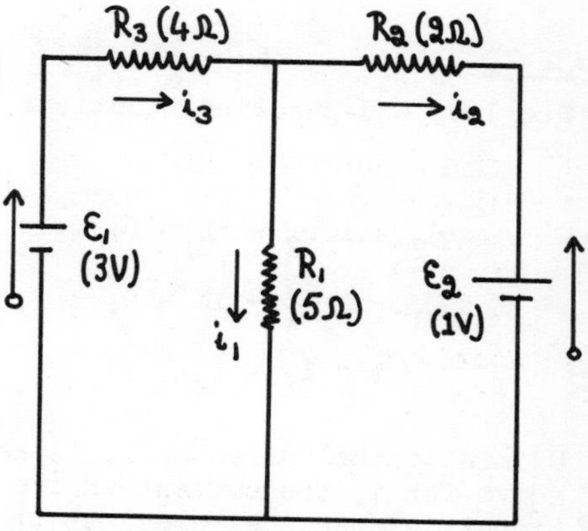

By the loop theorem, applied to the left and right hand branches respectively,

$$\mathcal{E}_1 - i_3R_3 - i_1R_1 = 0,$$

$$\mathcal{E}_2 + i_2R_2 - i_1R_1 = 0,$$

assuming the currents are directed as shown. The junction equation further provides that

$$i_3 = i_1 + i_2.$$

These three equations may be solved for the currents to yield (writing $r^2 = R_1R_2 + R_1R_3 + R_2R_3$),

$$r^2 i_1 = \mathcal{E}_1R_2 + \mathcal{E}_2R_3,$$

$$r^2 i_2 = \mathcal{E}_1R_1 - \mathcal{E}_2(R_1 + R_3),$$

$$r^2 i_3 = \mathcal{E}_1(R_1 + R_2) - \mathcal{E}_2R_1.$$

Numerically, these give

$$i_1 = \frac{5}{19}\text{ A};\ i_2 = \frac{3}{19}\text{ A};\ i_3 = \frac{8}{19}\text{ A}.$$

(a) The rates at which thermal energy appears in the resistors are

$$P_1 = i_1^2R_1 = 0.346\text{ W};\ P_2 = i_2^2R_2 = 0.050\text{ W};\ P_3 = i_3^2R_3 = 0.709\text{ W}.$$

(b) The powers supplied by the batteries are

$$P_{b1} = i_3\mathcal{E}_1 = 1.263\text{ W};\ P_{b2} = -i_2\mathcal{E}_2 = -0.158\text{ W};$$

the second is negative since the current i_2 flows through battery 2 in the direction opposite to the battery's emf.

(c) Energy is dissipated as heat in the resistors, supplied to the circuit by battery 1 and stored in battery 2; as expected, then, $1.263 \approx 0.346 + 0.050 + 0.709 + 0.158$, in watts.

32-29

The loop and junction equations are

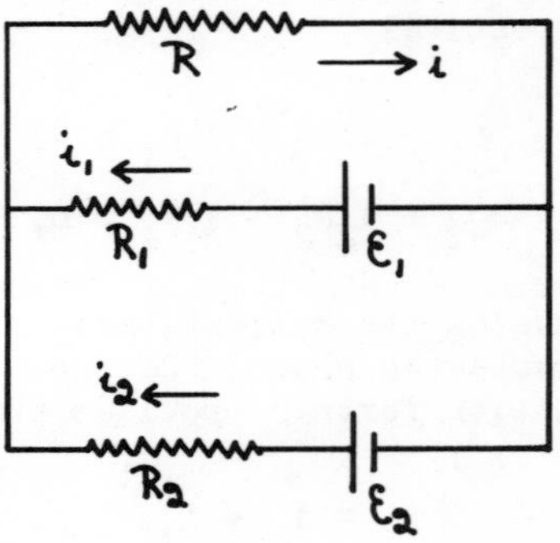

$$\mathcal{E}_2 - i_2r_2 - iR = 0,$$
$$\mathcal{E}_1 - i_1r_1 - iR = 0,$$
$$i = i_1 + i_2.$$

Eliminate the currents i_1, i_2 and solve for i, the current in the circuit elements external to the batteries:

$$i = \frac{r_1\mathcal{E}_2 + r_2\mathcal{E}_1}{Rr_2 + Rr_1 + r_1r_2}.$$

This, by definition, gives $\mathcal{E}_{eff}$: $i = \mathcal{E}_{eff}/R$, except that, clearly, the current i depends on R. To obtain an unambiguous value, set $R = 0$ after solving for $\mathcal{E}_{eff}$ to obtain

$$\mathcal{E}_{eff} = \frac{r_2\mathcal{E}_1 + r_1\mathcal{E}_2}{r_1 + r_2}$$

equivalent, in turn, to the expression given in the text.

32-31

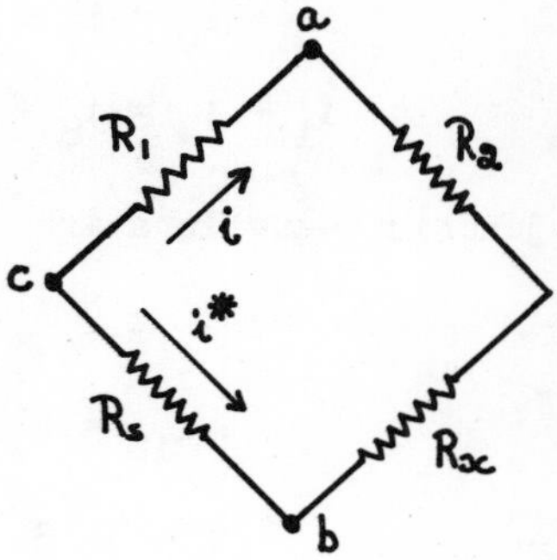

Ignore the battery and R_0 and focus on the square. By the loop theorem,

$$i(R_1 + R_2) = i^*(R_x + R_s).$$

Also, $V_{ac} = V_{bc}$ giving

$$iR_1 = i^*R_s.$$

Eliminate i, say, between these equations to get

$$i^*R_s(R_1 + R_2) = i^*R_1(R_x + R_s),$$
$$R_x = R_s(R_2/R_1).$$

32-35

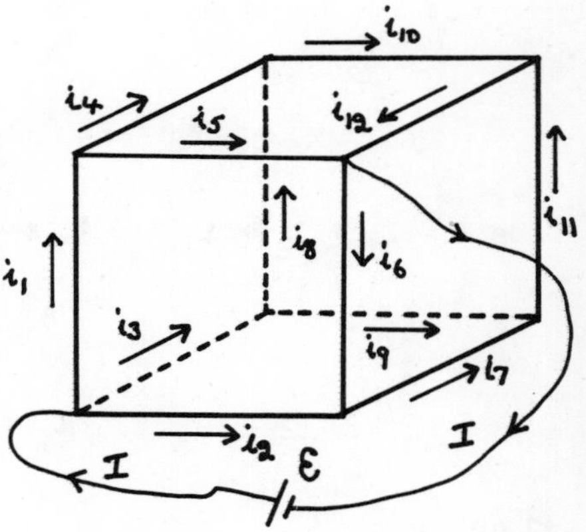

The method is illustrated by solving part (b) for the equivalent resistance of a face diagonal. Connect a battery between points B and C, between which the equivalent resistance is desired. This equivalent resistance R_e is defined by

$$\mathcal{E} = IR_e.$$

By the loop equation,

$$\mathcal{E} - i_1R - i_5R = 0,$$

and therefore

$$IR_e = (i_1 + i_5)R.$$

Now write down all the loop equations; since the resistances are identical the R's in each equation will drop out, leaving

$$i_1 + i_5 = i_2 + i_6, \quad i_7 + i_{11} = i_6 - i_{12},$$
$$i_3 + i_9 = i_7 + i_2, \quad i_8 + i_{10} = i_9 + i_{11},$$
$$i_1 + i_4 = i_8 + i_3, \quad i_4 + i_{10} = i_5 - i_{12}.$$

The junction equations are

$$i_1 = i_4 + i_5, \quad i_{10} + i_{11} = i_{12},$$
$$i_{10} = i_4 + i_8, \quad i_2 = i_6 + i_7,$$
$$i_3 = i_8 + i_9, \quad i_{11} = i_7 + i_9.$$

Solving these 12 equations for the 12 currents gives

$$i_1 = i_2 = i_5 = i_6 = i,$$
$$i_3 = i_{12} = \frac{2}{3}\, i,$$
$$i_4 = i_7 = 0,$$
$$i_8 = i_9 = i_{10} = i_{11} = \frac{1}{3}\, i.$$

Thus $I = i_1 + i_2 + i_3 = \frac{8}{3}\, i$, and $i_2 + i_5 = 2i$. Therefore,

$$\frac{8}{3} i R_e = 2iR,$$
$$R_e = \frac{3}{4}R \quad \underline{\text{Ans}}.$$

32-40

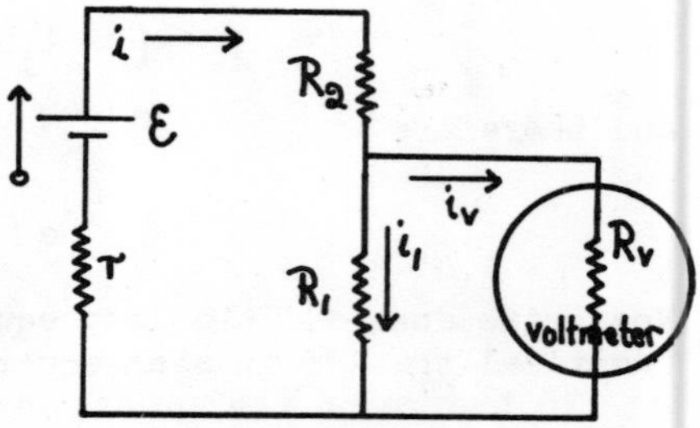

With $R_V = \infty$, $i_V = 0$ and

$$i = i_1 = \frac{\mathcal{E}}{r + R_1 + R_2} = 0.04545 \text{ A};$$

hence, the voltmeter reading is $iR_1 = 2.2727$ V. When R_V is finite $i_1R_1 = i_VR_V$ = voltmeter reading. But in this case also,

$$\mathcal{E} - (r + R_2)i - i_1R_1 = 0,$$

$$i = i_1 + i_V.$$

The three equations yield

$$i_1R_1 = \frac{R_1R_V\mathcal{E}}{rR_V + R_2R_V + R_1R_V + R_1R_2 + R_1r} = 2.2124 \text{ V}.$$

Thus, the error is

$$\frac{2.2727 - 2.2124}{2.2727} = 2.7\% \quad \underline{\text{Ans}}.$$

<u>32-44</u>

The charge on the capacitor is $q = q_0e^{-t/\tau}$ and therefore the potential difference across it is

$$V = q/C = (q_0/C)e^{-t/\tau} = V_0e^{-t/\tau}.$$

When $t = 0$, $V = 100$ volts so that, by the above equation, $V_0 = 100$ volts also; that is

$$V = 100e^{-t/\tau}.$$

Furthermore, at $t = 10$ s, $V = 1$ volt; substituting these gives

$$1 = 100e^{-10/\tau}.$$

(a) At $t = 20$ s,

$$V = 100e^{-20/\tau} = 100(e^{-10/\tau})^2 = 100(\tfrac{1}{100})^2 = 0.01 \text{ V} \quad \underline{\text{Ans}}.$$

(b) As shown above, $1 = 100e^{-t/\tau}$ and therefore

$$\ln(100) = 10/\tau,$$

$$\tau = 10/\ln(100) = 2.17 \text{ s} \quad \underline{\text{Ans}}.$$

<u>32-46</u>

(a) The charge on the capacitor and the energy stored within it are

$$q = q_0 e^{-t/\tau}, \; U_C = \tfrac{1}{2}q^2/C.$$

At $t = 0$, $U_C = 0.5$ J, so that $0.5 = \frac{1}{2}q_0^2/(10^{-6})$, giving $q_0 = 10^{-3}$ C.
(b) The current i is found by differentiating the charge q with respect to time t:

$$i = \frac{dq}{dt} = -(q_0/\tau)e^{-t/\tau}.$$

Therefore, disregarding the sign,

$$i(0) = q_0/\tau = q_0/RC = \frac{10^{-3}}{(10^6)(10^{-6})} = 10^{-3} \text{ A} \quad \underline{\text{Ans}}.$$

(c) The voltage across the capacitor is

$$V_C = q/C = (q_0/C)e^{-t/\tau}.$$

But, from (b), the time constant is evidently $\tau = RC = 1$ s, and $q_0/C = (10^{-3}\text{ C})/(10^{-6}\text{ F}) = 1000$ V. All this gives

$$V_C = 1000e^{-t} \text{ V} \quad \underline{\text{Ans}}.$$

From the loop theorem, $V_C + V_R = 0$ and thus $V_R = -1000e^{-t}$ volts.
(d) The rate of generation of thermal energy in the resistor is

$$P = i^2R = [(-q_0/\tau)e^{-t/\tau}]^2R = (q_0^2R/\tau^2)e^{-2t/\tau} = e^{-2t} \text{ W} \quad \underline{\text{Ans}},$$

using the numerical results from the other parts of the problem.

<u>32-47</u>
(a) The time constant is $\tau = RC = (3 \times 10^6\ \Omega)(10^{-6}\text{ F}) = 3$ s. Also,

$$q = \mathcal{E}C(1 - e^{-t/\tau}), \; i = (\mathcal{E}/R)e^{-t/\tau}.$$

Hence, for $t = 1$ s,

$$\frac{dq}{dt} = i = (4\text{ V})(e^{-1/3})/(3 \times 10^6\ \Omega) = 9.5538 \times 10^{-7} \text{ A} \quad \underline{\text{Ans}}.$$

(b) The rate at which energy is being stored in the capacitor is

$$dU_C/dt = d(q^2/2C)/dt = \frac{q}{C}\frac{dq}{dt} = iq/C.$$

Now $q(1) = (10^{-6}\ \text{F})(4\ \text{V})(1 - e^{-1/3}) = 1.1339 \times 10^{-6}$ C; i(1) was found in (a) and therefore

$$dU_C/dt = (1.1339 \times 10^{-6}\ \text{C})(9.5538 \times 10^{-7}\ \text{A})/(10^{-6}\ \text{F}),$$

$$dU_C/dt = 1.0833 \times 10^{-6}\ \text{W} \quad \underline{\text{Ans}}.$$

(c) For the resistor,

$$dU_R/dt = i^2R = (9.5538 \times 10^{-7}\ \text{A})^2(3 \times 10^6\ \Omega) = 2.7383 \times 10^{-6}\ \text{W}.$$

(d) Finally, the battery delivers energy at the rate

$$dU_B/dt = i\mathcal{E} = (9.5538 \times 10^{-7}\ \text{A})(4\ \text{V}) = 3.8215 \times 10^{-6}\ \text{W} \quad \underline{\text{Ans}}.$$

Notice that, within round-off error,

$$\frac{dU_B}{dt} = \frac{dU_C}{dt} + \frac{dU_R}{dt},$$

as required by the loop theorem.

32-49

(a) At t = 0 the capacitor exerts no influence and therefore

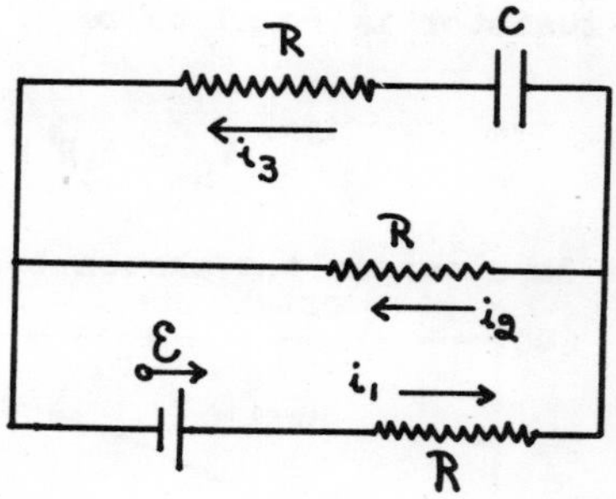

$$\mathcal{E} - i_1R - i_2R = 0,$$

$$-i_3R + i_2R = 0,$$

$$i_1 = i_2 + i_3.$$

Solving for the currents gives

$$i_2 = i_3 = \mathcal{E}/3R = 5.48 \times 10^{-4}\ \text{A} \quad \underline{\text{Ans}};$$

$$i_1 = 2\mathcal{E}/3R = 1.10 \times 10^{-3}\ \text{A} \quad \underline{\text{Ans}}.$$

At $t = \infty$ the capacitor prevents the flow of current in its branch of the circuit: $i_3 = 0$ Ans. In the other branch, now just a series circuit,

$$i_1 = i_2 = i = \mathcal{E}/2R = 8.22 \times 10^{-4}\ \text{A} \quad \underline{\text{Ans}}.$$

(b) For other values of t,

$$i_1 = i_2 + i_3,$$
$$\mathcal{E} - i_1 R - i_2 R = 0,$$
$$-\frac{q}{C} - i_3 R + i_2 R = 0.$$

Using the first two equations to eliminate i_2 in the third gives

$$\frac{3}{2}R\frac{dq}{dt} + \frac{q}{C} = \frac{1}{2}\mathcal{E},$$

since $i_3 = dq/dt$. The solution of this equation, for $q(0) = 0$, is

$$q = \frac{1}{2}\mathcal{E}C(1 - e^{-2t/3RC}).$$

Therefore,

$$i_3 = (\mathcal{E}/3R)e^{-2t/3RC}.$$

Using the third equation in (b) above, the potential drop across the resistor is found to be

$$V_R = i_2 R = \frac{\mathcal{E}}{6}(3 - e^{-2t/3RC}).$$

This is shown in the sketch below.

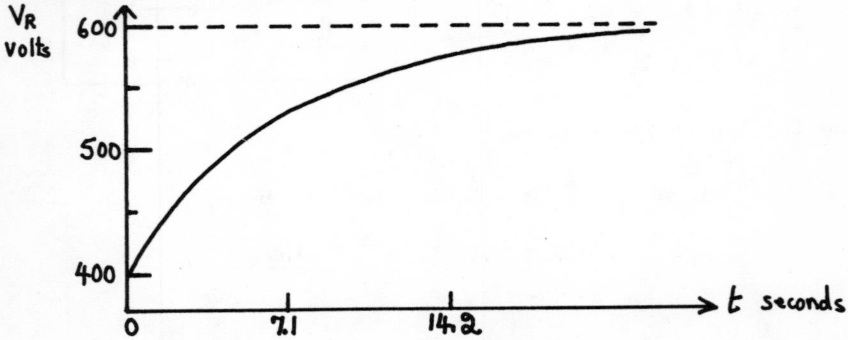

(c) From (b)

$$V_R(0) = \mathcal{E}/3 = 400\ \text{V};\ V_R(\infty) = \mathcal{E}/2 = 600\ \text{V}\quad \underline{\text{Ans}}.$$

(d) The time constant is

$$\tau = \frac{3}{2}RC = \frac{3}{2}(7.3 \times 10^5\ \Omega)(6.5 \times 10^{-6}\ \text{F}) = 7.1\ \text{s}.$$

After many time constants i_3 is close to zero, and at t"="∞ has fallen below measurable values.

33-2

(a) The charge on the electron is $q = -e$, with $e = 1.6 \times 10^{-19}$ C. Therefore, $F = q\vec{v} \times \vec{B} = -e\vec{v} \times \vec{B} = e\vec{B} \times \vec{v}$, and this points to the east Ans.

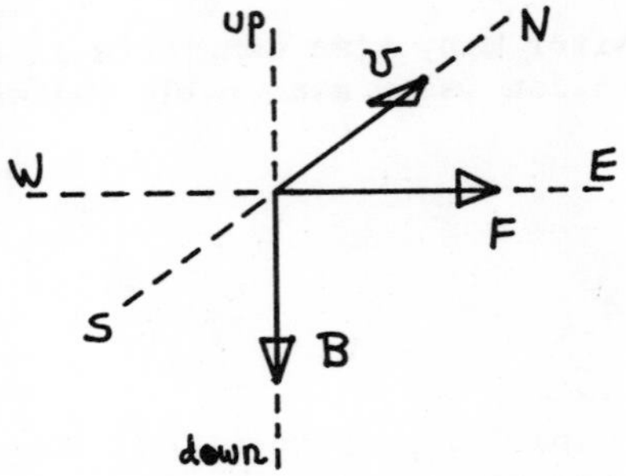

(b) Since the angle between $\vec{v}$ and $\vec{B}$ is 90°, the magnitude of $\vec{F}$ is

$$F = evB = ma,$$

$$a = (\frac{e}{m})vB.$$

But

$$v = (2K/m)^{\frac{1}{2}}$$

and since the kinetic energy $K = (12,000)(1.6 \times 10^{-19})$ J, the mass $m = 9.11 \times 10^{-31}$ kg, then $v = 6.492 \times 10^{7}$ m/s. Combining this with $e/m = 1.76 \times 10^{11}$ C/kg and $B = 55\ \mu T$ gives $a = 6.28 \times 10^{14}\ m/s^2$ Ans.

(c) The electrons follow a circular path of radius R at a constant speed v; for uniform circular motion,

$$mv^2/R = evB,$$

$$R = \frac{mv}{eB} = 6.7066 \text{ m}.$$

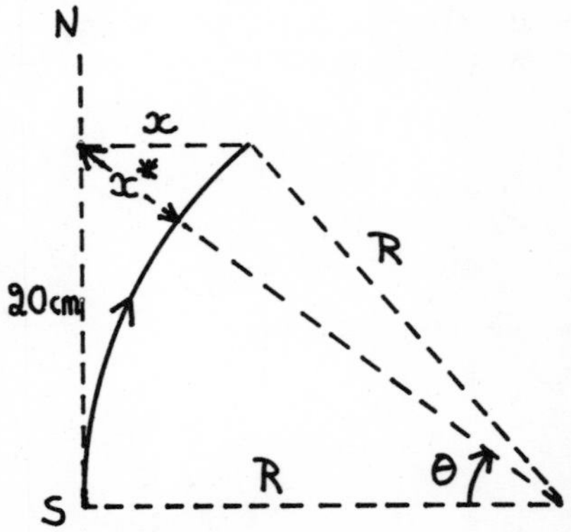

The deflection sought is x. From the figure,

$$\tan\theta = \frac{0.20 \text{ m}}{6.7066 \text{ m}} = 0.02982,$$

so that $\theta = 1°43'$. Also,

$$R^2 = (R + x^*)^2 + x^2 - 2x(R + x^*)\cos\theta.$$

Now $x^* = R(\sec\theta - 1)$, and therefore this equation becomes

$$x^2 - 2Rx + R^2\tan^2\theta = 0,$$

or numerically,

$$x^2 - 13.4132x + 0.039996 = 0,$$

$$x = 0.002983;\ 2R - 0.002983,$$

in meters. Thus the required deflection is about 3.0 mm. The second solution above is the deflection from the same N-S line to the other side of the upper semicircular path.

33-9

Clearly, to remove the tensions in the leads the force F exerted by the magnetic field on the wire of length L must equal the weight mg of the wire. Since $\vec{F} = i\vec{L} \times \vec{B}$, the current must pass from left to right in order that $\vec{F}$ point upward, opposite to the direction of the weight. Then,

$$F = iLB\sin 90° = iLB = mg,$$

$$i = \frac{mg}{LB} = \frac{(10^{-2}\ \text{kg})(9.8\ \text{m/s}^2)}{(0.6\ \text{m})\ (0.4\ \text{T})} = 0.41\ \text{A} \quad \underline{\text{Ans}}.$$

33-12

The impulse is $I = \int F dt = mv$, v the speed with which the wire is projected upward. If the wire reaches a height h,

$$v^2 = 2gh,$$

$$I^2 = 2ghm^2.$$

But, as the wire and the magnetic field are at right angles,

$$\int F dt = \int i\ell B dt = \ell B \int i dt = \ell Bq,$$

$$q = \frac{m}{\ell B}(2gh)^{\frac{1}{2}} = 3.8341\ \text{C} \quad \underline{\text{Ans}}.$$

33-16

If the cylinder rolls, the instantaneous axis of rotation will pass through P, the point on the rim in contact with the plane. Neither the normal force nor the force of rolling friction (not shown) will exert any torque about P as their lines of action pass through P. The torque due to gravity is

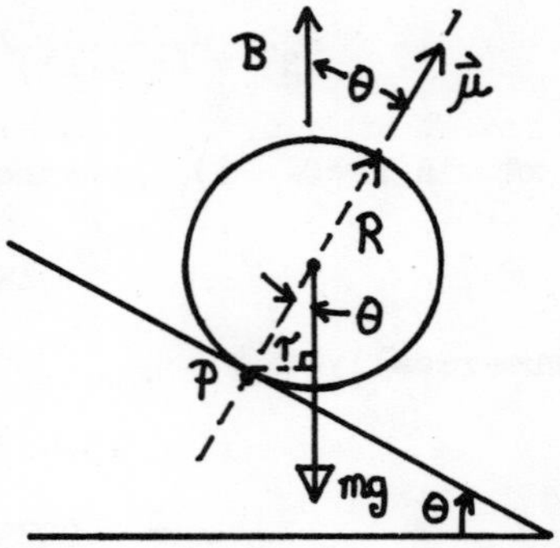

$$\tau_g = mgr = mgR\sin\theta,$$

directed into the page. The torque due to the magnetic field acting on the current loop is

$$\tau_m = \mu B\sin\theta = NiAB\sin\theta = Ni2R\ell B\sin\theta,$$

directed out of the page. The expression $\vec{\tau} = \vec{\mu} \times \vec{B}$ gives the magnetic torque about any axis, and hence about P, since the forces on the loop can be decomposed into couples. For no rotation the net torque must vanish, so that

$$Ni2R\ell B\sin\theta = mgR\sin\theta,$$

$$i = \frac{mg}{2NB\ell} = 2.45 \text{ A} \quad \underline{\text{Ans}}.$$

33-17

(a) The deflection of a galvanometer is proportional to the current passing through it. If the device obeys Ohm's law, this current is proportional to the voltage across the instrument. Hence, the deflection is proportional to the voltage also. Thus, if the total resistance of the galvanometer plus auxiliary resistor is R,

$$1.00 \text{ V} = (0.00162 \text{ A})R,$$

$$R = 617.3 \ \Omega.$$

Thus, connect the auxiliary resistor r in series with the galvanometer, choosing $r = 617.3 - 75.3 = 542 \ \Omega$ Ans.

(b) In this case attach the auxiliary resistor in parallel with the galvanometer so that a current $i = 0.00162$ A flows through the instrument when it is attached to a circuit branch in which the current is $I = 0.050$ A, as shown. For a parallel connection,

$$i_a r_a = i r_G;$$

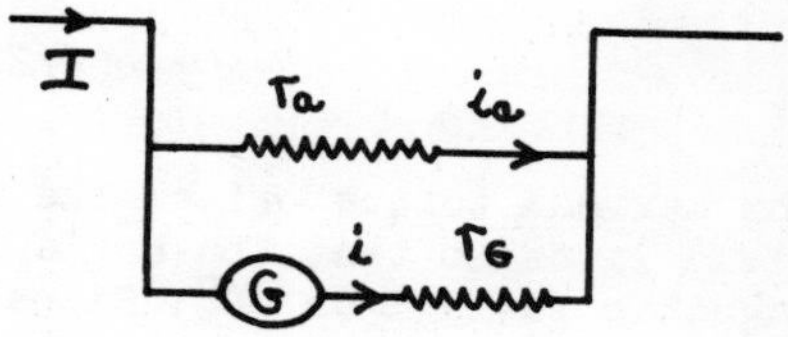

furthermore,

$$I = i + i_a.$$

Hence,

$$r_a = \frac{i}{I - i} r_G = (75.3\ \Omega)\frac{0.001620}{0.05 - 0.00162} = 2.52\ \Omega \quad \underline{\text{Ans}}.$$

<u>33-19</u>

Let the hinge lie along the z-axis. It is evident that the only force that can exert a torque along the hinge is the one exerted on the side of the rectangle opposite the hinge. (The forces on the other, 5 cm, sides are in the z-direction and, by the cross-product rule, their torques will be perpendicular to z.) The magnitude F of this force is

$$F = ibB\sin 90^\circ = ibB.$$

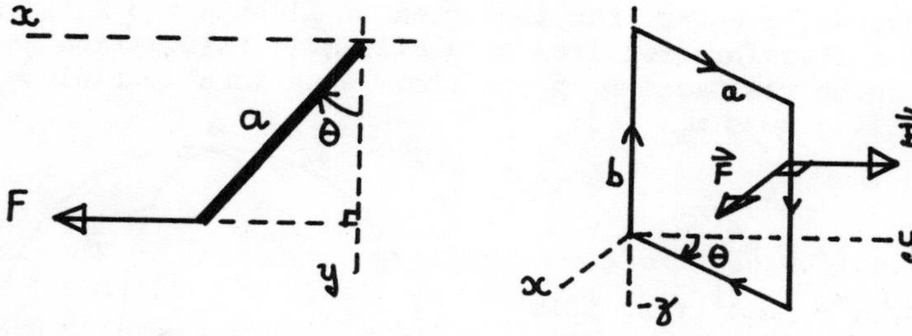

From the sketch the appropriate moment arm is seen to be $a\cos\theta$ so that the torque will be, for N loops,

$$\tau = N(ibB)(a\cos\theta) = 4.33 \times 10^{-3}\ \text{N·m} \quad \underline{\text{Ans}}.$$

From the right-hand rule, it is seen that this is directed in the -z-direction, i.e., down.

33-21

If N closed loops are formed from a wire of length ℓ, the circumference of each loop is ℓ/N, the radius of each is $\ell/2\pi N$ and therefore the area A of each is

$$A = \pi(\ell/2\pi N)^2 = \ell^2/4\pi N^2.$$

For maximum torque orient the plane of the loops parallel to the field lines so that the angle between the magnetic moment μ and the magnetic field B is 90°. Since there are N loops,

$$\tau = NiAB = Ni(\ell^2/4\pi N^2)B = i\ell^2 B/4\pi N.$$

Clearly, then, the maximum torque occurs for N = 1.

33-30

(a) At the equator the velocity will be perpendicular to the field B; hence $F = qvB = mv^2/R$ so that $v = qBR/m$; numerically,

$$v = \frac{(1.6 \times 10^{-19}\ C)(41 \times 10^{-6}\ T)(6.4 \times 10^{6}\ m)}{(1.67 \times 10^{-27}\ kg)} = 2.5 \times 10^{10}\ m/s.$$

This is, however, greater than the speed of light $c = 3 \times 10^8$ m/s; the situation therefore requires a relativistic calculation in which the proton rest mass m in the above equations is replaced by its relativistic mass m_r:

$$m_r = \frac{m}{(1 - \beta^2)^{\frac{1}{2}}},$$

where $\beta = v/c$. Substituting this for m in the equation for v gives

$$\beta = \frac{A}{(1 + A^2)^{\frac{1}{2}}}$$

with $A = RqB/mc = 83.8004$. This yields $v = \beta c = 2.9998 \times 10^8$ m/s.

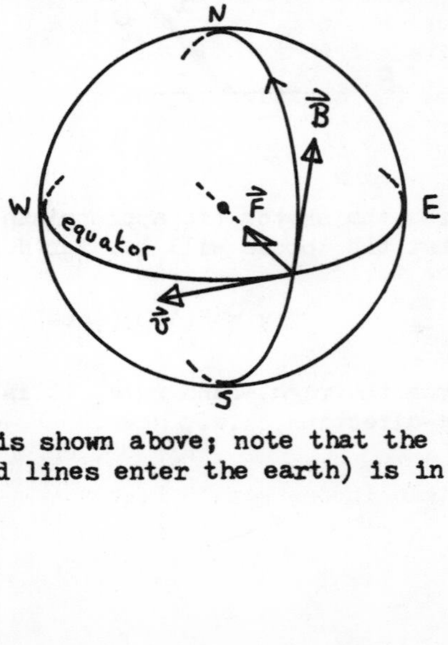

(b) The arrangement of the vectors is shown above; note that the magnetic SOUTH pole (where the field lines enter the earth) is in the geographic NORTHERN hemisphere.

33-36

The ion enters the spectrometer with a speed v related to the accelerating potential V by

$$\tfrac{1}{2}mv^2 = qV.$$

Inside the instrument the ion undergoes uniform circular motion, the speed v remaining unchanged; thus

$$mv^2/r = qvB.$$

But $v^2 = 2qV/m$ from the first equation and $r = \tfrac{1}{2}x$; substituting these gives

$$\frac{m(2qV/m)^{\frac{1}{2}}}{\tfrac{1}{2}x} = qB,$$

$$m = B^2qx^2/8V.$$

33-38

(a) The orientation of the vectors $\vec{v}$, $\vec{B}$, and $\vec{F} = q\vec{v} \times \vec{B}$ ($q > 0$) is shown in the sketch. Let $\vec{v} = v_y\vec{j} + v_z\vec{k}$; then $\vec{F} = v_yB\vec{i}$, independent of v_z. If $v_z = 0$, uniform circular motion results, with $v_y = v\sin\theta$ the governing speed. If $v_z \neq 0$, the additional uniform rectilinear motion along $\vec{B}$ stretches the circle into a helix.

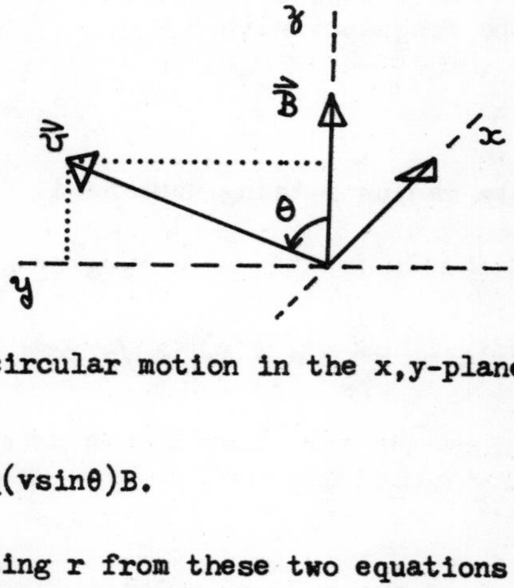

(b) The equation of motion for the circular motion in the x,y-plane is

$$\frac{m(v\sin\theta)^2}{r} = q(v\sin\theta)B.$$

The period $T = 2\pi r/(v\sin\theta)$; eliminating r from these two equations gives

$$T = \frac{2\pi}{B(q/m)} = 2\pi/(0.1)(1.76 \times 10^{11}) = 3.57 \times 10^{-10} \text{ s} \quad \underline{\text{Ans}}.$$

(c) The pitch p is the distance travelled parallel to the magnetic field in one period T: i.e.,

$$p = (v\cos\theta)T = (2K/m)^{\frac{1}{2}}T\cos\theta,$$

with the kinetic energy K = 2 keV = 3.2×10^{-16} J. Putting in the other numbers gives p = 0.17 mm Ans.

(d) From (a),

$$r = \frac{mv\sin\theta}{qB} = 1.5 \text{ mm} \quad \underline{\text{Ans}}.$$

33-41

With the plane of the electron's orbit perpendicular to B, the force evB due to the magnetic field will be directed either radially in toward the nucleus (antiparallel to E) or radially out (parallel to E), depending on the direction of the electron's motion about the proton. The net force on the electron will be

$$F = F_E \pm ev^*B = \frac{mv^{*2}}{r} = 4\pi^2 mr\nu^{*2},$$

ν^* the frequency of circulation with the magnetic field. If ν is the frequency with B = 0,

$$F_E = 4\pi^2 m\nu^2 r,$$

the radius r being unchanged. Thus,

$$4\pi^2 mr\nu^2 \pm ev^*B = 4\pi^2 mr\nu^{*2}.$$

Writing $\nu^* = \nu + \Delta\nu$, $\Delta\nu/\nu \ll 1$, this becomes

$$4\pi^2 mr\nu^2 \pm ev^*B = 4\pi^2 mr(\nu + \Delta\nu)^2 \approx 4\pi^2 mr\nu^2(1 + 2\,\frac{\Delta\nu}{\nu}),$$

$$\pm\, ev^*B = 8\pi^2 mr\nu\Delta\nu.$$

But $v^* \approx v = 2\pi r\nu$; substituting this and solving for $\Delta\nu$ gives

$$\pm\, e(2\pi r\nu)B = 8\pi^2 mr\nu\Delta\nu,$$

$$\Delta\nu = \frac{eB}{4\pi m} \quad \underline{\text{Ans}}.$$

33-47

From Example 7, the final deuteron energy is 17 MeV and the dee radius is 0.53 m. Thus, the number of revolutions made by a deuteron is

$$\frac{17\ \text{MeV}}{2(80\ \text{keV})} = 106$$

since there are two accelerations per revolution. Over one of these accelerations, as the particle passes from one dee to the other, the radius of the semicircular path jumps from r to r* where

$$\tfrac{1}{2}m(r\omega)^2 + qV = 2\pi^2 m\nu^2 r^2 + qV = 2\pi^2 m\nu^2 r^{*2}.$$

If $r^* = r + \Delta r$, $\Delta r/r \ll 1$ so that $r^{*2} \approx r^2(1 + 2\Delta r/r)$, then

$$\Delta r = \frac{qV}{4\pi^2 m\nu^2}\,\frac{1}{r}.$$

Thus the deuteron spends more of the 106 revolutions at radii greater than $\tfrac{1}{2}(0.53\ \text{m})$ than at smaller radii; hence, as an average radius, choose $(0.53\ \text{m})(\tfrac{1}{2}\sqrt{2})$, say, for which the circumference of the circle is $2\pi(0.37\ \text{m}) = 2.3\ \text{m}$. Since the deuteron makes 106 revolutions, the path traveled must be about $(106)(2.3\ \text{m}) = 240\ \text{m}$.

33-49

If the electron travels in a straight line, the net force on it must be zero (assuming constant speed), and therefore

$$evB = eE,$$

$$B = \frac{E}{v} = \frac{\mathcal{E}}{vd}.$$

But $\tfrac{1}{2}mv^2 = eV$ so that

$$B = \frac{\mathcal{E}}{d}\left(\frac{m}{2eV}\right)^{\frac{1}{2}}.$$

Putting $\mathcal{E} = 100$ volts, $V = 1000$ volts and $d = 0.02$ m gives $B = 2.67 \times 10^{-4}$ T <u>Ans.</u>

CHAPTER 34

34-2

Apply Ampere's law by integrating around the dashed rectangle shown. Since there is no current cutting the area bounded by the path of integration, it is expected that

$$\oint \vec{B}\cdot\vec{d\ell} = 0.$$

On section 2, $\int \vec{B}\cdot\vec{d\ell} = 0$ since $B = 0$ there. On sections 1 and 3, the integral is zero because either $B = 0$ or $\vec{B}$ is perpendicular to $\vec{d\ell}$. On section 4, however,

$$\int \vec{B}\cdot\vec{d\ell} = B_4 b.$$

Thus $\oint \vec{B}\cdot\vec{d\ell} = B_4 b \neq 0$, contradicting Ampere's law. Hence, the geometry assumed for the field lines must be in error.

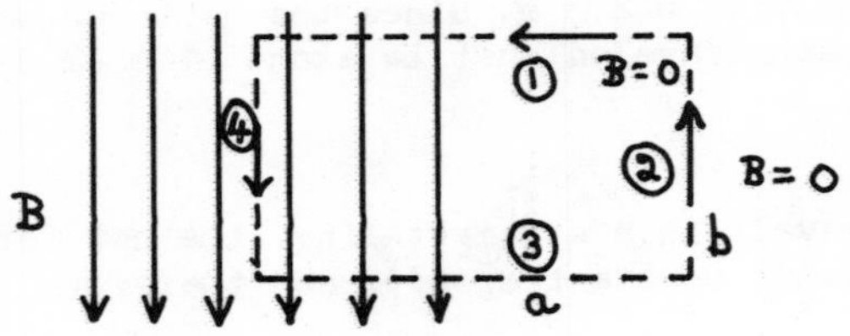

34-13

Consider a large number of long wires carrying identical currents and arranged in a plane, as in the sketch. By the right-hand rule, the fields above the plane all are directed to the left and below the plane all are directed to the right. Between the wires the fields due to neighboring wires are oppositely pointing. It seems reasonable that as the wires are brought closer together the already opposing fields between them will vanish (cancel), yielding the pattern shown. Selecting as a path of integration a rectangle of length L and width W oriented as shown, perpendicular to the current sheet, it is clear that $\vec{B}$ is at right angles to $\vec{d\ell}$ along the sides W of the rectangle, so the line integral in Ampere's law gives nil on these sides, while along the top side L,

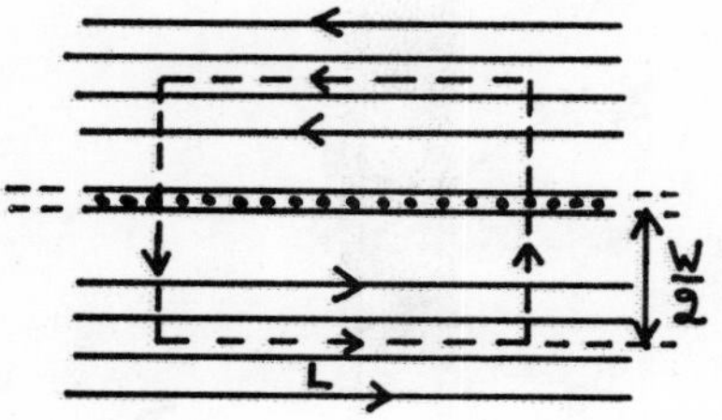

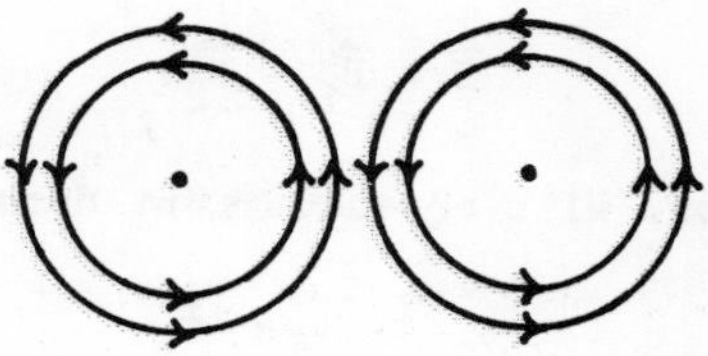

$$\int \vec{B}\cdot d\vec{\ell} = \int B d\ell \cos 0° = B\int d\ell = BL.$$

Since both of these sides (length L, parallel to the sheet) are at the same distance from the sheet, B is the same at each of them, so that Ampere's law gives

$$\oint \vec{B}\cdot d\vec{\ell} = 2BL = \mu_0 i_{enc} = \mu_0(niL),$$

$$B = \tfrac{1}{2}\mu_0 ni.$$

To obtain this result from the field a distance R above a flat strip of width a, carrying a total current i, for which

$$B = (\mu_0 i/\pi a)\tan^{-1}(\frac{a}{2R}),$$

make the substitutions

$$a \text{ "=" } \infty,\ \frac{i}{a} = ni;$$

that is, the strip width becomes very large, and the current per unit width of the strip becomes the current per unit length of the sheet. Then,

$$B = (\mu_0 ni/\pi)(\pi/2) = \tfrac{1}{2}\mu_0 ni,$$

the same result as obtained directly above.

<u>34-14</u>

Let P represent a point in the hole. By superposition, the field $\vec{B}$ at P = (field $\vec{B}_1$ at P due to a wire of radius R, current I, that contains no hole) - (field $\vec{B}_2$ at P due to the hole *if* it carried a uniform current of current density $I/\pi R^2$ in the same direction as i). That is

$$\vec{B} = \vec{B}_1 - \vec{B}_2.$$

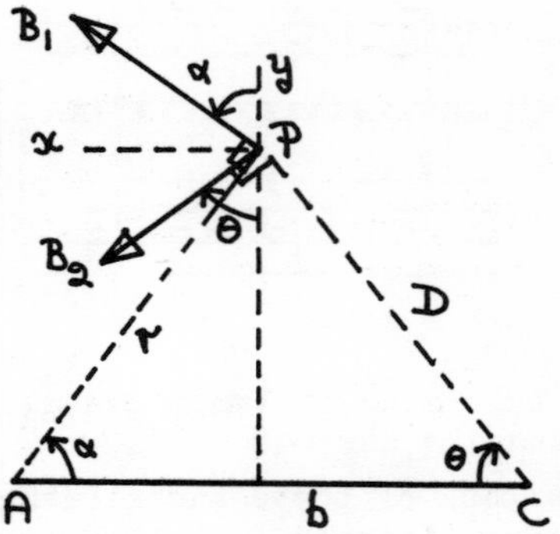

Now, with equal current densities

$$\frac{i}{\pi R^2 - \pi a^2} = \frac{I}{\pi R^2}.$$

Also,

$$B_1 = \frac{\mu_0 I}{2\pi R^2} r, \; B_2 = \frac{\mu_0 I}{2\pi R^2} D.$$

Let A be the center of the wire and C the center of the hole; then,

$$B_y = B_1 \cos\alpha + B_2 \cos\theta = \frac{\mu_0 I}{2\pi R^2}(r\cos\alpha + D\cos\theta) = \frac{\mu_0 I}{2\pi R^2} b;$$

$$B_x = B_1 \sin\alpha - B_2 \sin\theta = \frac{\mu_0 I}{2\pi R^2}(r\sin\alpha - D\sin\theta) = 0.$$

Therefore $B = B_y$; using the current density relation to replace I in the expression for B_y yields

$$B = \frac{\mu_0 i}{2\pi} \frac{b}{R^2 - a^2},$$

perpendicular to AC, independent of the location of P in the hole.

<u>34-17</u>

Let the current in the wire be i_w and the current in the rectangle i_r. The force $\vec{F}$ on the rectangle is the vector sum of the forces on the four sides of the rectangle, labelled 1, 2, 3, 4. Then,

$$\vec{F}_1 = i_r \vec{\ell} \times \vec{B},$$

$\vec{\ell}$ pointing in the direction of the current in side 1. From Ampere's law, the field B due to the wire is

$$B = \mu_0 i_w / 2\pi a,$$

directed into the paper; thus

$$F_1 = \mu_0 i_r i_w \ell / 2\pi a, \text{ to the left.}$$

Similarly,

$$F_3 = \mu_0 i_r i_w \ell / 2\pi(a + b), \text{ to the right,}$$

since this side of the rectangle is at a distance (a + b) from the wire and its current is directed oppositely from the current in side 1. F_2 is more difficult to compute since different parts of

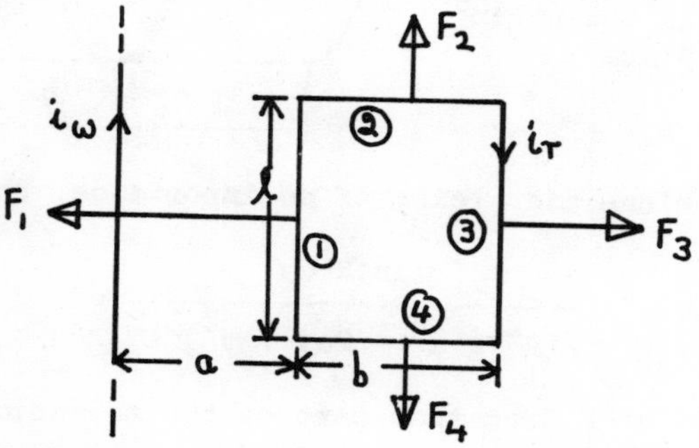

side 2 are at different distances from the wire:

$$F_2 = \int dF_2 = \int i_r B(r)dr = \int_a^{a+b} i_r(\mu_0 i_w / 2\pi r)dr = \frac{\mu_0 i_r i_w}{2\pi} \ln(1 + \frac{b}{a}),$$

pointing "upward". But clearly $\vec{F}_4 = -\vec{F}_2$ so that, with $\vec{F} = \vec{F}_1 + \vec{F}_2 + \vec{F}_3 + \vec{F}_4$,

$$F = i_r i_w \mu_0 \ell / 2\pi a - i_r i_w \mu_0 \ell / 2\pi(a + b) = \frac{\mu_0 i_r i_w \ell b}{2\pi a(a + b)},$$

$$F = \frac{(20\text{ A})(30\text{ A})(4\pi \times 10^{-7}\text{ T}\cdot\text{m/A})(0.3\text{ m})(0.08\text{ m})}{2\pi(0.01\text{ m})(0.09\text{ m})}$$

$$F = 3.2 \times 10^{-3}\text{ N} \quad \underline{\text{Ans}},$$

to the left.

34-24

To find the field at P, compute the contributions B_r and B_l due to the segments of the solenoid lying to the right and to the left of P; since the current in the turns flows in the same direction, the fields are parallel and the net field is $B = B_r + B_l$. Due to the strip of width dx shown in the sketch, the field is

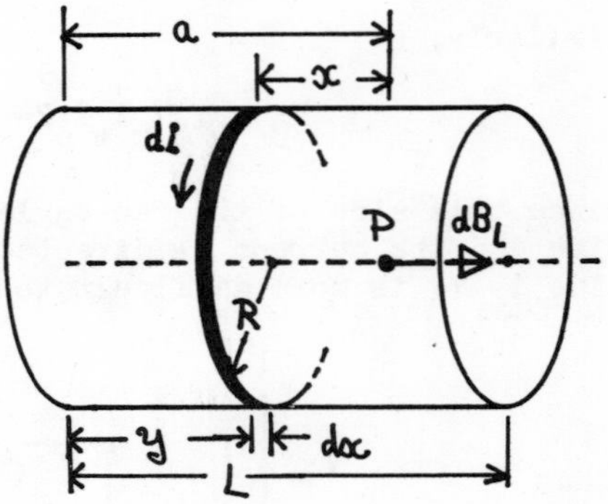

$$dB_l = \frac{\mu_0 R^2}{2(R^2 + x^2)^{3/2}}\, ni\, dx,$$

n the number of turns per unit length of the solenoid, each turn carrying a current i; there are ndx turns of wire in the strip being considered. Since

$$y + x = a,$$

then dy = dx, the minus sign being of no importance. Then,

$$dB_l = \frac{\mu_0 niR^2 dy}{2(R^2 + a^2 - 2ay + y^2)^{3/2}}$$

and the total field at P from that part of the solenoid to the left of P is

$$B_l = \tfrac{1}{2}\mu_0 niR^2 \int_0^a \frac{dy}{(R^2 + a^2 - 2ay + y^2)^{3/2}} = \tfrac{1}{2}\mu_0 ni\, \frac{a}{(R^2 + a^2)^{\frac{1}{2}}}.$$

From this, it follows directly that

$$B_r = \tfrac{1}{2}\mu_0 ni\, \frac{L - a}{[R^2 + (L - a)^2]^{\frac{1}{2}}}$$

the segment on the right having a length L - a. But for an ideal solenoid $L \gg R$ and also $a \gg R$; ignoring the terms in R and adding gives

$$B = B_r + B_l = \tfrac{1}{2}\mu_0 ni + \tfrac{1}{2}\mu_0 ni = \mu_0 ni$$

34-25

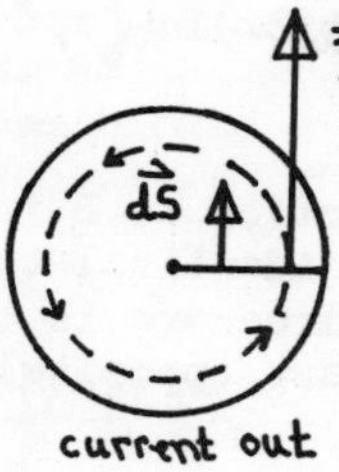

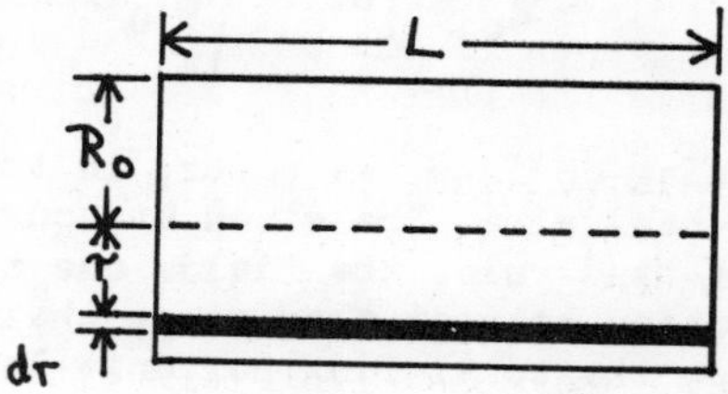

Choose $\vec{dS}$, always perpendicular to the surface over which the flux is to be computed, in the same direction as $\vec{B}$ as it cuts the surface. From Example 34-1

$$B = \mu_0 i_0 r/2\pi R_0^2.$$

Thus,

$$\Phi_B = \int \vec{B}\cdot\vec{dS} = \int B(dS)\cos 0^\circ = \int B dS.$$

Now B varies across the surface (r direction) but not along the narrow strip, since all parts of the strip are at the same distance r from the axis of the wire. Hence,

$$\Phi_B = \int_0^{R_0} (\mu_0 i_0 r/2\pi R_0^2) L dr = \mu_0 i_0 L/4\pi;$$

therefore, the flux through one meter is $\mu_0(10\ A)/4\pi = 10^{-6}\ T\cdot m^2$.

34-26

(a) Call the radius of each wire b and let a be the distance their axes are apart. Consider a single wire. The flux inside the wire is

$$\Phi_i = \mu_0 i L/4\pi,$$

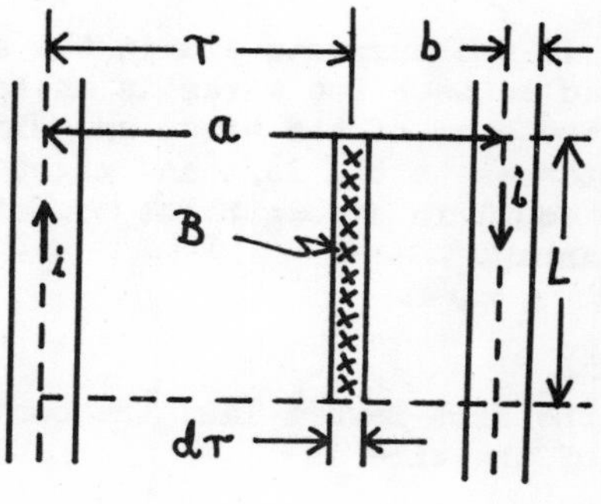

by Problem 34-25. The flux outside out to a distance r = a from the axis is

$$\Phi_o = \int BdS = \int_b^a (\mu_0 i/2\pi r)Ldr = (\mu_0 iL/2\pi)\ln(\frac{a}{b}),$$

choosing $\vec{dS}$ into the paper, in the same direction as $\vec{B}$ from the left-hand wire. The wires and currents being identical then, by the right-hand rule, the fields due to the two wires are in the same direction between them since their currents are oppositely directed. Hence, the total flux per unit length is

$$\Phi = (2\mu_0 i/4\pi)[1 + \ln(\frac{a}{b})^2].$$

Numerically, $a/b = 20/1.27$, $i = 10$ A and $\mu_0 = 4\pi \times 10^{-7}$ Wb/A·m giving $\Phi = 13.027 \times 10^{-6}$ Wb/m Ans.

(b) Consider the left-hand wire, say. Part of its flux for the region $b \leq r \leq a$ lies inside the other wire; this flux is

$$\Phi^* = \int BdS = (\mu_0 iL/2\pi)\ln(\frac{a}{a - b}).$$

A similar flux lies inside the left-hand wire due to the right-hand wire. Hence, the total flux inside the wires is

$$2\Phi_i + 2\Phi^* = (2\mu_0 iL/4\pi)[1 + \ln(\frac{a}{a - b})^2].$$

Numerically, per unit length this is 2.262×10^{-6} Wb/m. Therefore, the fraction desired is

$$\frac{2.2624}{13.027} = 17.4\% \text{ Ans.}$$

(c) If the currents are in the same direction, the net magnetic field between the wires is antisymmetrical about a line parallel to the axes of the wires and midway between them. That is, at equal distances to the left and right of this line, the magnetic fields are equal in strength but oppositely directed. The net flux, then, vanishes.

34-29

By the Biot-Savart law, the contribution to B from a short section $d\ell$ of the wire is

$$\vec{dB} = \frac{\mu_0 i}{4\pi}\frac{\vec{dl} \times \vec{r}}{r^3}.$$

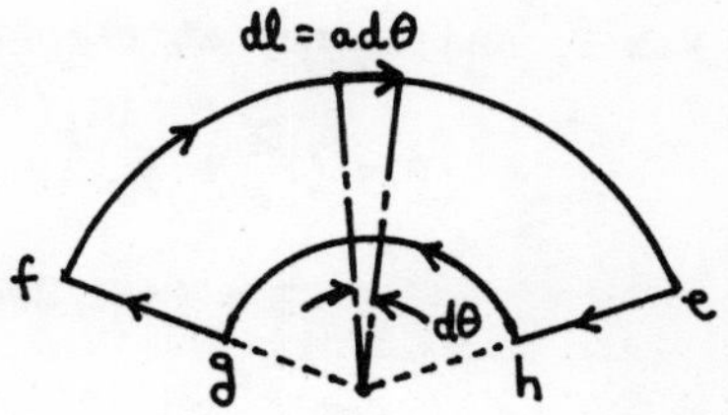

Now sections he and fg give nothing since $\vec{dl} \times \vec{r} = 0$ on these pieces. Along section ef,

$$B_2 = \frac{\mu_0 i}{4\pi}\int\frac{r dl \sin(\vec{r},\vec{dl})}{r^3},$$

$$B_2 = \frac{\mu_0 i}{4\pi}\int\frac{a(ad\theta)\sin(\pi/2)}{a^3} = \frac{\mu_0 i}{4\pi a}\int_0^\theta d\theta = \frac{\mu_0 i\theta}{4\pi a}.$$

Clearly, the contribution due to gh is $B_1 = (\mu_0 i\theta/4\pi b)$. Also, $B_1 > B_2$ since $b < a$, and B_1 is out of, B_2 into, the page. Thus,

$$B = \frac{\mu_0 i\theta}{4\pi}\left(\frac{1}{b} - \frac{1}{a}\right),$$

out of the page.

34-34

The field $\vec{B}$ at C = field $\vec{B}_w$ due to an infinite straight wire + field $\vec{B}_c$ due to a circle of wire; these are

$$B_w = \mu_0 i/2\pi R,$$
$$B_c = \mu_0 i/2R,$$

the former from Eq. 34-4 and the latter from Eq. 34-11 with x = 0.

(a) Here $\vec{B}_w$ and $\vec{B}_c$ both are directed normal to and out of the page; thus,

$$B = B_c + B_w = (\mu_0 i/2R)\left(1 + \frac{1}{\pi}\right),$$

out of the page.

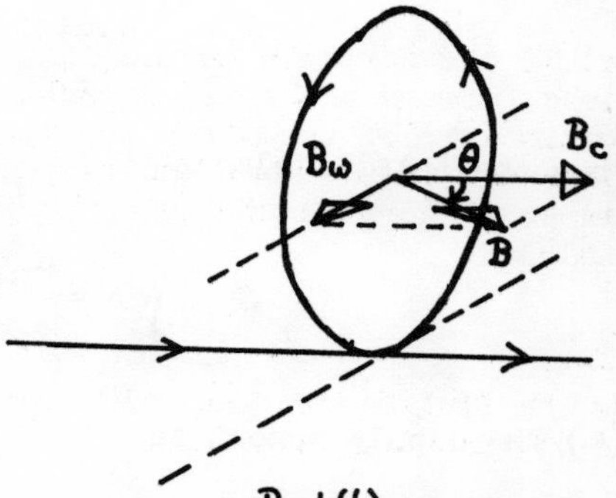

Part(b)

(b) Now $\vec{B}_c$ and $\vec{B}_w$ are at right angles and therefore

$$B = (B_w^2 + B_c^2)^{\frac{1}{2}} = (\mu_0 i/2R)[1^2 + (\frac{1}{\pi})^2]^{\frac{1}{2}},$$

$$B = (\mu_0 i/2\pi R)(1 + \pi^2)^{\frac{1}{2}} \quad \underline{\text{Ans}}.$$

The direction of $\vec{B}$ is at an angle θ out of the page, where

$$\theta = \tan^{-1}(\frac{B_w}{B_c}) = \tan^{-1}(\frac{1}{\pi}) = 18° \quad \underline{\text{Ans}}.$$

<u>34-37</u>

(a) Consider a narrow ring of radius r and width dr; it carries a charge

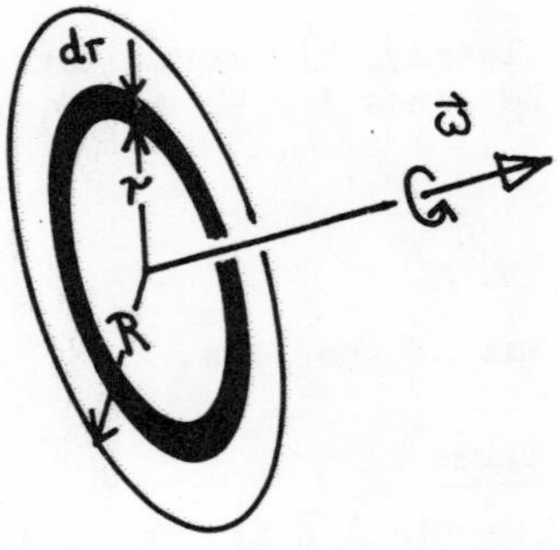

$$dq = \frac{q}{\pi R^2}(2\pi r dr),$$

i.e., the charge per unit area of the disc times the area of the ring. In time $2\pi/\omega$ all of this charge passes any fixed point near the ring, so that the equivalent current is

$$di = \frac{dq}{\tau} = \frac{2\pi q r \, dr/\pi R^2}{2\pi/\omega} = q\omega r \, dr/\pi R^2.$$

By Eq. 34-11 (x = 0) this ring sets up a field dB at the center of the disc given by

$$dB = \frac{\mu_0 di}{2r} = \frac{\mu_0}{2r}(q\omega r dr/\pi R^2).$$

Thus the total field is

$$B = \int dB = \frac{\mu_0 q\omega}{2\pi R^2}\int_0^R dr = \mu_0 q\omega/2\pi R.$$

(b) The dipole moment is

$$\mu = \int A di = \int_0^R (\pi r^2)\frac{q\omega r dr}{\pi R^2} = \frac{q\omega}{R^2}\int_0^R r^3 dr = \tfrac{1}{4}q\omega R^2.$$

34-39

(a) Let the wire rest along the x-axis, its midpoint at the origin. Since the wire is of finite length, the Biot-Savart law is called for. All elements of the wire give rise to a field directed into the paper, the magnitude of the total field being

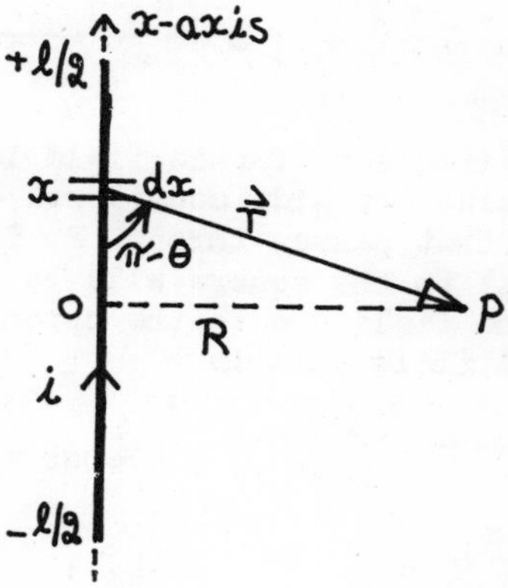

$$B = (\mu_0 i/4\pi)\int_{x=-\frac{1}{2}\ell}^{x=\frac{1}{2}\ell} \frac{(d\ell)(r)\sin\theta}{r^3}.$$

But,

$$\sin\theta = \frac{R}{r},\ r^2 = x^2 + R^2.$$

Therefore,

$$B = (\mu_0 iR/4\pi)\int_{-\frac{1}{2}\ell}^{\frac{1}{2}\ell} (x^2 + R^2)^{-3/2}dx = \frac{\mu_0 i}{2\pi R}\frac{\ell}{(\ell^2 + 4R^2)^{\frac{1}{2}}}.$$

(b) If $R \ll \ell$, ignore R^2 in the denominator to obtain $B = \mu_0 i/2\pi R$, the field due to a very long wire: from points extremely close to the original wire, the wire appears to be very long.

34-41

The field due to the square is the vector sum of the fields due to the four sides of the square. Consider, then, one side. The point at which the field is desired lies along the perpendicular bisector to that side, at a distance R given by

$$R = \tfrac{1}{2}(4x^2 + a^2)^{\frac{1}{2}}.$$

Hence, using the result of Problem 34-39, the field from one side is

$$B = (\mu_0 i/2\pi R) \frac{a}{(a^2 + 4R^2)^{\frac{1}{2}}},$$

putting a for ℓ as the length of the wire section. Substituting for R from the first equation gives

$$B = \frac{\mu_0 i}{\pi} \frac{a}{(4x^2 + a^2)^{\frac{1}{2}}(4x^2 + 2a^2)^{\frac{1}{2}}}.$$

The direction of this field is perpendicular to the plane that contains the side considered and the perpendicular bisector of that side that passes through P. The component perpendicular to the normal to the square will be cancelled by the analogous component of the field due to the opposite side of the square. Thus, the total field will be

$$B_T = 4B\cos\theta = 4B \frac{a/2}{R} = \frac{4aB}{(4x^2 + a^2)^{\frac{1}{2}}},$$

$$B_T = \frac{4\mu_0 i}{\pi} \frac{a^2}{(4x^2 + a^2)(4x^2 + 2a^2)^{\frac{1}{2}}}.$$

As expected, for x = 0 (center of square) this reduces to the result of Problem 34-40.

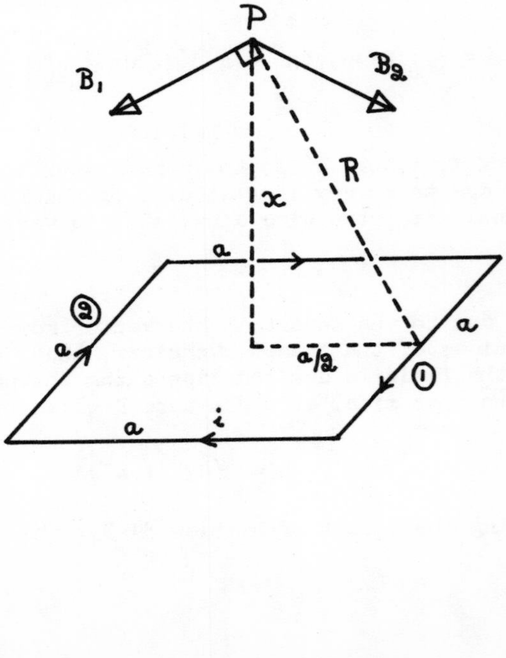

CHAPTER 35

35-6

(a) By Faraday's law, $\mathcal{E} = -N\cdot d\Phi_B/dt$ and therefore the flux through the rectangular loop at any time t must be computed first. This is

$$\Phi_B = \int\vec{B}\cdot\vec{dS} = \int B(dS)\cos\theta = B\cos\theta\int dS = abB\cos\theta,$$

since B is uniform and θ, the angle between $\vec{B}$ and $\vec{dS}$, has at any instant the same value at all points "on" the rectangle. If the loop rotates at the constant rate ν, then $\theta = 2\pi\nu t$ and

$$d\Phi_B/dt = d(abB\cos 2\pi\nu t)/dt = -2\pi\nu abB\sin 2\pi\nu t,$$

and therefore $\mathcal{E} = 2\pi\nu NabB\sin 2\pi\nu t = \mathcal{E}_0\sin 2\pi\nu t$.

(b) If $\mathcal{E}_0 = 150$ V, $\nu = 60$ Hz and B = 0.50 T, then

$$2\pi Nab = \mathcal{E}_0/\nu B = 5\ m^2$$

and any loop built according to the specification $Nab = 5/2\pi\ m^2$ will produce the desired effect.

35-7

(a) By Faraday's law the current i induced in the coil is

$$i = \frac{\mathcal{E}}{R} = \frac{N}{R}\,d\Phi_B/dt$$

in which N = 1, R = 5 Ω. Now $\Phi_B = \int\vec{B}\cdot\vec{dS}$, the integral taken over the cross-section of the coil, or any area bounded by the coil, and B is the field due to the solenoid. Since $B = \mu_0 n i_s$ inside the solenoid (i_s = current in the solenoid) and B = 0 outside the solenoid, the integral for the flux over the area between the coil and solenoid will be zero and

$$\Phi_B = \int_{\substack{\text{area of}\\ \text{solenoid}}} (\mu_0 n i_s)dS + 0 = \mu_0 n i_s A_s,$$

where A_s is the cross-sectional area of the solenoid. Thus,

$$i = \frac{N}{R} d(\mu_0 n i_s A_s)/dt = \frac{N}{R}\mu_0 n A_s d(i_s)/dt = \frac{N}{R}\mu_0 n A_s \frac{2i_0}{t}.$$

Numerically, $N = 1$, $R = 5\ \Omega$, $\mu_0 = 4\pi \times 10^{-7}$ Wb/A·m, $n = 20{,}000\ \text{m}^{-1}$, $A_s = \pi(0.015\ \text{m})^2$, $i_0 = 1.5$ A, $t = 0.05$ s; these give $i = 2.1 \times 10^{-4}$ amperes Ans.

(b) The magnetic flux is entirely confined to the internal volume of the solenoid only if the current in the solenoid is constant; if the current increases, say, additional lines of flux must snake into the solenoid, to reflect the increased field strength, cutting the coil as they do so.

35-11

Consider a conduction electron in the rod. As a result of the rod's motion, the electron too is moving to the left with the same speed v. Hence, the magnetic field exerts a force $\vec{F} = q\vec{v} \times \vec{B} = (-e)\vec{v} \times \vec{B}$ on the electron. This is directed away from the long wire, as shown. From Ampere's law

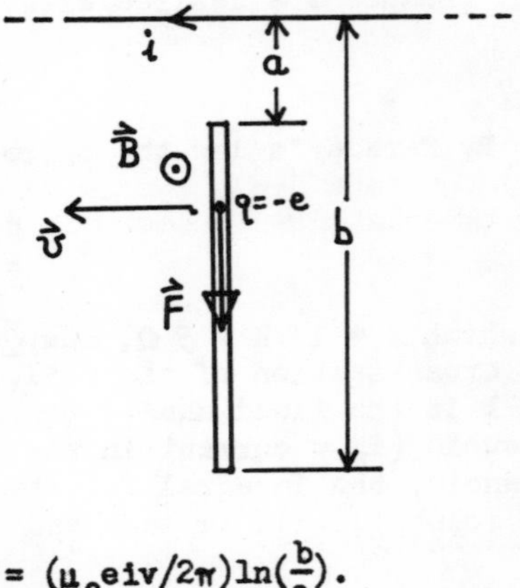

$$B = \mu_0 i/2\pi r,$$

so that

$$F = \mu_0 eiv/2\pi r.$$

If the electron started from the end of the rod closest to the wire, the total work W done by the magnetic force in moving the electron to the other end is

$$W = \int F dr = (\mu_0 eiv/2\pi)\int_a^b \frac{dr}{r} = (\mu_0 eiv/2\pi)\ln(\frac{b}{a}).$$

But $W = e\mathcal{E}$, and therefore

$$\mathcal{E} = \frac{\mu_0 iv}{2\pi}\ln\left(\frac{b}{a}\right) = 3.00 \times 10^{-4}\ \text{V} \quad \underline{\text{Ans}}.$$

35-15

As the small area a^2 leaves the magnetic field, the emf induced will be

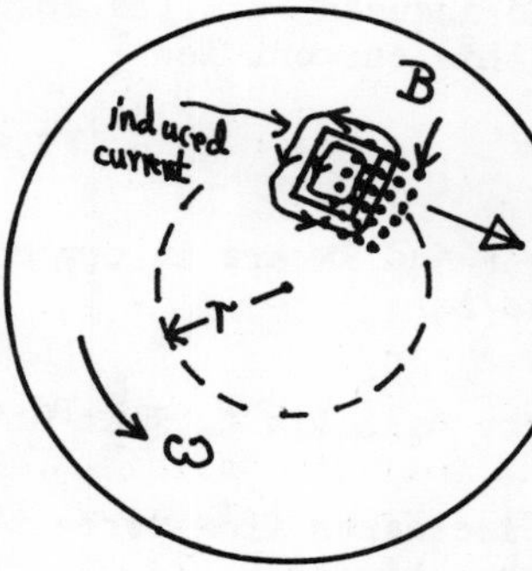

$$\mathcal{E} = d\Phi_B/dt = Bav = Bar\omega.$$

The eddy current induced is $\mathcal{E}/R$, R the resistance of the current path. This is

$$R = \rho\frac{L}{A} \approx \rho\frac{4a}{at} = \frac{1}{\sigma}\frac{4}{t}.$$

Thus,

$$i = \frac{\mathcal{E}}{R} = \frac{Bar\omega}{4/\sigma t} = \tfrac{1}{4}Bar\omega\sigma t$$

The force opposing the motion is $F \approx iaB$ and is felt twice each revolution, as the area (or the one ahead of it) enters or leaves the field; hence,

$$\tau = 2rF \approx \tfrac{1}{2}B^2a^2r^2\sigma\omega t \quad \underline{\text{Ans}}.$$

35-17

(a) The force on the wire is $F = iBd$, a constant since i is taken to be constant; by Newton's second law,

$$a = \frac{F}{m} = \frac{iBd}{m} = \frac{dv}{dt},$$

$$v = \frac{iBd}{m}\cdot t,$$

since $v(0) = 0$. By the right-hand rule the force is directed away from the generator and , with $v(0) = 0$, $\vec{v}$ is in this direction.

(b) The battery sets up a current in the wire, and the wire will begin to move under a force

$$F = iBd = \frac{\mathcal{E}}{R}\cdot Bd.$$

But now a current, flowing oppositely to the first, is induced and is given by

$$i^* = \mathcal{E}^*/R = \frac{Bvd}{R},$$

$\mathcal{E}^*$ the induced emf. The force opposing the motion and associated with this current is

$$F^* = i^*Bd = B^2d^2v/R.$$

Since F and F* are in opposite directions, the resultant of these forces is

$$F_t = \frac{\mathcal{E}}{R}\cdot Bd - \frac{1}{R}\cdot B^2d^2v = \frac{Bd}{R}(\mathcal{E} - Bvd).$$

As v increases from zero it eventually reaches a value making F_t vanish; this is

$$(\mathcal{E} - Bvd) = 0,$$

$$v_t = \mathcal{E}/Bd \quad \underline{\text{Ans}}.$$

(c) When the terminal velocity v_t is reached the net current i_n must be zero, for if it is not a force i_nBd would act on the wire, causing it to accelerate.

<u>35-22</u>

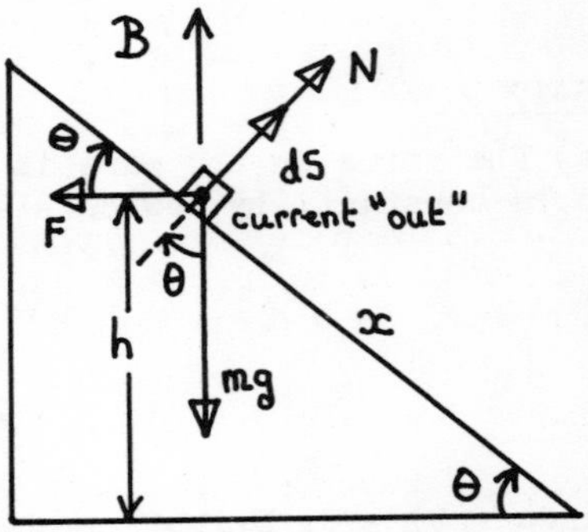

(a) The forces acting on the wire are its weight mg, the normal force N, and the force F exerted by the magnetic field on the current induced in the wire by virtue of its motion induced by gravity. The equation of motion of the wire is

$$mg\sin\theta - F\cos\theta = ma.$$

But $F = i\ell B = (\mathcal{E}/R)\ell B$, ℓ the length of the wire. Since the angle between $\vec{dS}$ (along the normal to the inclined plane) and $\vec{B}$ is θ, the angle of the incline, $\mathcal{E} = B\ell v\cos\theta$ and therefore

$$mg\sin\theta - (B^2\ell^2 v/R)\cos^2\theta = ma.$$

Initially v = 0; as the wire accelerates down the incline v will increase until it reaches a value v_t given by

$$v_t = \frac{mgR\sin\theta}{B^2\ell^2\cos^2\theta},$$

at which point ma = 0 and the wire stops accelerating, as the net force on it now is zero. Subsequently, then, the wire slides down with constant speed v_t.

(b) The kinetic and gravitational potential energies are $K = \frac{1}{2}mv^2$ and U = mgh. Let x be the distance of the wire from the bottom of the incline measured along the incline. Then $h = x\sin\theta$. By the conservation of energy it is expected that

$$\frac{dU}{dt} = \frac{dK}{dt} + P,$$

the last term being the rate of generation of thermal energy. As v = dx/dt and a = dv/dt, this equation becomes

$$mgv\sin\theta = mv\frac{dv}{dt} + i^2R,$$

$$mg\sin\theta = ma + (B^2\ell^2 v/R)\cos^2\theta,$$

since $i = \mathcal{E}/R = B\ell v\cos\theta/R$. But this last equation agrees with (a); hence the result in (a) is consistent with energy conservation. After the rod has reached its terminal speed, dK/dt = 0 so that P = dU/dt.

(c) If B is directed down instead of up the induced current will flow in the direction opposite from that in (a), but $\vec{F} = (-i)\vec{\ell} \times (-\vec{B})$ will be in the same direction as in (a), so the motion of the wire will be unchanged.

35-27

(a) By Faraday's law $\mathcal{E} = -d\Phi_B/dt$ so that $i = \mathcal{E}/R = (-1/R)d\Phi_B/dt$. But i = dq/dt also, and therefore

$$\frac{dq}{dt} = -\frac{1}{R}\,d\Phi_B/dt,$$

$$\int_0^q \frac{dq}{dt}\,dt = -\frac{1}{R}\int_{\Phi_B(0)}^{\Phi_B(t)} (d\Phi_B/dt)dt,$$

$$q = -\frac{1}{R}[\Phi_B(t) - \Phi_B(0)],$$

where the limits on the integrals are the values of charge and flux at times t and 0.

(b) If the flux at time t and the flux at t = 0 both are zero, the equation derived in (a) implies that the net charge through the circuit is zero; if the flux changed during the time interval, this indicates that equal amounts of charge flowed clockwise through the circuit for part of the time, say, and counterclockwise for the remaining part of the time interval during which charge flowed, giving a net charge flow of zero. This means that induced currents of varying senses could have existed during the entire time period.

35-28

(a) Far away from the larger loop the field may be taken, across the area of the small loop, as approximately uniform and equal to the value B on the axis. By Example 34-8, this is

$$B = \mu_0 iR^2/2x^3,$$

so that the flux through the small loop is

$$\Phi = B(\pi r^2) = \mu_0 iR^2\pi r^2/2x^3 \quad \underline{\text{Ans}}.$$

(b) For a single loop the emf is

$$\mathcal{E} = d\Phi/dt = (\mu_0 iR^2\pi r^2/2)\,\frac{dx^{-3}}{dt}.$$

But $dx^{-3}/dt = -3x^{-4}dx/dt = -3v/x^4$. Hence, except for sign,

$$\mathcal{E} = \frac{3}{2}\mu_0 i\pi R^2 r^2 v/x^4 \quad \underline{\text{Ans}}.$$

(c) The field due to i in the larger loop, near its axis, points away from the large loop toward the small loop. Therefore, as the small loop moves away from the larger, the small loop sees a steadily decreasing upward directed flux cutting through it. To oppose this, the induced emf will seek to set up an induced current the magnetic field of which will be directed upward also, in the area bounded by the small loop. By the right-hand rule, this will necessitate a counterclockwise current as seen looking down from above the small loop (i.e., in the same direction as i in the large loop).

35-32

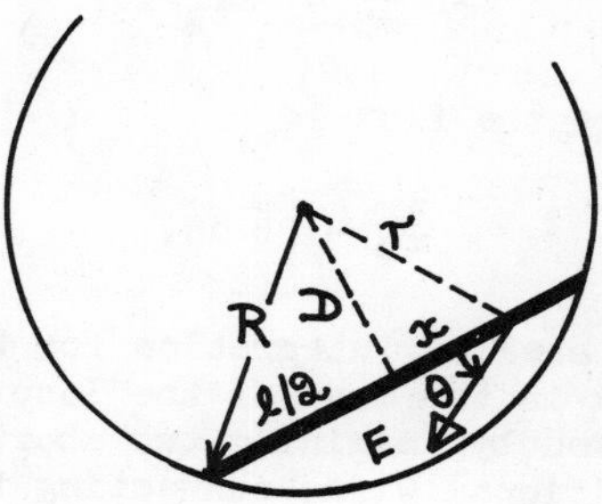

By Example 35-4, the induced electric field is

$$E = \tfrac{1}{2}r\,\frac{dB}{dt}$$

at points a distance r from the center of the cylindrical region; the direction of E is perpendicular to the radius, normal to the axis. Thus, the emf induced in a length $d\ell$ of the rod is

$$d\mathcal{E} = \vec{E}\cdot\vec{d\ell} = E(dx)\cos\theta = (\tfrac{1}{2}r\,\frac{dB}{dt})(dx)(\frac{D}{r}),$$

$$d\mathcal{E} = \tfrac{1}{2}D\,\frac{dB}{dt}\,dx.$$

Hence, the emf induced between the ends of the rod is

$$\mathcal{E} = \frac{D}{2}\frac{dB}{dt}\int dx = \frac{\ell}{2}\,D\,\frac{dB}{dt} = \frac{\ell}{2}\frac{dB}{dt}\,[R^2 - (\tfrac{1}{2}\ell)^2]^{\frac{1}{2}}.$$

CHAPTER 36

36-9

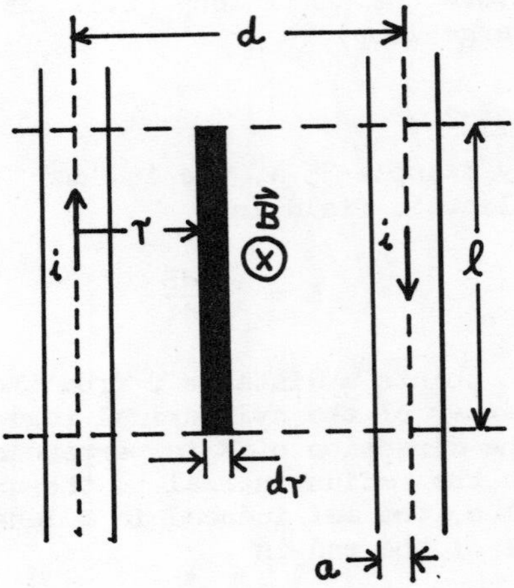

The inductance L is defined by

$$-L\frac{di}{dt} = -d\Phi_B/dt,$$

where the flux is

$$\Phi_B = \int \vec{B}\cdot d\vec{S}.$$

The area of integration for the flux is the area of the loop formed by imagining two short additional wires connecting those given, to form a closed circuit. The lengths of these new wires being very small compared to the original wires, their fluxes may be ignored. Then the field $\vec{B}$ is the sum of the fields set up by the two wires given: both of these are into the paper and therefore, by Ampere's law,

$$B = \frac{\mu_0 i}{2\pi r} + \frac{\mu_0 i}{2\pi(d - r)}.$$

B does not vary in a direction parallel to the wires and therefore for dS take a skinny rectangle of length ℓ and width dr; choose the direction of $d\vec{S}$ as into the paper (the same as $\vec{B}$). Then,

$$\Phi_B = \int B(dS)\cos 0^\circ = \int_a^{d-a}\left[\frac{\mu_0 i}{2\pi r} + \frac{\mu_0 i}{2\pi(d - r)}\right]\ell dr = \frac{\mu_0 i\ell}{\pi}\ln\left(\frac{d - a}{a}\right).$$

Hence,

$$d\Phi_B/dt = \frac{\mu_0\ell}{\pi}\frac{di}{dt}\ln\left(\frac{d - a}{a}\right) = L\frac{di}{dt},$$

and therefore, with the flux within the wires themselves ignored,

$$L = \frac{\mu_0 \ell}{\pi}\ln\left(\frac{d - a}{a}\right).$$

36-10

The area through which the flux is calculated is a rectangle of length ℓ lying between the cylinders and whose plane contains the common axis of those cylinders. By Problem 34-9,

$$B = \frac{\mu_0 i}{2\pi r},$$

for $a \le r \le b$ (i.e., over the rectangle). The flux through this area is

$$\Phi_B = \int \vec{B}\cdot d\vec{S} = \int_a^b \frac{\mu_0 i}{2\pi r}(\ell dr) = \frac{\mu_0 i \ell}{2\pi}\ln\left(\frac{b}{a}\right),$$

and the inductance L follows directly:

$$L = \frac{d\Phi_B/dt}{di/dt} = \frac{\mu_0 \ell}{2\pi}\ln\left(\frac{b}{a}\right) \text{ Ans.}$$

36-13

The current in the circuit is

$$i = \frac{\mathcal{E}}{R}(1 - e^{-t/\tau})$$

where $\tau = L/R$ is the inductive time constant. The equilibrium value of the current is $\mathcal{E}/R$, the value reached "after a long time". Therefore, if t is the time being sought,

$$0.999\,\frac{\mathcal{E}}{R} = \frac{\mathcal{E}}{R}(1 - e^{-t/\tau}),$$

$$0.999 = 1 - e^{-t/\tau},$$

$$e^{-t/\tau} = 1 - 0.999 = 0.001,$$

$$\ln(e^{-t/\tau}) = \ln(0.001),$$

$$-t/\tau = -6.908,$$

$$t = 6.908\tau,$$

or that 6.9 time constants, approximately, must elapse.

36-18

The current in an LR circuit drops exponentially:

$$i = i_0 e^{-t/\tau},$$

where i_0 is the current at $t = 0$ and $\tau = L/R$. At $t = 0$ the current is 1 A, so that $i_0 = 1$ A. At any subsequent time, then,

$$i = e^{-t/\tau}.$$

When $t = 1$ s, the current $i = 0.01$ A. Hence,

$$0.010 = e^{-1/\tau},$$

$$\ln(0.01) = -1/\tau,$$

$$-\ln(100) = -R/L,$$

$$R = L \cdot \ln(100) = 46\ \Omega \quad \text{Ans.}$$

36-19

(I) With the switch S just closed, the inductor effectively breaks the right-hand branch: $i_2 = 0$.

(a) $i_1 = \mathcal{E}/R_1 = (10\text{ V})/(5\ \Omega) = 2.0$ A Ans.

(b) $i_2 = 0$ Ans.

(c) $i = i_1 + i_2 = 2 + 0 = 2.0$ A Ans.

(d) $V_2 = i_2R_2 = (0)(10\ \Omega) = 0$ Ans.

(e) By the loop theorem, $V_L + V_2 + V_1 = 0$ giving $V_L = -i_1R_1 + i_2R_2$ (going clockwise around the branch). Hence, $V_L = -(2\text{ A})(5\ \Omega) + 0 = -10$ V Ans.

(f) In absolute value, $V_L = L\, di_2/dt$, so that $10 = 5\, di_2/dt$, or $di_2/dt = 2.0$ A/s Ans.

(II) After a long time the inductor has no effect and the circuit reduces to two resistors connected in parallel across a battery, and all currents are independent of time.

(a) $i_1 = \mathcal{E}/R_1 = 2.0$ A <u>Ans</u>.

(b) $i_2 = \mathcal{E}/R_2 = 1.0$ A <u>Ans</u>.

(c) $i = i_1 + i_2 = 3.0$ A <u>Ans</u>.

(d) $V_2 = i_2R_2 = 10$ V <u>Ans</u>.

(e) $V_L = -L\, di_2/dt = 0$ <u>Ans</u>.

(f) $di_2/dt = 0$ <u>Ans</u>.

<u>36-20</u>

(a) The inductor "breaks" the right-hand branch; $i_1 = i_2 = i$. But

$$i = \frac{\mathcal{E}}{R_1 + R_2} = \frac{100}{10 + 20} = 3.33 \text{ A} \quad \underline{\text{Ans}}.$$

(b) Now the inductor has no effect and therefore

$$\mathcal{E} - i_1R_1 - i_1R_e = 0,$$

where

$$\frac{1}{R_e} = \frac{1}{R_2} + \frac{1}{R_3}$$

giving $R_e = 12\ \Omega$. Thus $i_1 = (100 \text{ V})/(22\ \Omega) = 4.55$ A <u>Ans</u>. Also,

$$\mathcal{E} - i_1R_1 - i_2R_2 = 0,$$

$$100 = (5.545)(10) + i_2(20),$$

$$i_2 = 2.73 \text{ A} \quad \underline{\text{Ans}}.$$

(c) The left-hand branch is broken now, so that $i_1 = 0$. The current through R_2 equals the current through R_3 since the elements remaining form a series circuit. The initial value of this current equals the current through R_3 a long time after S originally was closed. From (b), this is $i_2 = 4.545 - 2.727 = 1.82$ A.

(d) Now there are no sources of emf in the circuit and therefore all currents vanish.

36-25

(a) The current is

$$i = \frac{\mathcal{E}}{R}(1 - e^{-t/\tau}) = \frac{50}{10,000}(1 - e^{-t/\tau}).$$

At t = 5 ms, i = 2 mA; hence,

$$0.002 = 0.005(1 - e^{-0.005/\tau}),$$

$$0.4 = 1 - e^{-0.005/\tau},$$

$$e^{-0.005/\tau} = 0.6,$$

$$\tau = -\frac{0.005}{\ln(0.6)} = 9.79 \times 10^{-3} = L/R,$$

$$L = (9.79 \times 10^{-3})(10^4) = 97.9\ \text{H} \quad \underline{\text{Ans}}.$$

(b) The energy stored is $U = \frac{1}{2}Li^2 = \frac{1}{2}(97.9)(0.002)^2 = 1.96 \times 10^{-4}$ in joules.

36-28

Suppose the switch to have been in position a for a time T before being thrown to b. The energy stored in the inductor the instant the switch is thrown is

$$U_B(T) = \frac{1}{2}Li_T^2$$

where $i_T = \frac{\mathcal{E}}{R}(1 - e^{-T/\tau})$ and $\tau = L/R$. With the switch in position b the current in the circuit becomes

$$i = i_T e^{-(t - T)/\tau},$$

time still being measured from the instant the switch was closed on a. The energy dissipated in the resistor after the switch is thrown to b over all subsequent time is

$$E = \int_T^{\infty} i^2R\ dt = i_T^2R\int_T^{\infty} e^{-2(t - T)/\tau}dt = i_T^2R\tau/2,$$

$$E = i_T^2 RL/2R = \tfrac{1}{2}Li_T^2,$$

proving the assertion.

36-33

(a) The energy density is given by

$$u = \tfrac{1}{2}B^2/\mu_0 = (\mu_0 iN/2\pi r)^2/2\mu_0 = \mu_0 i^2N^2/8\pi^2 r^2 \quad \underline{\text{Ans}}.$$

(b) Since u depends on r, take as a volume element the volume between two coaxial circular cylinders, radii r and r + dr, the axes of which coincide with the axis of the toroid. That is, dV = $2\pi rh\,dr$; thus, the stored energy is

$$U = \int u dV = \int_a^b (\mu_0 i^2N^2/8\pi^2 r^2)2\pi rh dr = (\mu_0 i^2N^2h/4\pi)\int_a^b \frac{dr}{r},$$

$$U = \frac{\mu_0 i^2N^2h}{4\pi}\ln\left(\frac{b}{a}\right) = 1.73 \times 10^{-4}\ \text{J} \quad \underline{\text{Ans}}.$$

(c) The energy U is also

$$U = \tfrac{1}{2}Li^2 = \tfrac{1}{2}\left[\frac{\mu_0 N^2h}{2\pi}\ln\left(\frac{b}{a}\right)\right]i^2$$

and this is seen to agree with (b).

CHAPTER 37

37-6

In a time $2\pi/\omega$, one period of rotation, all of the charge q on the ring passes any fixed point near the ring; thus the equivalent current i is

$$i = \frac{q}{2\pi/\omega} = \frac{\omega q}{2\pi}.$$

Therefore,

$$\mu = NiA = (1)\left(\frac{\omega q}{2\pi}\right)(\pi r^2) = \tfrac{1}{2}\omega q r^2 \quad \underline{\text{Ans}}.$$

By the right-hand rule, the magnetic moment vector is parallel to the angular velocity $\vec{\omega}$.

37-8

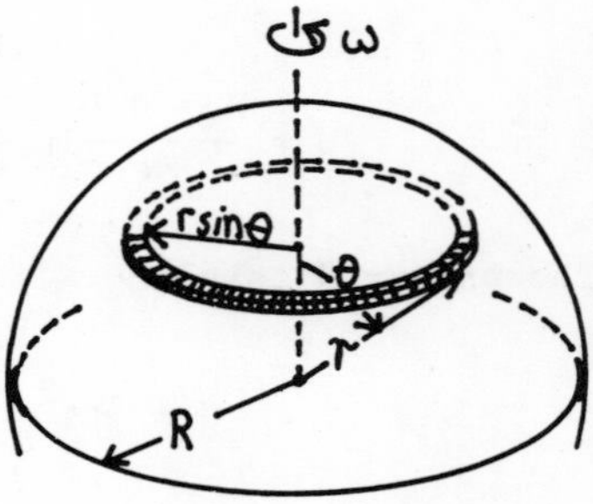

Consider the ring shown; it contains a charge

$$dq = \frac{e}{4\pi R^3/3}(2\pi r\sin\theta)(r\,d\theta)(dr),$$

the product of the charge per unit volume of the electron and the volume of the ring. The magnetic moment of this ring is

$$d\mu = A\,di = \pi(r\sin\theta)^2\frac{dq}{2\pi/\omega} = \frac{3e\omega}{4R^3}r^4\sin^3\theta\,dr\,d\theta.$$

Thus the magnetic moment of the entire electron is

$$\mu = \int d\mu = \frac{3e\omega}{4R^3}\int_0^R\int_0^\pi r^4\sin^3\theta\,d\theta\,dr = \frac{1}{5}\,e\omega R^2.$$

The spin angular momentum is just

$$L = I\omega = \frac{2}{5} mR^2\omega,$$

and therefore

$$\frac{e}{m} = 2\frac{\mu}{L}.$$

The actual spin g-factor is very slightly greater than two.

37-11

(a) From Example 2,

$$\mu_\ell = L_\ell \frac{e}{2m} = (1)(1.05 \times 10^{-34}\ \text{J}\cdot\text{s})\tfrac{1}{2}(1.76 \times 10^{11}\ \text{C/kg}),$$

$$\mu_\ell = 0.924 \times 10^{-23}\ \text{A}\cdot\text{m}^2 \quad \underline{\text{Ans}}.$$

(b) Since $U = -\vec{\mu}\cdot\vec{B}$,

$$\Delta U = \mu B - (-\mu B) = 2\mu B = 2(0.924 \times 10^{-23}\ \text{A}\cdot\text{m}^2)(1.2\ \text{T}),$$

$$\Delta U = 2.218 \times 10^{-23}\ \text{J} \quad \underline{\text{Ans}}.$$

(c) Setting the two energies considered equal,

$$2\mu B = \tfrac{1}{2}kT,$$

$$T = 3.2\ \text{K} \quad \underline{\text{Ans}},$$

using (b) and $k = 1.38 \times 10^{-23}$ J/K.

37-15

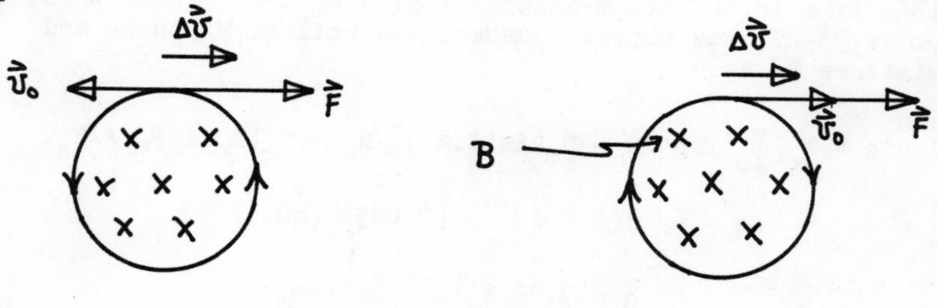

As the magnetic field is introduced, B increases and a counter-clockwise electric field is induced; by Example 35-4 this field is $E = \frac{1}{2}r dB/dt$ and the electrons feel an electrical force, directed as shown. Suppose the magnetic field increases by an amount B in time T. Then, for each electron,

$$\Delta v = aT = \frac{F}{m}T = \frac{eE}{m}T = \frac{e}{m}\left(\frac{r}{2}\frac{B}{T}\right)T = \frac{erB}{2m}.$$

The new velocities are

$$v = v_0 \pm \frac{erB}{2m},$$

+ for the clockwise circulating electron and - for the other. Dividing by r, assumed to remain constant, gives

$$\omega = \omega_0 \pm \frac{eB}{2m};$$

the changed angular speed leads to an increase or decrease in the orbital magnetic moment.

37-17

(a) The field in a toroid is $B = (\mu_0 iN)/2\pi r$, where N is the total number of turns. This is a non-uniform field, but consider it to be approximately uniform and equal to the actual value at the center of the toroidal tube. It is desired that this field B_0 equal 2×10^{-4} T:

$$2 \times 10^{-4} = (4\pi \times 10^{-7})(i)(400)/2\pi(0.055),$$

$$i = 0.14 \text{ A} \quad \underline{\text{Ans.}}$$

(b) With the iron present inside the toroid the field is $B_M + B_0 = 801B_0$. If A is the cross-sectional area of the toroid, then, by Problem 35-27, the charge induced in a coil of N_c turns and resistance R_c is

$$q = N_c[\Phi_B(\text{final}) - \Phi_B(\text{initial})]/R_c = N_c(B_0 + B_M)A/R_c,$$

$$q = (50)(2 \times 10^{-4})\pi(0.005)^2(801)/(8),$$

$$q = 7.86 \times 10^{-5} \text{ C} \quad \underline{\text{Ans.}}$$

37-22

Evaluate

$$\oint \vec{H} \cdot \vec{d\ell} = iN$$

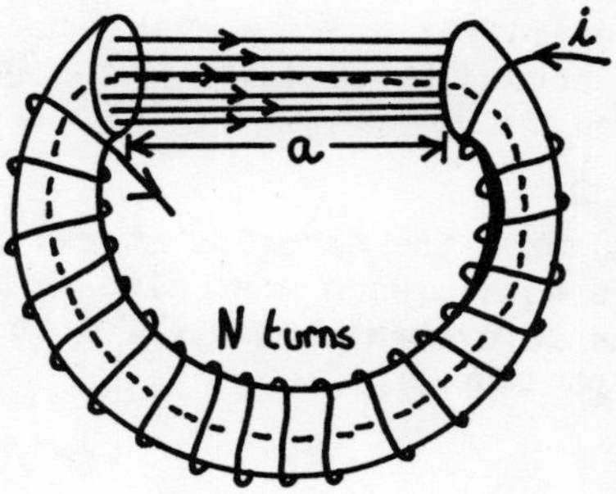

where the integration is along the dotted path, a length a of which is in the air and a length b in the iron of the magnet. Then,

$$(H_a a + H_i b) = iN.$$

In the air M = 0 and $H_a = B_a/\mu_0$; in the iron $M \neq 0$ and $H_i = \frac{B_i}{k_m \mu_0}$. Also $B_a \approx B_i = B$. Thus,

$$B\left(\frac{a}{\mu_0} + \frac{b}{k_m \mu_0}\right) = Ni,$$

$$i = \frac{B}{\mu_0 N}\left(a + \frac{b}{k_m}\right) = 29.2 \text{ A} \quad \underline{\text{Ans}}.$$

CHAPTER 38

38-5

(a) When the energies stored in the electric and magnetic fields are equal, each must also equal half the total energy present in the LC system; hence, if Q is the maximum charge to appear on the capacitor,

$$\tfrac{1}{2}q^2/C = \tfrac{1}{2}(\tfrac{1}{2}Q^2/C),$$

$$q = \frac{Q}{\sqrt{2}} \quad \underline{\text{Ans}}.$$

(b) Since $q = Q\sin(\omega t + \phi)$, if $q = Q/\sqrt{2}$ when $t = t'$, then

$$\sin(\omega t' + \phi) = 1/\sqrt{2},$$

$$\omega t' + \phi = \pi/4,\ 3\pi/4.$$

Let $q = Q$ when $t = t''$; then $\omega t'' + \phi = \pi/2$ and therefore

$$\omega t' + (\frac{\pi}{2} - \omega t'') = \pi/4,\ 3\pi/4,$$

$$(t' - t'') = \pm\frac{\pi}{4\omega} = \pm\frac{\pi}{4}\frac{T}{2\pi} = \pm\frac{T}{8};$$

thus the desired time interval is T/8 before and after the instant when $q = Q$.

38-12

(a) The frequencies that can be tuned are $\nu = 1/2\pi(LC)^{\frac{1}{2}}$; thus,

$$\frac{\nu_{max}}{\nu_{min}} = \frac{(LC_{min})^{-\frac{1}{2}}}{(LC_{max})^{-\frac{1}{2}}} = (C_{max}/C_{min})^{\frac{1}{2}} = (365/10)^{\frac{1}{2}} = 6.04 \quad \underline{\text{Ans}}.$$

(b) Let C be the added capacitance. For capacitors in parallel the equivalent capacitance C* is the sum of the individual capacitances and therefore C* ranges between C + 10 and C + 365 pF. Hence,

$$\frac{1.60}{0.54} = \left(\frac{365 + C}{10 + C}\right)^{\frac{1}{2}},$$

$$C = 35.63 \text{ pF} \quad \underline{\text{Ans}}.$$

The inductance must be given by

$$\nu = \frac{\omega}{2\pi} = \frac{1}{2\pi}[(L)(C + 10)]^{-\frac{1}{2}} = 1.6 \times 10^6,$$

$$L = 2.2 \times 10^{-4} \text{ H} \quad \underline{\text{Ans}},$$

using the value of C (expressed in farads) found in the above.

<u>38-17</u>

(a) The charge q as a function of time is

$$q = q_m \sin\omega t,$$

choosing sine rather than cosine so that $q(0) = 0$ with a phase constant of zero. The current corresponding to the above is

$$i = \frac{dq}{dt} = \omega q_m \cos\omega t.$$

Thus, the maximum current is $i_m = \omega q_m$ and therefore

$$q_m = i_m/\omega = i_m(LC)^{\frac{1}{2}} = (2)[(3 \times 10^{-3})(2.7 \times 10^{-6})]^{\frac{1}{2}},$$

$$q_m = 1.8 \times 10^{-4} \text{ C} \quad \underline{\text{Ans}}.$$

(b) The energy stored in the capacitor at any time t is

$$U = \tfrac{1}{2}q^2/C = (\tfrac{1}{2}q_m^2/C)\sin^2\omega t.$$

The rate of increase is

$$\frac{dU}{dt} = (\tfrac{1}{2}q_m^2/C)(2\omega)\sin\omega t\cdot\cos\omega t = (\tfrac{1}{2}\omega q_m^2/C)\sin 2\omega t.$$

This reaches its greatest value when $\sin 2\omega t = 1$, or

$$2\omega t = \frac{\pi}{2},$$

$$2\left(\frac{2\pi}{T}\right)t = \frac{\pi}{2},$$

$$t = T/8 \quad \underline{\text{Ans}}.$$

(c) The greatest rate of increase is, from (b),

$$\left(\frac{dU}{dt}\right)_{max} = \tfrac{1}{2}\omega q_m^2/C.$$

Numerically, $\omega = (LC)^{-\frac{1}{2}} = (9 \times 10^{-5})^{-1}$, and $q_m = 1.8 \times 10^{-4}$ C, from (a). Also, $C = 2.7 \times 10^{-6}$ F. These give $(dU/dt)_{max} = 66.7$ W.

38-18

The charge q on the capacitor as a function of time t in a damped LC, or LCR, circuit is, by Eq.38-10,

$$q = q_m e^{-Rt/2L}\cos\omega' t = Q(t)\cos\omega' t$$

where q_m is the charge at $t = 0$. The energy present over one oscillation, assuming $2\pi/\omega' \ll 2L/R$ (small damping) is

$$U = \tfrac{1}{2}Q^2/C = (\tfrac{1}{2}q_m^2/C)e^{-Rt/L} = [U(0)]e^{-Rt/L}.$$

The required time is found by setting $U = \frac{1}{2}U(0)$:

$$\tfrac{1}{2}U(0) = [U(0)]e^{-Rt/L},$$

$$\ln 2 = Rt/L,$$

$$t = \frac{L}{R}\ln 2 \quad \underline{\text{Ans}}.$$

38-21

The envelope of the charge oscillations is given by

$$q = q_m e^{-Rt/2L},$$

and thus the time t* required for q to fall to $\frac{1}{2}q_m$ is found by setting

$$\tfrac{1}{2}q_m = q_m e^{-Rt^*/2L},$$

$$t^* = \frac{2L}{R}\cdot \ln 2.$$

The number n of oscillations occuring during this time satisfies the relation

$$n\cdot\frac{2\pi}{\omega'} = t^*,$$

$$n = \frac{L\ln 2}{\pi R}\,\omega' \approx \frac{L\ln 2}{\pi R}\,\omega,$$

$$n^2 = \frac{L(\ln 2)^2}{\pi^2 R^2 C},$$

since $\omega^2 = 1/LC$. But,

$$\omega' = \left[\frac{1}{LC} - \left(\frac{R}{2L}\right)^2\right]^{\frac{1}{2}} = \left[\frac{1}{LC}\left(1 - \frac{R^2C}{4L}\right)\right]^{\frac{1}{2}} \approx \left(\frac{1}{LC}\right)^{\frac{1}{2}}\left(1 - \frac{R^2C}{8L}\right),$$

$$\omega' = \omega\left(1 - \frac{R^2C}{8L}\right).$$

Thus,

$$\frac{\omega - \omega'}{\omega} = \frac{R^2C}{8L} = \frac{(\ln 2)^2}{8\pi^2 n^2} \approx \frac{0.0061}{n^2}.$$

38-22

The charge q(t) on the capacitor is

$$q(t) = q_m e^{-Rt/2L}\cos\omega' t,$$

$$\omega' = \left[\frac{1}{LC} - \left(\frac{R}{2L}\right)^2\right]^{\frac{1}{2}}.$$

This function may be considered a harmonic oscillation with a time-dependent amplitude: that is, as

$$q(t) = Q(t)\cos\omega' t.$$

If the damping is small, then just as for the undamped LC system, the total energy in the system during one oscillation is

$$U = \tfrac{1}{2}Q^2/C,$$

to a high degree of approximation. The energy loss per cycle is

$$\Delta U \approx \frac{dU}{dt}(\Delta t),$$

where $\Delta t = 2\pi/\omega' \approx 2\pi/\omega$. But $\omega = 1/(LC)^{\frac{1}{2}}$ so that

$$\frac{\Delta U}{U} \approx \frac{\frac{d}{dt}(\frac{1}{2}Q^2/C)}{\frac{1}{2}Q^2/C}\left(\frac{2\pi}{\omega}\right) = \frac{2\pi R}{\omega L},$$

using $Q = q_m e^{-Rt/2L}$.

CHAPTER 39

39-5

(a) The capacitative reactance is $X_C = 1/\omega C$; hence,

$$X_C = \frac{1}{\omega(10 \times 10^{-6})} = \frac{10^5}{\omega}.$$

For the inductor,

$$X_L = \omega L = \omega(6 \times 10^{-3}).$$

Setting these equal gives

$$\frac{10^5}{\omega} = \omega(6 \times 10^{-3}),$$

$$\omega^2 = \frac{10^5}{6 \times 10^{-3}} = \frac{1}{6} \times 10^8,$$

$$\nu = \frac{\omega}{2\pi} = 650 \text{ Hz} \quad \underline{\text{Ans}}.$$

(b) The reactance, calculated for the inductor say, is $X_L = (2\pi)\cdot(650)(0.006) = 24.5\ \Omega$ Ans.

(c) Using the formulae for X_C and X_L in (a) and setting $X_L = X_C$ yields

$$\omega L = \frac{1}{\omega C},$$

$$\omega = (LC)^{-\frac{1}{2}}$$

which equals the frequency of an LC circuit.

39-11

The average power dissipated is

$$P_{av} = \mathcal{E}_{rms} i_{rms} \cos\phi = \tfrac{1}{2}\mathcal{E}_m i_m \cos\phi = \tfrac{1}{2}\mathcal{E}_m\left(\frac{\mathcal{E}_m}{Z}\right)\left(\frac{R}{Z}\right) = \frac{1}{2}\,\frac{\mathcal{E}_m^2 R}{Z^2},$$

$$P_{av} = \tfrac{1}{2}\mathcal{E}_m^2 R \frac{1}{R^2 + (\omega L - 1/\omega C)^2}.$$

To find the desired values of C set $dP_{av}/dC = 0$; this gives

$$-\frac{\mathcal{E}_m^2 R}{Z^4}(\omega L - \frac{1}{\omega C})(1/\omega C^2) = 0.$$

Two values of C satisfy this equation and they are:

(a)

$$C = 1/L\omega^2 = 1/4\pi^2\nu^2 L = 117\ \mu F \quad \underline{\text{Ans}};$$

(b)

$$C = \infty \quad \underline{\text{Ans}}.$$

(c) Substituting the values of C from (a) and (b) into P_{av} gives

$$P_{av} = \tfrac{1}{2}\mathcal{E}_m^2/R = 9000\ W \quad \underline{\text{Ans}};$$

$$P_{av} = \tfrac{1}{2}\mathcal{E}_m^2 R \frac{1}{R^2 + \omega^2 L^2} = 419\ W \quad \underline{\text{Ans}}.$$

For $C = 1/L\omega^2$, the impedance $Z = R$ and therefore $\cos\phi = R/Z = 1$, so that $\phi = 0$.
On the other hand, for C "=" ∞, $Z = (R^2 + \omega^2 L^2)^{\frac{1}{2}}$, and $\cos\phi = R/Z = 0.22$, giving $\phi = 78°$.
The power factors are just $\cos\phi = 1$; 0.22.

(d) There is no value of C that will make $\cos\phi$ negative. Also, $P_{av} = \tfrac{1}{2}\mathcal{E}_m^2 R/Z$; the dependence on C enters through Z, and $Z \geq 0$ always.

<u>39-16</u>

(a) The power factor is $\cos\phi$, by definition.

(b) Since

$$\tan\phi = (X_C - X_L)/R,$$

if $\phi > 0$, then $X_C > X_L$ and the element is capacitative.

(c) If $\mathcal{E} = \mathcal{E}_m \sin\omega t$ and $i = i_m \sin(\omega t + \phi)$ and $\phi > 0$, the emf lags the current.

(d) The average power is

$$P_{av} = \tfrac{1}{2}\mathcal{E}_m i_m \cos\phi = \tfrac{1}{2}(12\ \text{A})(750\ \text{V})\cos\phi = (4500\ \text{W})\cos\phi.$$

(e) The power is an average over many cycles and $\overline{\sin\omega t \sin(\omega t + \phi)} = \tfrac{1}{2}\cos\phi$, independent of ω when the average is so taken.

(f) From (b), if $\phi = 0$, its value at resonance, $X_C = X_L$. Since the box is assumed capacitative, an inductive element must be added to reach resonance.

(g) At resonance $\phi = 0$ and therefore, from (d), $P_{av} = 4500$ W.

39-17

The maximum value of i_m is $\mathcal{E}_m/R$. To find the half-width, then, set $i_m = \tfrac{1}{2}\mathcal{E}_m/R$:

$$\tfrac{1}{2}\mathcal{E}_m/R = \frac{\mathcal{E}_m}{[R^2 + (\omega L - 1/\omega C)^2]^{\frac{1}{2}}}.$$

Squaring to solve for the term in parenthesis gives

$$\omega L - \frac{1}{\omega C} = \sqrt{3}R.$$

Now set $\omega = \omega_0 + \delta\omega = \omega_0(1 + \delta\omega/\omega_0)$; this should yield an accurate result if the half-width is small compared to the resonant frequency (sharp peak). To this same degree of approximation,

$$\frac{1}{\omega} \approx \frac{1}{\omega_0}(1 - \frac{\delta\omega}{\omega_0}).$$

The previous equation now becomes

$$\omega_0(1 + \frac{\delta\omega}{\omega_0})L - \frac{1}{\omega_0 C}(1 - \frac{\delta\omega}{\omega_0}) = \sqrt{3}R,$$

$$(\omega_0 L - \frac{1}{\omega_0 C}) + \frac{\delta\omega}{\omega_0}(\omega_0 L + \frac{1}{\omega_0 C}) = \sqrt{3}R.$$

But $\omega_0^2 = 1/LC$; hence the first term vanishes and

$$\omega_0 L + \frac{1}{\omega_0 C} = \frac{\omega_0^2 LC + 1}{\omega_0 C} = \frac{2}{\omega_0 C} = 2\left(\frac{L}{C}\right)^{\frac{1}{2}}.$$

Thus,

$$\frac{\delta\omega}{\omega_0} = \tfrac{1}{2}\sqrt{3}\left(\frac{C}{L}\right)^{\frac{1}{2}}R.$$

The half width $\Delta\omega = 2\delta\omega$, so that

$$\frac{\Delta\omega}{\omega_0} = \left(\frac{3C}{L}\right)^{\frac{1}{2}}R = \frac{\sqrt{3}R}{\omega_0 L}.$$

39-23

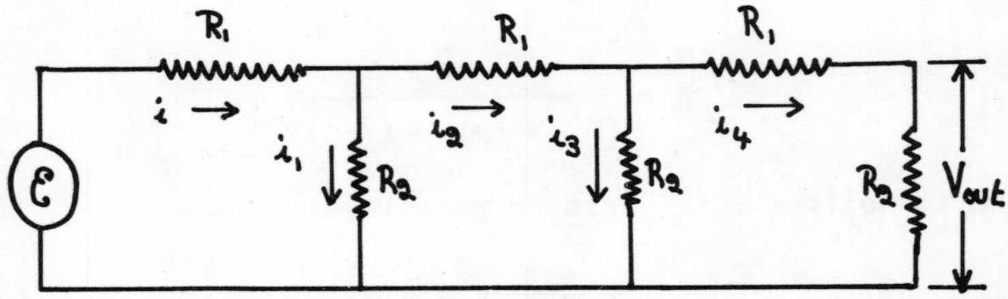

(a) The three-stage filter resembles three filters in series, with the output from one stage serving as the input for the subsequent stage. Since for any stage,

$$V_{out}/V_{in} = R_2/R_1$$

for the dc equivalent, it is anticipated that for three stages,

$$V_{out}/V_{in} = (R_2/R_1)^3.$$

(b) For the equivalent dc circuit shown, the loop theorem and the junction equations are

$$i = i_1 + i_2,$$

$$i_2 = i_3 + i_4,$$

$$\mathcal{E} - iR_1 - i_1R_2 = 0,$$
$$- i_2R_1 - i_3R_2 + i_1R_2 = 0,$$
$$- i_4R_1 - i_4R_2 + i_3R_2 = 0.$$

These may be solved for the five unknown currents. Since $V_{out} = i_4R_2$, only i_4 is of immediate interest and this is found to be

$$i_4 = \mathcal{E}R_2^2/(R_1^3 + 5R_1^2R_2 + 6R_1R_2^2 + R_2^3).$$

With $R_1 \gg R_2$, all terms in the denominator save the first may be ignored: that is,

$$i_4 \approx \mathcal{E}R_2^2/R_1^3.$$

Thus the attentuation factor becomes

$$V_{out}/V_{in} = i_4R_2/\mathcal{E} = (R_2/R_1)^3 \quad \underline{\text{Ans}}.$$

39-26

The subscript "rms" is dropped for brevity.

(a) The ratio of the voltages equals the ratio in the numbers of windings; that is,

$$V_2/V_1 = N_2/N_1,$$
$$V_2 = (120)\frac{10}{500} = 2.4 \text{ V} \quad \underline{\text{Ans}}.$$

(b) The current ratio is

$$i_2/i_1 = (N_1/N_2)\cos\phi.$$

With $\phi = 0$, $N_1 = 500$, $N_2 = 10$ this becomes $i_2 = 50i_1$. But,

$$i_2 = V_2/R = (2.4)/(15) = 0.16 \text{ A} \quad \underline{\text{Ans}}.$$

Thus $i_1 = i_2/50 = 0.0032 \text{ A} = 3.2 \text{ mA}$ Ans.

39-29

The power output at the primary is

$$P_{av} = V_1 i_1 \cos\phi.$$

However, if Φ_1 and Φ_2 are the fluxes in the core due to the primary and secondary windings, then

$$\cos\phi = (\Phi_2/\Phi_1) = (N_2 i_2)/(N_1 i_1).$$

But $V_1/V_2 = N_1/N_2$ and therefore

$$P_{av} = V_1 i_1 (N_2/N_1)(i_2/i_1) = V_1 i_1 (V_2/V_1)(i_2/i_1) = V_2 i_2;$$

hence,

$$V_1 i_1 \cos\phi = V_2 i_2,$$

$$i_1 = \frac{N_2}{N_1} i_2/\cos\phi.$$

With $i_2 = V_2/R = (N_2 V_1/N_1)/R$ this becomes

$$i_1 = (N_2^2 V_1/N_1^2)/(R\cos\phi).$$

If $\phi = 0$ this can be written

$$i_1 = \frac{V_1}{(N_1/N_2)^2 R}$$

proving the assertion.

39-31

The resistance added to the amplifier by virtue of the connected transformer is, by Problem 39-29,

$$r_t = (N_1/N_2)^2 R$$

and thus the total resistance of the amplifier is $r + r_t$ in effect.

The average power delivered to R is

$$P_{av} = i^2_{2,rms}R = (\frac{N_1}{N_2} i_{1,rms})^2 R = (\frac{N_1}{N_2})^2 \frac{\mathcal{E}^2_m \sin^2 \omega t}{[r + (N_1/N_2)^2 R]^2} R.$$

If $x = (N_1/N_2)^2$ then, as far as dependence on x is concerned,

$$P_{av} = \frac{x}{(r + xR)^2}.$$

To find the value of x leading to maximum P_{av}, calculate dP_{av}/dx:

$$dP_{av}/dx = \frac{r - xR}{(r + xR)^3} = 0,$$

$$x = \frac{r}{R} = (N_1/N_2)^2 = \frac{1000\ \Omega}{10\ \Omega},$$

$$N_1/N_2 = 10 \quad \underline{\text{Ans}};$$

that is, the coil must have a turn ratio of 10 (amplifier) to 1 (speaker).

CHAPTER 40

40-3

The displacement current i_d is, by definition,

$$i_d = \epsilon_0 \frac{d\Phi_E}{dt} = \epsilon_0 \frac{d(EA)}{dt} = \epsilon_0 A \frac{dE}{dt}.$$

Now let x be the separation between the plates; thus $E = V/x$, V the potential difference across the plates. Then, if the plates are fixed in position so that x does not change,

$$i_d = \epsilon_0 A \frac{d(V/x)}{dt} = \frac{\epsilon_0 A}{x} \frac{dV}{dt}.$$

But for a parallel-plate capacitor ($k = 1$), $C = \epsilon_0 A/x$ giving

$$i_d = C \frac{dV}{dt}.$$

40-8

(a) The displacement current i_d in the gap between the plates is equal to the conduction current i in the wires; hence,

$$i_{max} = i_{d,max} = 8.9 \times 10^{-6}\ \text{A} \quad \underline{\text{Ans}}.$$

(b) By definition,

$$i_d = \epsilon_0 (d\Phi_E/dt),$$

so that

$$(d\Phi_E/dt)_{max} = i_{d,max}/\epsilon_0 = 8.9 \times 10^{-6}/8.85 \times 10^{-12},$$

$$(d\Phi_E/dt)_{max} = 1.0056 \times 10^6\ \text{V·m/s} \quad \underline{\text{Ans}}.$$

(c) By Problem 40-3, and noting that the potential across the capacitor, except for sign, is equal to the battery emf (by the loop theorem),

$$i_d = \frac{\epsilon_0 A}{d}\frac{dV}{dt} = \frac{\epsilon_0 A}{d}\frac{d\mathcal{E}}{dt} = \frac{\epsilon_0 A}{d}\mathcal{E}_m \omega \cos\omega t.$$

The plate separation, then, must be

$$d = \epsilon_0 A\mathcal{E}_m\omega/i_{d,max} = \frac{(8.85 \times 10^{-12})(0.1)(200)(100)}{8.90 \times 10^{-6}},$$

$$d = 1.99 \times 10^{-3}\ \text{m} = 1.99\ \text{mm}\quad \underline{\text{Ans}}.$$

(d) In the gap between the plates the conduction current $i = 0$. Ampere's law reduces to

$$\oint \vec{B}\cdot\vec{d\ell} = \mu_0 I_d,$$

where I_d is the displacement current flowing across the area bounded by the circular path of integration of radius r. It is easy to verify that $r < R$, the radius of the plates. Then, if the current density j_d is uniform,

$$I_d = j_d(\pi r^2) = \frac{i_d}{A}(\pi r^2).$$

Returning to Ampere's law,

$$B(2\pi r) = \mu_0 I_d,$$

$$B_{max}(2\pi r) = \mu_0 I_{d,max} = \mu_0(i_{d,max}\pi r^2/A),$$

$$B_{max} = \tfrac{1}{2}\mu_0 R i_{d,max}/A = \tfrac{1}{2}(4\pi \times 10^{-7})(0.1)(8.9 \times 10^{-6})/(0.1),$$

$$B_{max} = 5.59 \times 10^{-12}\ \text{T}\quad \underline{\text{Ans}}.$$

<u>40-13</u>

(a) Since $i = dq/dt$, the charge $q(t)$ on the faces will be

$$q = \int i\,dt = \int \alpha t\,dt = \tfrac{1}{2}\alpha t^2\quad \text{Ans};$$

[with $q(0) = 0$ the integration constant is zero].

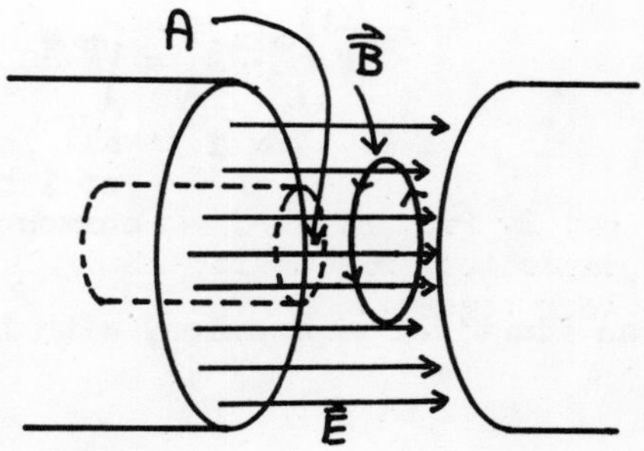

(b) Use the pill-box Gaussian surface shown. The charges on the rod faces do not contribute to the electric field in the rod, so set E = 0 there. $\vec{E}$ is parallel to the x-axis in the gap and Gauss's law under these circumstances gives

$$\epsilon_0 EA = qA/\pi a^2,$$

$$E = q/\pi\epsilon_0 a^2 = \tfrac{1}{2}\alpha t^2/\pi\epsilon_0 a^2 \quad \underline{\text{Ans}},$$

using the result from (a).

(c) From an examination of Figure 40-1, the lines of B are concentric circles, centered on and perpendicular to the x-axis.

(d) Choosing as path of integration a circle of radius r coincident with a line of B,

$$\oint \vec{B}\cdot d\vec{\ell} = B(2\pi r).$$

In the gap i = 0; hence,

$$2\pi r B = \mu_0\epsilon_0 \frac{d\Phi_E}{dt} = \mu_0\epsilon_0 \frac{d}{dt}(ES) = \mu_0\epsilon_0(\pi r^2)\frac{dE}{dt} = \mu_0\epsilon_0 \pi r^2\left(\frac{\alpha t}{\pi\epsilon_0 a^2}\right),$$

$$B = \frac{\frac{1}{2}\mu_0 r\alpha t}{\pi a^2} \quad \underline{\text{Ans}}.$$

(e) The displacement current in the gap equals the conduction current in the rod; thus the magnetic field in the gap and rod are the same, by Ampere's law.

40-17

(a) Let box 1 contain charge q_1 and box 2 a charge q_2. Applying Gauss's law to boxes 1 and 2 in turn,

$$\underset{\text{box 1}}{\oint \vec{E}\cdot d\vec{S}_1} = \underset{\text{all faces of 1 save common face}}{\int \vec{E}\cdot d\vec{S}_1} + \underset{\text{common face}}{\int \vec{E}\cdot d\vec{S}_1} = q_1/\epsilon_0;$$

an identical expression, with 1 replaced with 2, holds for box 2.

$$\oint \vec{E}\cdot\vec{dS} + \int \vec{E}\cdot(\vec{dS}_1 + \vec{dS}_2) = q/\epsilon_0,$$

enveloping surface   common face

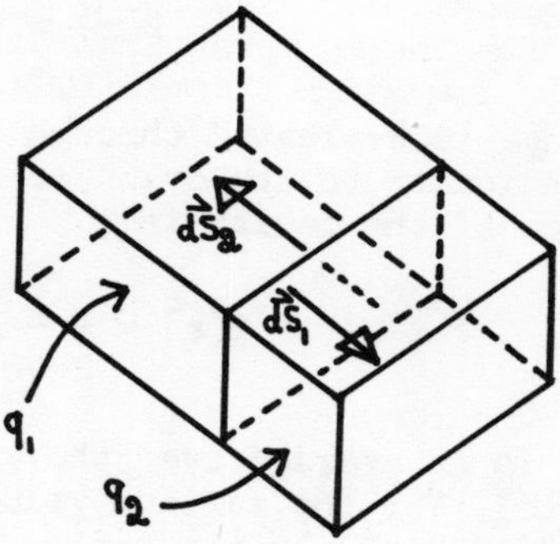

where $q = q_1 + q_2$, the total charge inside the enveloping surface. Now, along the common face,

$$\vec{dS}_1 = -\vec{dS}_2,$$

and therefore the integral over the common face is nil, leaving

$$\oint_{\text{enveloping surface}} \vec{E}\cdot\vec{dS} = q/\epsilon_0.$$

(b) This is formally identical to (a): just set $q_1 = q_2 = 0$ and replace E with B.

<u>40-18</u>

(a) Since $E = E_m \sin\omega t$, the rate at which E changes is

$$\frac{dE}{dt} = E_m \omega\cos\omega t,$$

$$\left(\frac{dE}{dt}\right)_{max} = \omega E_m.$$

The angular frequency for the mode of oscillation shown in the figure is $\omega = 2.41c/a$ (see Section 38-5); hence,

$$\left(\frac{dE}{dt}\right)_{max} = \frac{2.41c}{a}E_m = \frac{(2.41)(3 \times 10^8)}{0.050}(10^4),$$

$$\left(\frac{dE}{dt}\right)_{max} = 1.45 \times 10^{14}\ \text{V/m·s} \quad \underline{\text{Ans}}.$$

(b) The magnetic field is given by

$$B(r) = \left(\frac{\mu_0\epsilon_0}{2\pi r}\right)d\Phi_E/dt,$$

where Φ_E is evaluated through a circular area of radius r oriented perpendicular to the cylinder axis. Hence, the field at the surface of the cavity is

$$B(a) = \frac{1}{2\pi ac^2}\frac{d\bar{E}}{dt}\pi a^2 = \frac{a}{2c^2}\frac{d\bar{E}}{dt};$$

$d\bar{E}/dt$ is an average over the area through which the flux is computed. For the maximum value of B use the largest possible value of $d\bar{E}/dt$ which, by assumption, is half the value found in (a). Then,

$$B_{max}(a) = \frac{0.025}{2(3 \times 10^8)^2}\tfrac{1}{2}(1.45 \times 10^{14}),$$

$$B_{max}(a) = 1.0 \times 10^{-5}\ \text{T} \quad \underline{\text{Ans.}}$$

CHAPTER 41

41-15

The displacement current is

$$i_d = \epsilon_0 d\Phi_E/dt = \epsilon_0 \frac{d}{dt}\int \vec{E}\cdot\vec{dS}.$$

Since in Figs. 41-13, 41-14 the field $\vec{E}$ is parallel to the z-axis, choose the shaded rectangle in Fig. 41-14b as the area over which to compute the flux. This rectangle is thin enough to consider E(x) essentially constant within it. Then,

$$d\Phi_E = EdS = E_m\sin(kx - \omega t)(h\ dx),$$

and the displacement current through the rectangle is

$$di_d = -\epsilon_0\omega E_m\cos(kx - \omega t)(h\ dx).$$

Hence, the displacement current density is

$$j_d = di_d/dS = -\epsilon_0\omega E_m\cos(kx - \omega t) \quad \underline{\text{Ans}}.$$

41-16

The equations alluded to are

Eq. 41-10: $\frac{\partial E}{\partial x} = -\frac{\partial B}{\partial t}$; Eq. 41-13: $-\frac{\partial B}{\partial x} = \mu_0\epsilon_0\frac{\partial E}{\partial t}$.

(a) Take $\partial/\partial x$ of the first equation and $\partial/\partial t$ of the second to get,

$$\frac{\partial^2 E}{\partial x^2} = -\frac{\partial^2 B}{\partial x\partial t}, \qquad -\frac{\partial^2 B}{\partial t\partial x} = \mu_0\epsilon_0\frac{\partial^2 E}{\partial t^2}.$$

But it is anticipated that $\partial^2 B/\partial t\partial x = \partial^2 B/\partial x\partial t$ and if this is satisfied, then

$$\frac{\partial^2 E}{\partial x^2} = \mu_0 \epsilon_0 \frac{\partial^2 E}{\partial t^2},$$

which is the wave equation.

(b) As for B, take $\partial/\partial t$ of the first equation and $\partial/\partial x$ of the second to get

$$\frac{\partial^2 E}{\partial t \partial x} = -\frac{\partial^2 B}{\partial t^2}, \qquad -\frac{\partial^2 B}{\partial x^2} = \mu_0 \epsilon_0 \frac{\partial^2 E}{\partial x \partial t}.$$

For the same reason as in (a), these imply that

$$\frac{\partial^2 B}{\partial t^2} = \frac{1}{\mu_0 \epsilon_0} \frac{\partial^2 B}{\partial x^2},$$

again, a wave equation.

41-26

(a) By Ampere's law,

$$B = \frac{\mu_0 i}{2\pi r} = \frac{\mu_0 (\mathcal{E}/R)}{2\pi r} = \frac{\mu_0 \mathcal{E}}{2\pi R r} \quad \underline{\text{Ans}}.$$

If λ is the charge per unit length on the inner cable, $E = \lambda/2\pi\epsilon_0 r$ and the potential difference across the cable is

$$V_a - V_b = \mathcal{E} = \int_a^b \frac{\lambda}{2\pi\epsilon_0 r}\, dr = \frac{\lambda}{2\pi\epsilon_0} \ln\left(\frac{b}{a}\right),$$

$$\frac{\lambda}{2\pi\epsilon_0} = \frac{\mathcal{E}}{\ln(b/a)};$$

hence,

$$E = \frac{\mathcal{E}}{r \ln(b/a)} \quad \underline{\text{Ans}}.$$

(b) Since $\vec{E}$ is perpendicular to $\vec{B}$ ($\vec{E}$ is radially out from the cable and $\vec{B}$ is transverse to it), the magnitude of the Poynting vector (directed along the cable) will be

$$S = EB/\mu_0 = \frac{\mathcal{E}^2}{2\pi R r^2 \ln(b/a)} \quad \underline{\text{Ans}}.$$

(c) The power flowing across the area is

$$P = \int_a^b S(2\pi r\, dr) = \frac{2\pi\mathcal{E}^2}{2\pi R \ln(b/a)}\int_a^b \frac{dr}{r} = \frac{\mathcal{E}^2}{R}.$$

If $i = \mathcal{E}/R$ this becomes $P = i\mathcal{E}$, as expected.

(d) If the battery is reversed, $\vec{E}$ and $\vec{B}$ reverse directions, but

$$\vec{S} = \frac{\vec{E} \times \vec{B}}{\mu_0} = \frac{(-\vec{E}) \times (-\vec{B})}{\mu_0}$$

does not.

41-27

(a) The received intensity is the Poynting vector; since $\vec{E}$ and $\vec{B}$ are perpendicular (the receiver seeing practically a plane wave), and $E = cB$,

$$10 \times 10^{-6} = \overline{S} = \frac{1}{\mu_0}\overline{EB} = \frac{1}{\mu_0}\frac{\overline{EE}}{c}.$$

If $\overline{E} = (\overline{EE})^{\frac{1}{2}}$, then

$$\overline{E} = (\overline{S}\mu_0 c)^{\frac{1}{2}} = [(10^{-5})(4\pi \times 10^{-7})(3 \times 10^8)]^{\frac{1}{2}},$$

$$\overline{E} = 0.0614 \text{ V/m} \quad \underline{\text{Ans}}.$$

(b) Since $E = cB$,

$$\overline{B} = \frac{\overline{E}}{c} = \frac{0.0614}{3 \times 10^8} = 2.05 \times 10^{-10} \text{ T} \quad \underline{\text{Ans}}.$$

(c) With the transmitter radiating spherical waves,

$$\overline{P} = \overline{S}\cdot 4\pi r^2 = (10^{-5})(4\pi)(10^4)^2 = 12.6 \text{ kW} \quad \underline{\text{Ans}}.$$

41-28

(a) As $\vec{S} = (\vec{E} \times \vec{B})/\mu_0$, $\vec{S}$ is directed in the negative z-direction. Thus, the only faces across which S cuts are the two that are parallel to the x,y-plane. Energy at the rate EBa^2/μ_0 "enters" the top face and "leaves" via the bottom face (a^2 the area of each face), according to the Poynting vector picture. If the fields E and B are static, however, this model, being dynamic energetically, does not really make sense.

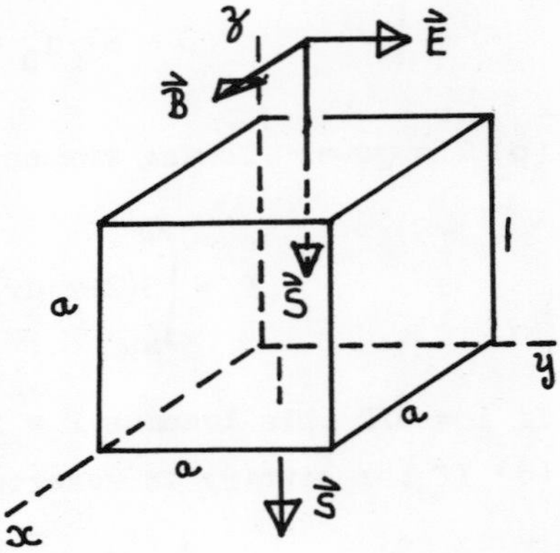

(b) Since energy appears to enter and leave at the same rate, the energy in the cube does not change, even in the Poynting point of view. It appears, then, that the Poynting vector picture makes sense when the energy flow through a closed surface is examined, but sometimes not if only part of such a surface is considered.

41-31

(a) Sighting along the resistor in the direction of the current flow, $\vec{E}$ points directly away from the observer, while $\vec{B}$ is directed transverse and at right angles to the resistor axis in the clockwise sense. Then, by the right-hand rule, $\vec{E} \times \vec{B}$ and therefore $\vec{S}$ are pointed radially inward.

(b) Since $\vec{E}$ and $\vec{B}$ are at right angles,

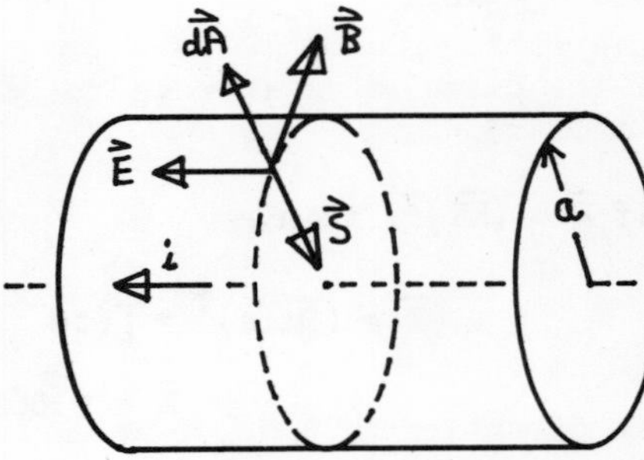

$$S = \frac{EB}{\mu_0} = \frac{(V/\ell)B}{\mu_0} = \frac{i}{\mu_0\ell}(R)B = \frac{i}{\mu_0\ell}(\rho\ell/\pi a^2)(\mu_0 i/2\pi a) = \frac{i^2\rho}{2\pi^2 a^3}.$$

The integral desired is $\int\vec{S}\cdot d\vec{A}$ over the cylindrical surface of the wire, considering a piece of length ℓ. The angle between $\vec{S}$ and the element of area $d\vec{A}$ (called $d\vec{S}$ in the previous chapters, a notation that now would conflict with that for the Poynting vector) is 180° and the magnitude of $\vec{S}$ is the same all over the surface. Thus,

$$\int \vec{S}\cdot d\vec{A} = -SA = -(\frac{i^2\rho}{2\pi^2 a^3})(2\pi a \ell) = -i^2(\rho \ell/\pi a^2) = -i^2 R.$$

41-33

(a) Treating the rotating cylinder as a solenoid, $B = \mu_0 ni$ with ni identified as

$$ni = \frac{ni\ell}{\ell} = \frac{\text{charge on cylinder/period of rotation}}{\ell},$$

$$ni = \frac{\sigma(2\pi R\ell)/(2\pi/\omega)}{\ell}.$$

Therefore,

$$B = \sigma R\mu_0\omega \quad \underline{\text{Ans}}.$$

(b) The field due to the static charge distribution is zero inside the cylinder. However, $\omega = \alpha t$ so that $B = B(t)$, and therefore an induced electric field E is present by the law of induction:

$$\oint \vec{E}\cdot d\vec{\ell} = -\, d\Phi_B/dt.$$

Integrating along the perimeter of the circular cross-section of the cylinder,

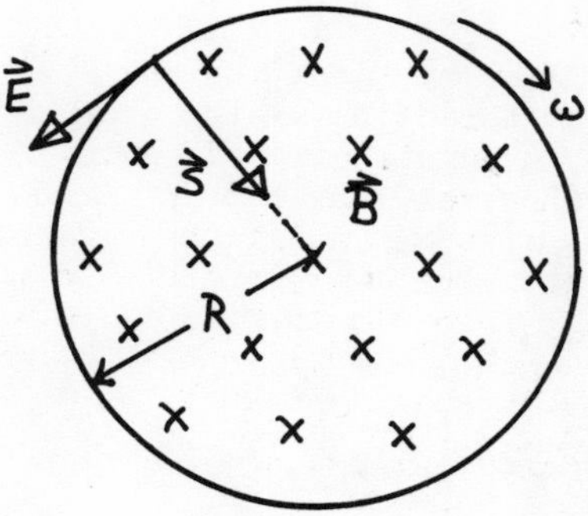

$$E(2\pi R) = \pi R^2 \frac{dB}{dt} = \pi R^2(\mu_0\sigma R\alpha),$$

$$E = \tfrac{1}{2}R^2\mu_0\alpha\sigma \quad \underline{\text{Ans}}.$$

The direction of $\vec{E}$ is shown in the sketch: $\vec{B}$ is into the paper and increasing; a current flowing counterclockwise will generate a magnetic field in the cylinder pointing out of the paper, as required by Lenz's law.

(c) Using the results from (a) and (b) and that $\vec{E}$ and $\vec{B}$ are at right-angles, it follows that

$$S = \frac{EB}{\mu_0} = \tfrac{1}{2}R^3\mu_0\sigma^2\alpha^2 t \quad \underline{\text{Ans}},$$

directed into the cylinder.

(d) The flux cutting the cylinder's surface is

$$\Phi_S = \int \vec{S}\cdot d\vec{A} = -\,S2\pi R\ell = -\,\mu_0\pi\sigma^2R^4\alpha^2\ell t.$$

From (a),

$$\frac{dB^2}{dt} = 2\mu_0^2\sigma^2R^2\alpha^2t,$$

and therefore

$$\Phi_S = \frac{d}{dt}\left(\pi R^2\ell\cdot\frac{B^2}{2\mu_0}\right);$$

that is,

$$\Phi_S = \frac{d}{dt}(\text{magnetic field energy inside cylinder}).$$

CHAPTER 42

42-4

Let f be the fraction of the incident beam energy that is reflected. The radiation pressure due to the part of the beam energy that is absorbed is

$$P_a = \frac{S_a}{c} = \frac{(1 - f)S}{c},$$

S the magnitude of the Poynting vector of the incident beam. The radiation pressure due to the reflected part of the incident beam is

$$P_r = \frac{2S_r}{c} = \frac{2(fS)}{c}.$$

(The factor of two occurs because, on reflection, the momentum p of the beam is reversed, in which case $\Delta p = 2p$.) The total radiation pressure is, then,

$$P = P_a + P_r = \frac{(1 + f)S}{c}.$$

For a plane wave with energy flux S, an amount of energy SAt crosses an area A normal to the beam in time t. But in this same time the wave travels a distance ct. Hence, SAt is the energy contained in a rectangular volume of base area A and length ct, so that the energy density u in the wave is

$$u = \frac{SAt}{Act} = \frac{S}{c}.$$

In terms of the energy density u, then, the radiation pressure P found above is

$$P = (1 + f)u = u + fu.$$

The first term on the right is the energy density of the incident beam and the second is the energy density of the reflected beam. Since this energy is a scalar, the total radiation energy density just outside the surface is u + fu, rather than u - fu, even though the incident and reflected beams are oppositely directed. Thus, the assertion of the problem has been established.

42-6

The light being absorbed, all of its momentum p is transferred to the absorbing object, for which A is the area perpendicular to the beam. In a time t,

$$p = \frac{U}{c} = \frac{SAt}{c},$$

U the energy absorbed in time t and S the magnitude of the beam's Poynting vector (energy carried per unit area per unit time). The force F exerted on the object will be

$$F = \frac{dp}{dt} = \frac{SA}{c},$$

so that the radiation pressure P is

$$P = \frac{F}{A} = \frac{S}{c}.$$

42-8

(a) Let r = radius, ρ = density of the particle a distance x from the sun, which has a mass M and a luminosity (rate of energy output) L. The gravitational force on the particle is

$$F_{grav} = G\frac{Mm}{x^2} = G\frac{M(4\pi r^3\rho/3)}{x^2}.$$

The force due to radiation pressure is, by Problem 42-6,

$$F_{rad} = \frac{SA}{c} = \frac{1}{c}(L/4\pi x^2)(\pi r^2),$$

since the area A perpendicular to the beam is the sphere's cross-sectional area πr^2. The critical radius r = R_0 occurs when these forces balance:

$$G\frac{M(4\pi R_0^3\rho/3)}{x^2} = \frac{1}{c}(L/4\pi x^2)\pi R_0^2,$$

$$R_0 = \frac{3L}{16\pi cG\rho M} \quad \underline{\text{Ans}}.$$

(b) Putting $L = 4 \times 10^{26}$ W, $\rho = 1000$ kg/m^3 and $M = 2 \times 10^{30}$ kg gives $R_0 = 600$ nm <u>Ans</u>.

(c) R_0 is independent of x since both forces, gravitation and that due to radiation pressure, are proportional to $1/x^2$; hence the x disappears when they are set equal.

<u>42-10</u>

The momentum carried off by the beam in time t is

$$p = \frac{U}{c},$$

where U is the energy carried by the beam that is emitted in the same time interval. If P = 10 kW is the power of the beam, then U = Pt and therefore

$$p = \frac{Pt}{c}$$

so that the force will be

$$F = \frac{dp}{dt} = \frac{P}{c} = ma.$$

Hence, the speed v reached in time t, assuming the spaceship started from rest is, since v = at,

$$v = \frac{Pt}{mc} = \frac{(10^4 \text{ W})(86400 \text{ s})}{(1500 \text{ kg})(3 \times 10^8 \text{ m/s})} = 1.92 \times 10^{-3} \text{ m/s},$$

$$v = 1.92 \text{ mm/s} \quad \underline{\text{Ans}}.$$

<u>42-12</u>

(a) The frequency ν is

$$\nu = \frac{c}{\lambda} = \frac{3 \times 10^8 \text{ m/s}}{3.0 \text{ m}} = 100 \text{ MHz} \quad \underline{\text{Ans}}.$$

(b) $\vec{E} \times \vec{B} = \mu_0 \vec{S}$ must be in the +x-direction. By the right-hand rule, if $\vec{E}$ lies along the y-axis, $\vec{B}$ must be directed along the z-axis. Also,

$$B_m = E_m/c = \frac{300 \text{ V/m}}{3 \times 10^8 \text{ m/s}} = 10^{-6} \text{ T} \quad \underline{\text{Ans}}.$$

(c) The angular frequency $\omega = 2\pi\nu = 6.28 \times 10^8$ rad/s; the wave number is $k = 2\pi/\lambda = 2.1$ m-1.

(d) The time-averaged Poynting vector may be written

$$\bar{S} = \overline{E^2}/\mu_0 c = E_m^2/2\mu_0 c = 120 \text{ W/m}^2 \quad \underline{\text{Ans.}}$$

(e) If p is the momentum delivered to the sheet of area A,

$$p = \frac{SA}{c} = 4 \times 10^{-7}(A) \text{ kg·m/s}$$

each second. The radiation pressure P is independent of the area A:

$$P = \frac{SA/c}{A} = 4 \times 10^{-7} \text{ Pa} \quad \underline{\text{Ans.}}$$

<u>42-20</u>

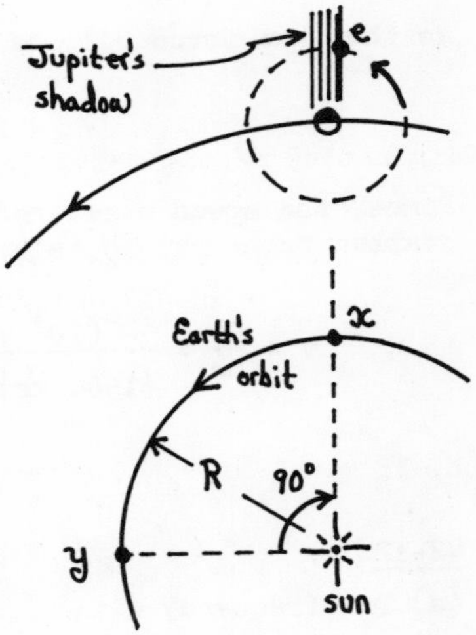

(a) Let the true period of revolution of one of Jupiter's satellites be T and the observed period T*. The latter may be determined by measuring, for example, the time between two successive passages of the satellite through point e, into eclipse. If this observation is made when the earth is near x, it is expected that T*(x) = T, since the earth, moving at a tangent to the earth-Jupiter line does not materially alter its distance to Jupiter in the time T. However, when the earth is near y it is moving almost directly away from Jupiter, so that T*(y) will differ from T by vT/c approximately, where v is the speed of the earth in its orbit. Hence, as the earth moves from point x to y, T* will increase steadily.

(b) With T determined (when the earth was near x) a prediction can be made, on the basis of the presumed infinite speed of light, of the number of ingresses of the satellite into Jupiter's shadow that will be observed before the earth reaches y; also the time of the ingress that occurs closest to the moment the earth reaches y can be computed. However, this particular ingress will in fact be observed at a time R/c, approximately, later than predicted. Hence, if R (the radius of the earth's orbit about the sun) is known, c can be calculated.

42-24

(a) Let the car be approaching, say, the radar trap. Also let

ν = frequency of waves leaving radar set, in radar set's frame;

ν' = frequency of waves as received and reflected by the car, in the car's frame;

ν'' = frequency of waves as received back at the radar set, in the radar set's frame.

Then, if v is the speed of the car,

$$\nu' = \nu \frac{1 + v/c}{(1 - v^2/c^2)^{\frac{1}{2}}},$$

$$\nu'' = \nu' \frac{1 + v/c}{(1 - v^2/c^2)^{\frac{1}{2}}}.$$

Eliminating ν' and retaining only linear terms in v/c,

$$\nu'' = \nu \frac{(1 + v/c)^2}{1 - v^2/c^2} = \nu \frac{1 + v/c}{1 - v/c} \approx \nu(1 + 2v/c),$$

$$\frac{\nu'' - \nu}{\nu} = \frac{\Delta\nu}{\nu} = 2\frac{v}{c}.$$

(b) If $v = 1$ mi/h $= 0.447$ m/s, this gives

$$\Delta\nu = 2\frac{0.447 \text{ m/s}}{3 \times 10^8 \text{ m/s}}(2.450 \times 10^9 \text{ s}^{-1}) = 7.3 \text{ Hz} \quad \underline{\text{Ans}}.$$

42-28

The Doppler formula is

$$\nu' = \nu \frac{1 \pm u/c}{(1 - u^2/c^2)^{\frac{1}{2}}}.$$

Since $\nu = c/\lambda$, this Doppler shift can be expressed in terms of the wavelengths as

$$\frac{1}{\lambda'} = \frac{1}{\lambda} \frac{1 \pm u/c}{(1 - u^2/c^2)^{\frac{1}{2}}}.$$

Expand the denominator in a Taylor series and, for $u/c \ll 1$, retain the first two terms only: to wit,

$$\lambda \approx \lambda'(1 \pm u/c)(1 + \tfrac{1}{2}u^2/c^2).$$

Thus, if now only linear terms are retained,

$$\lambda = \lambda'(1 \pm u/c).$$

If the source and observer are separating, the negative sign is appropriate; for this case $\Delta\lambda = \lambda' - \lambda$. On the other hand, if the source and observer are approaching use the positive sign and let $\Delta\lambda = \lambda - \lambda'$. In either case,

$$\Delta\lambda = \lambda'(\frac{u}{c}).$$

But, for $u \ll c$, $\lambda \approx \lambda'$ and therefore

$$\frac{\Delta\lambda}{\lambda} = \frac{u}{c}.$$

CHAPTER 43

43-5

(a) For normal incidence there is no refraction at the water surface.

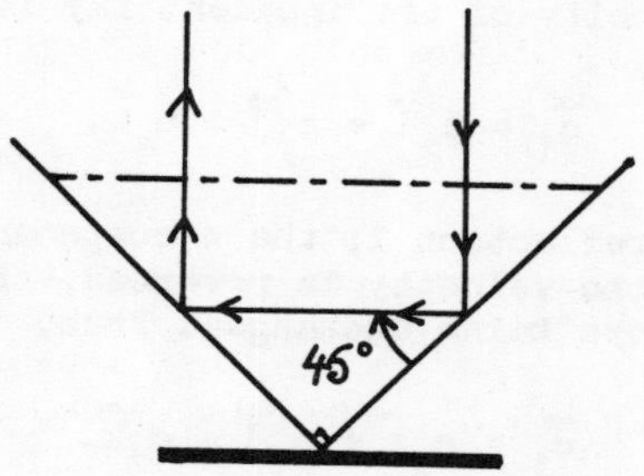

(b) By application of the law of reflection, together with the property of plane triangles that the sum of the angles is 180°, it can be seen that the incident and emergent rays are parallel. The relation between the angles θ and ϕ (i.e., the law of refraction) is not required).

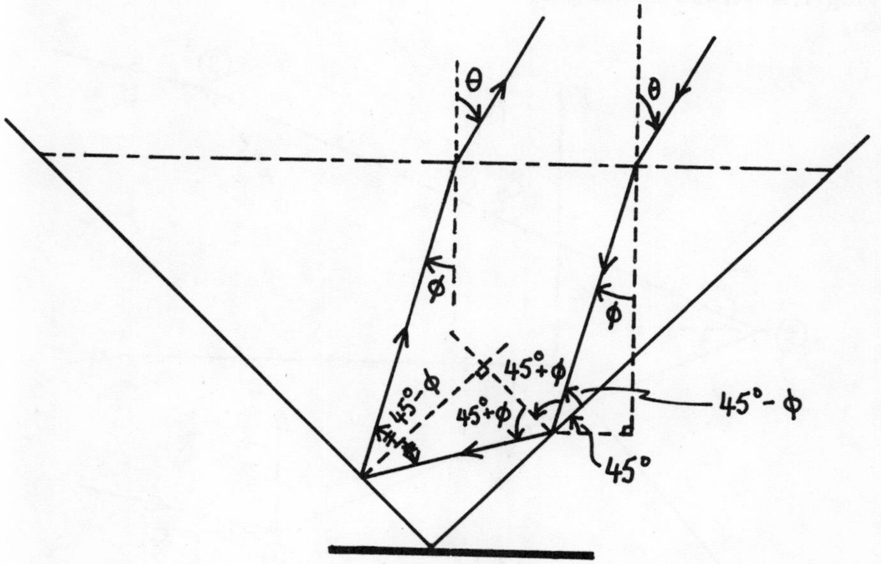

(c) Theorem: Three perpendicular mirrors form the sides of a vessel filled with water, as shown. A light ray entering the vessel and striking all three mirrors will emerge parallel to the incoming ray.

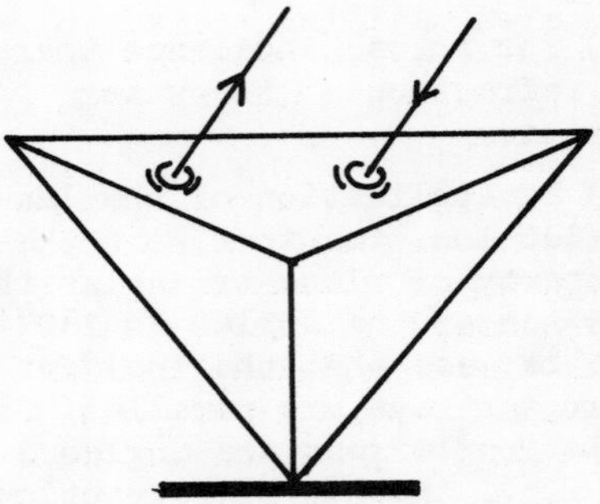

First, prove the theorem for an empty vessel; allow the mirrors to form coordinate planes. The velocity of the incident ray is

$$\vec{c}_i = c_x\vec{i} + c_y\vec{j} + c_z\vec{k}.$$

At reflection 1, the z-component of the velocity is reversed, the others being unchanged. Thus,

$$\vec{c}_1 = c_x\vec{i} + c_y\vec{j} - c_z\vec{k}.$$

At reflections 2 and 3, the y and x-components are reversed, also leaving the others unchanged; hence,

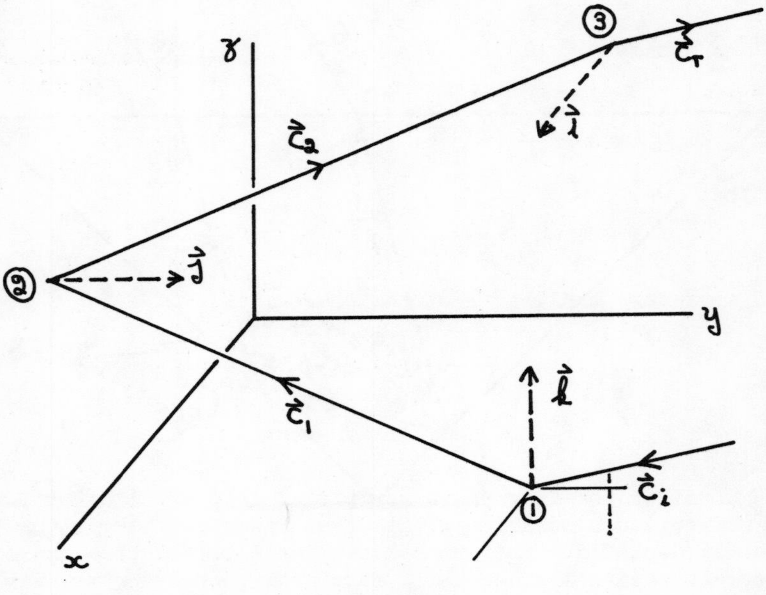

$$\vec{c}_2 = c_x\vec{i} - c_y\vec{j} - c_z\vec{k},$$

$$\vec{c}_r = -c_x\vec{i} - c_y\vec{j} - c_z\vec{k} = -\vec{c}_i.$$

Thus the ray emerges in the opposite direction from which it entered. Filling the vessel with water does not change this result since rays striking the water surface at the same angle are refracted through the same angle and thus remain parallel or antiparallel.

43-7

The length of the shadow is $L + \frac{1}{2}$ meters. By the law of refraction,

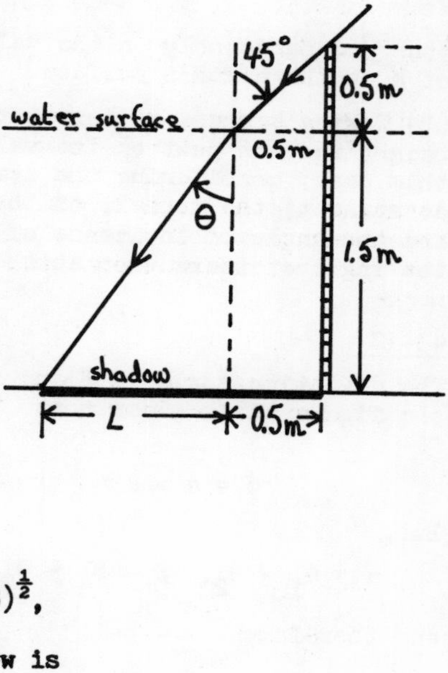

$$\sin 45^\circ = n\cdot\sin\theta = \frac{4}{3}\sin\theta,$$

$$\sin\theta = \frac{3}{8}\cdot\sqrt{2}.$$

Hence,

$$\tan\theta = \frac{\sin\theta}{\cos\theta} = \frac{\sin\theta}{(1 - \sin^2\theta)^{\frac{1}{2}}},$$

$$\tan\theta = 3/(23)^{\frac{1}{2}}.$$

Also,

$$\tan\theta = \frac{L}{1.5},$$

so that

$$L = 4.5/(23)^{\frac{1}{2}},$$

and therefore the length of the shadow is

$$L + \tfrac{1}{2} = \frac{4.5}{(23)^{\frac{1}{2}}} + 0.5 = 1.44 \text{ m} \quad \underline{\text{Ans.}}$$

43-10

(a) Consider the atmosphere stratified into N + 1 layers, the index of refraction n constant within each layer and larger for layers of greater depth into the atmosphere. Let θ be the true and ϕ the apparent zenith distance of a star. Then, applying the law of refraction from the top on down,

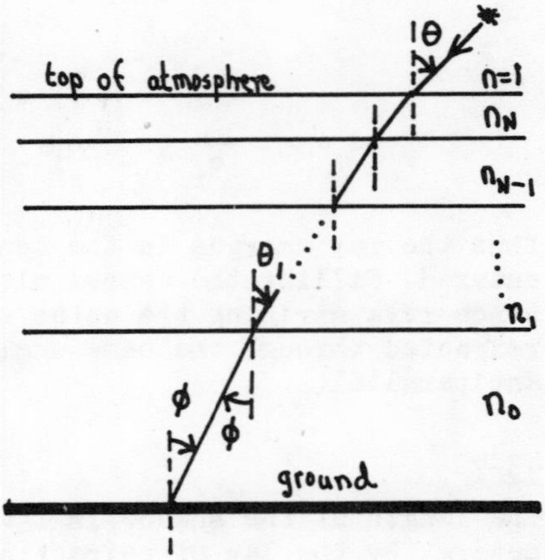

$$\sin\theta = n_N \sin\theta_N,$$

$$n_N \sin\theta_N = n_{N-1} \sin\theta_{N-1},$$

$$\vdots$$

$$n_1 \sin\theta_1 = n_0 \sin\phi.$$

Therefore,

$$\sin\phi = \frac{1}{n_0} \sin\theta;$$

thus ϕ depends only on the value of n at the earth's surface.

(b) Taking account of curvature, ϕ will depend on the run of n with height and not just on the value at the earth's surface, for in this case, considering the atmosphere stratified into shells of constant n, the normals of the different layers are not parallel and the angle of incidence of the ray at one shell is not equal to the angle of refraction at the boundary directly above.

<u>43-12</u>

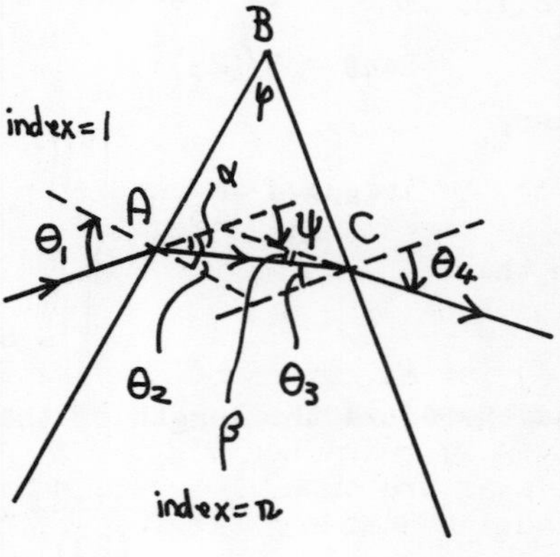

The deviation angle is ψ and from the figure it is clear that

$$\psi = \alpha + \beta.$$

But,

$$\alpha = \theta_1 - \theta_2, \quad \beta = \theta_4 - \theta_3,$$

and therefore

$$\psi = \theta_1 + \theta_4 - \theta_2 - \theta_3.$$

The angles of triangle ABC must add up to π rad.; hence

$$\pi = \phi + (\frac{\pi}{2} - \theta_2) + (\frac{\pi}{2} - \theta_3),$$

$$\phi = \theta_2 + \theta_3.$$

Also, from the law of refraction,

$$\sin\theta_1 = n\cdot\sin\theta_2, \quad n\cdot\sin\theta_3 = \sin\theta_4;$$

when these angles are small, the expressions become

$$\theta_1 = n\theta_2, \quad \theta_4 = n\theta_3.$$

Finally, these results give

$$\phi = n\theta_2 + n\theta_3 - \theta_2 - \theta_3 = (n - 1)(\theta_2 + \theta_3) = (n - 1)\phi.$$

43-22

(a) Under the assumptions of the problem, if θ is the critical angle for total internal reflection, the rays emitted at angles with respect to the vertical greater than θ are reflected back into the water and do not escape. Only those rays emitted into directions lying inside a cone of semiangle θ about the vertical will escape. Imagine the source surrounded by a sphere of radius $R < h$. If A is the area intercepted by the cone on the sphere, the desired ratio $f = A/4\pi R^2$ since the source radiates isotropically. But $A = 2\pi R^2(1 - \cos\theta)$, a "well known" result, and since $n\cdot\sin\theta = 1$ (condition for onset of total internal reflection), the ratio sought is

$$f = \tfrac{1}{2}[1 - (1 - 1/n^2)^{\frac{1}{2}}].$$

(b) The index of refraction n of water is 1.33, so the ratio in this case is 0.17 Ans.

43-25

(a) Let AB represent an edge of the cube, length a. The spot S is seen by refracted light that leaves the cube. However, rays leaving S at angles greater than

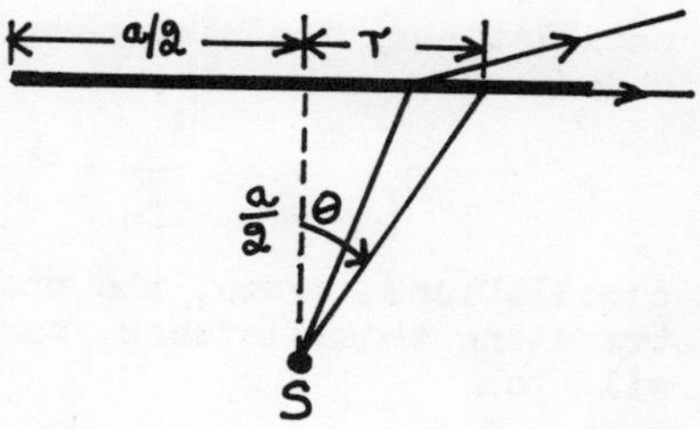

$$\theta = \sin^{-1}(\frac{1}{n})$$

with the vertical undergo total internal reflection and, if their subsequent behaviour is ignored, do not exit the cube. Hence, rays will emerge from a circular area, the radius r of which is given from

$$\tan\theta = \frac{r}{\frac{1}{2}a},$$

$$r = \frac{a}{2}\tan\theta = \tfrac{1}{2}a(n^2 - 1)^{-\frac{1}{2}}.$$

Thus, if an opaque circular disc is centered on each face of the cube, the spot will not be seen if the radius of the discs equal r found above. For a = 1.0 cm and n = 1.5, the radius r = 0.45 cm.

(b) The fraction f of the cube surface covered is

$$f = \frac{6\pi r^2}{6a^2} = \frac{\pi}{4(n^2 - 1)} = 0.63 \quad \underline{\text{Ans}}.$$

<u>43-28</u>

(a) If L is the geometrical path length, the optical path length is

$$D = nL = (1.5)(1.6 \times 10^{-4}\ \text{cm}) = 2.4 \times 10^{-4}\ \text{cm} \quad \underline{\text{Ans}}.$$

(b) The wavelength in the medium is

$$\lambda_n = \frac{\lambda}{n} = (6000\ \text{A})/(1.5) = 4000\ \text{A} \quad \underline{\text{Ans}}.$$

(c) The wave traveling a distance L in free space undergoes

$$\frac{L}{\lambda} = \frac{1.6 \times 10^{-4}\ \text{cm}}{6000\ \text{A}} = 266.67$$

oscillations. The same wave traveling this distance in the medium undergoes

$$\frac{L}{\lambda_n} = \frac{1.6 \times 10^{-4}\ \text{cm}}{4000\ \text{A}} = 400$$

oscillations. Hence, the phase difference δ between the waves after traveling this distance, assuming they were in phase at the start, will be

$$\delta = 0.67(2\pi) = \frac{4\pi}{3}\ \text{rad} = 240^\circ \quad \underline{\text{Ans}}.$$

Note that $2\pi - 4\pi/3 = 2\pi/3$ rad is also correct.

CHAPTER 44

44-4

In the diagram below are shown the three virtual images obtained even if the object does not lie on the bisector of the two mirrors.

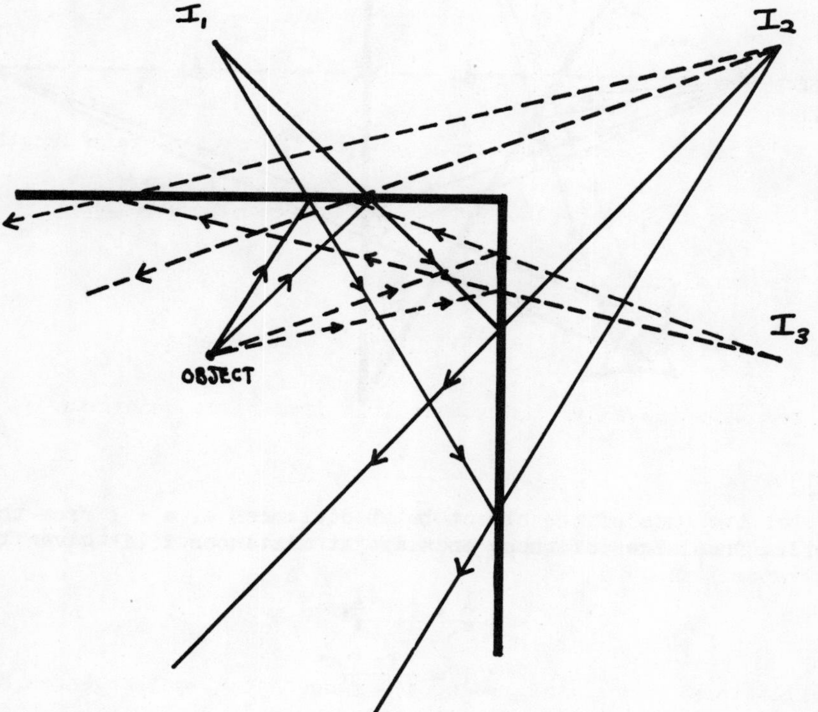

44-8

The line from image to object is perpendicular to, and bisected by, the plane mirror. If the mirror is rotated with the object held fixed, the image must move to maintain this geometrical relation. The eye will always be able to see this image (assuming perfect peripheral vision) since the situation is identical, for all practical purposes, is if the mirror be held fixed and the eye free to move about.

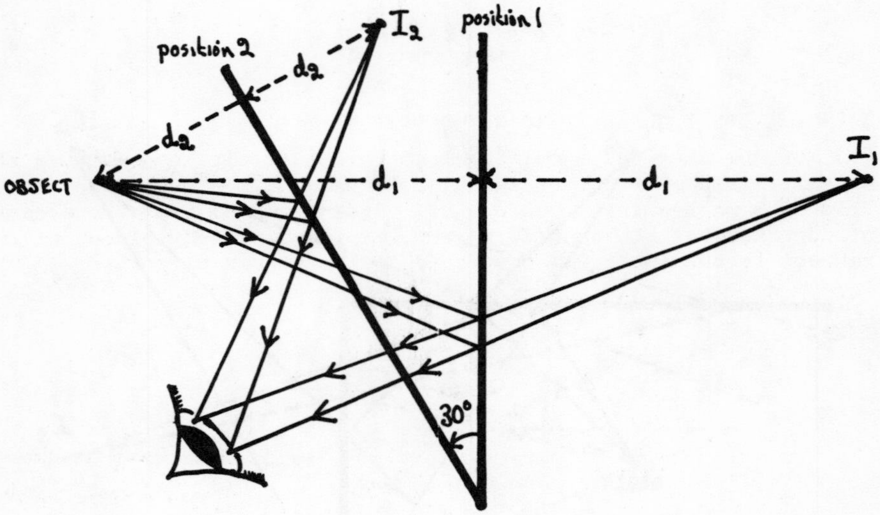

44-13

(a) Let the ends of the object be at distances a, a + ℓ from the mirror. The images of these ends are at distances i, i* given by

$$\frac{1}{a} + \frac{1}{i} = \frac{1}{f},$$

$$i = \frac{fa}{a - f},$$

$$\frac{1}{\ell + a} + \frac{1}{i^*} = \frac{1}{f},$$

$$i^* = \frac{f\ell + fa}{\ell + a - f}.$$

Thus the length ℓ^* of the image is

$$\ell^* = i - i^* = \frac{f^2\ell}{(\ell + a - f)(a - f)}.$$

The object being short, in the denominator set $\ell + a \approx a \approx o$, the object distance; then,

$$\ell^* = \ell\left(\frac{f}{o - f}\right)^2.$$

(b) The lateral magnification is $m = i/o$, ignoring sign. From (a),

$$m = \frac{f}{o - f}.$$

Hence, if $m^* = \ell^*/\ell$, it is clear that $m^* = m^2$.

(c) For the image of a cube to be a cube also, it is necessary that $m^* = m$. Since $m^* = m^2$, this will hold only if $m = m^2$, or $m = +1$. This in turn requires that $o = 2f$, placing the object at the center of curvature of the concave mirror, or symmetrically placed with respect to the center of curvature in the convex mirror.

44-15

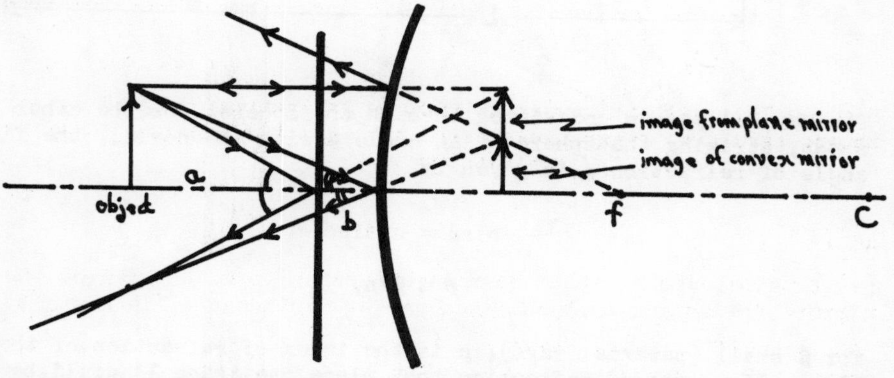

The image formed by the plane mirror lies a distance a behind this mirror, or at a distance b - a from the convex mirror. Thus, for this image to coincide in position with the image formed by the convex mirror, the latter image must be at an image distance of b - a from the convex mirror also. Since the source is at an object distance of b + a from the convex mirror,

$$\frac{1}{o} + \frac{1}{i} = \frac{1}{f},$$

$$\frac{1}{b + a} + \frac{1}{b - a} = \frac{1}{f},$$

$$a^2 = b(b - 2f).$$

Putting b = 7.5 cm, f = - 4b = -30 cm yields a = 3b = 22.5 cm Ans.

44-23

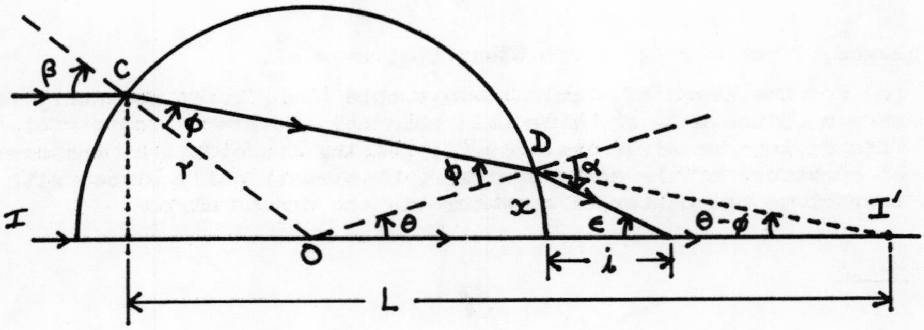

The ray II' passes undeviated through the sphere. For the other ray shown, striking the sphere at an angle β with the normal, the first angle of refraction ϕ is given by

$$\sin\beta = n\cdot\sin\phi,$$

$$\phi \approx \beta/n,$$

for β small (paraxial rays); n is the index of refraction of the glass. If no second refraction took place, an image I' would be formed a distance L behind the forward surface, with L determined from

$$\tan(\beta - \phi) = \frac{r \sin\beta}{L},$$

$$\beta - \phi \approx \frac{r\beta}{L},$$

$$L = \frac{r\beta}{\beta - \phi} = \frac{n}{n - 1}\, r.$$

For glass n = 1.5 approximately, so that $L \approx 3r$, and therefore I' lies outside the glass sphere. At the second refraction,

$$n \cdot \sin\phi = \sin\alpha,$$

$$\alpha = \beta.$$

The angles of triangle COD must total π rad:

$$2\phi + (\pi - \theta - \beta) = \pi,$$

$$\theta = 2\phi - \beta = \frac{2 - n}{n}\beta.$$

Hence,

$$x = r\theta = r\beta\,\frac{2 - n}{n}.$$

The image distance i is given by

$$i = \frac{x}{\tan\epsilon} = \frac{x}{\tan(\beta - \theta)} \approx \frac{x}{\beta - \theta} = \frac{r\beta\,(2 - n)/n}{\beta - (2 - n)\beta/n},$$

$$i = \frac{2 - n}{2(n - 1)}\,r \quad \underline{\text{Ans}}.$$

44-26

The lens maker's equation is

$$\frac{1}{f} = (n - 1)\left(\frac{1}{r'} - \frac{1}{r''}\right)$$

where n is the index of refraction of the glass ($n > 1$); r', r" = radius of curvature of the first and the second surface struck by the light. If $f > 0$ the lens is converging and if $f < 0$ it is diverging. Now, the centers of curvature of the surfaces shown in the figures all are to the right of each lens: hence, with light assumed incident from the left of each lens, r', r" > 0 since the centers of curvature lie on the R-side of each lens. (The same results, converging vs. diverging, are obtained if the light is imagined incident from the right since then r' and r" are interchanged, but they also change signs as the centers of curvature will then be on the V-side of each lens.)

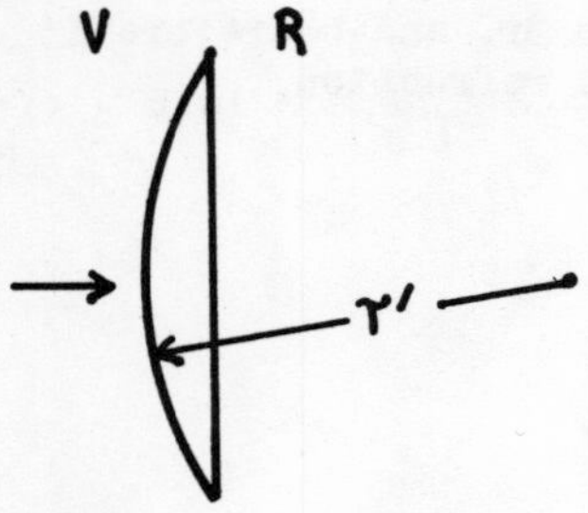

(a) $r' > 0$, $r'' = \infty$; $f > 0$, lens is converging.

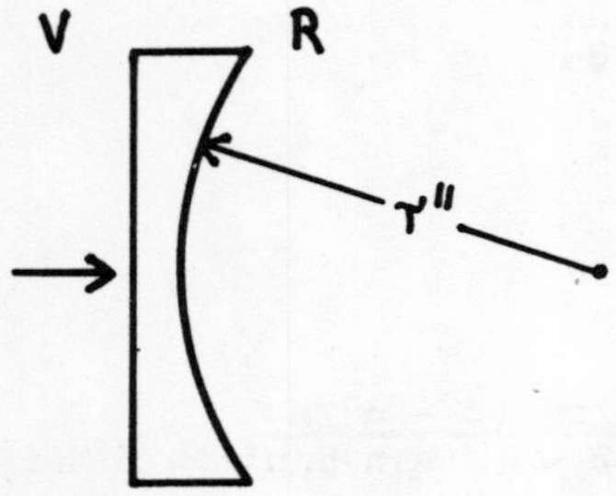

(b) $r' = \infty$, $r'' > 0$; $f < 0$, lens is diverging.

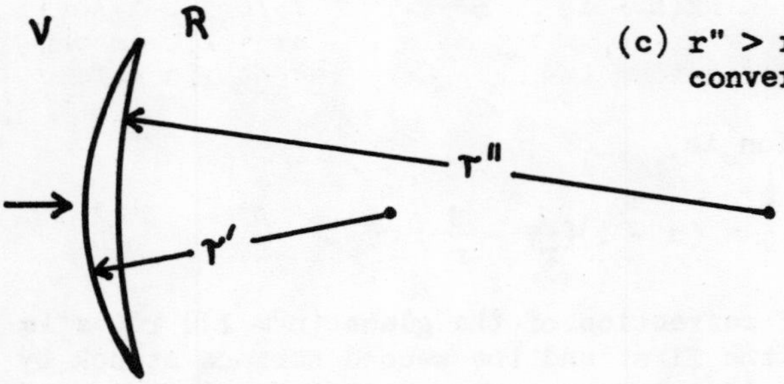

(c) $r'' > r'$; $f > 0$, lens is converging.

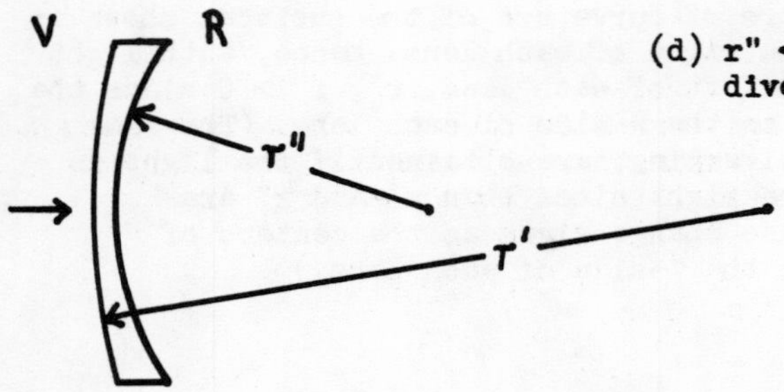

(d) $r'' < r'$; $f < 0$, lens is diverging.

44-31

(a) If the object distance is x, then the image distance is D - x; hence,

$$\frac{1}{o} + \frac{1}{i} = \frac{1}{f},$$

$$\frac{1}{x} + \frac{1}{D - x} = \frac{1}{f},$$

$$x^2 - Dx + Df = 0,$$

of which the solutions are

$$x_1 = \tfrac{1}{2}(D - d), \quad x_2 = \tfrac{1}{2}(D + d)$$

with

$$d = (D^2 - 4Df)^{\frac{1}{2}}.$$

Thus the separation in the two positions of the lens (object fixed) is $x_2 - x_1 = d$.

(b) The ratio of the magnifications is $m_2/m_1 = (-i_2/o_2)/(-i_1/o_1)$. But $o_1 = x_1$, $o_2 = x_2$, $i_1 = D - x_1 = x_2$, $i_2 = D - x_2 = x_1$, as can be verified from the expressions for x_1 and x_2 in (a). Therefore

$$m_2/m_1 = (x_1/x_2)^2 = \left(\frac{D - d}{D + d}\right)^2.$$

44-33

Let light be incident first on lens f_1 and assume $f_1 > 0$ (lens is converging). Then,

$$\frac{1}{o_1} + \frac{1}{i_1} = \frac{1}{f_1},$$

i_1 the distance of the image formed by lens f_1. Since $f_1 > 0$ light converges toward the image which thereby forms a virtual object for lens f_2. Hence $o_2 < 0$ and $i_1 = -o_2$. But

$$\frac{1}{o_2} + \frac{1}{i_2} = \frac{1}{f_2},$$

$$-\frac{1}{i_1} + \frac{1}{i_2} = \frac{1}{f_2}.$$

By definition of f,

$$\frac{1}{o_1} + \frac{1}{i_2} = \frac{1}{f}.$$

The first and third equations yield

$$\left(\frac{1}{o_1} - \frac{1}{f_1}\right) + \frac{1}{i_2} = \frac{1}{f_2}.$$

Comparing this with the preceding equation leads to

$$\frac{1}{f} - \frac{1}{f_1} = \frac{1}{f_2},$$

$$f = \frac{f_1 f_2}{f_1 + f_2}.$$

This holds regardless of whether f_2 be greater or less than zero. The only case not covered, then, is that of two diverging lenses. Here lens f_1 forms a real object for lens f_2. Thus, although $o_2 > 0$, $i_1 = -o_2$, as before. The rest of the proof is the same as above.

44-35

The Newtonian form applies to converging lenses ($f > 0$). To derive it, start with the Gaussian form

$$\frac{1}{o} + \frac{1}{i} = \frac{1}{f}.$$

The object and image distances in terms of x, x' are

$$o = f \pm x, \quad i = f \pm x',$$

the upper sign for a real image ($o > f$, $i > 0$), the lower sign for a virtual image ($o < f$, $i < 0$). Putting these into the first equation gives directly

$$\frac{1}{f \pm x} + \frac{1}{f \pm x'} = \frac{1}{f},$$

$$xx' = f^2,$$

in which $x, x' > 0$.

44-42

(a) The lens forms a real image at a distance i behind the lens, where i is found from the lens formula

$$\frac{1}{1} + \frac{1}{i} = \frac{1}{0.5},$$

giving i = 1.0 m. This image lies 2 - 1 = 1 m in front of the plane mirror which forms a virtual image of it 1 m behind, or 1 + 2 = 3 m behind the lens. Hence the final image is formed a distance I in front of the lens, with

$$\frac{1}{I} + \frac{1}{3} = \frac{1}{0.5},$$

$$I = 0.60 \text{ m} \quad \underline{\text{Ans}}.$$

(b) This image is real since the final object distance (3 m) is greater than the focal length (0.5 m) of the lens.
(c),(d) The magnification of the first image formed by the lens is

$$m_1 = -\frac{i}{o} = -\frac{1}{1} = -1.$$

The plane mirror does not disturb this magnification. The second magnification by the lens is

$$m_2 = -\frac{0.6}{3.0} = -0.2.$$

Compared to the original object the magnification is

$$m = m_1 m_2 = +0.20 \quad \underline{\text{Ans}}.$$

Since m is positive, the image must be erect.

44-45

(a) The image distance i is the focal length of the lens when the eye is focused at infinity: i = 2.5 cm. Hence the new focal length f* is given by

$$\frac{1}{40} + \frac{1}{2.5} = \frac{1}{f^*},$$

$$f^* = 2.35 \text{ cm} \quad \underline{\text{Ans}}.$$

(b) By the lens maker's formula

$$\frac{1}{f} = (n - 1)\left(\frac{1}{r'} - \frac{1}{r''}\right),$$

to diminish f, decrease r' alone or decrease both r' and r" keeping r" > r', but do not decrease r" alone.

<u>44-48</u>

(a) The mirror M may be considered replaced with a lens of the same focal length, the only difference introduced is that now the image is behind the lens rather than in front of the mirror. The mirror flat M' is of no import as far as magnification is concerned, for it only redirects the light. Hence,

$$m_0 = - f_{ob}/f_{eye},$$

as before, the lens also inverting the image.

(b) Considering the mirror to be spherical rather than paraboloidal gives

$$\frac{1}{2000} + \frac{1}{i} = \frac{1}{16.8},$$

$$i = 16.94 \text{ m} \quad \underline{\text{Ans}}.$$

(c) The focal length $f = r/2 = 5.0$ m. Thus,

$$200 = 5/f_{eye},$$

$$f_{eye} = 2.5 \text{ cm} \quad \underline{\text{Ans}}.$$

CHAPTER 45

45-10

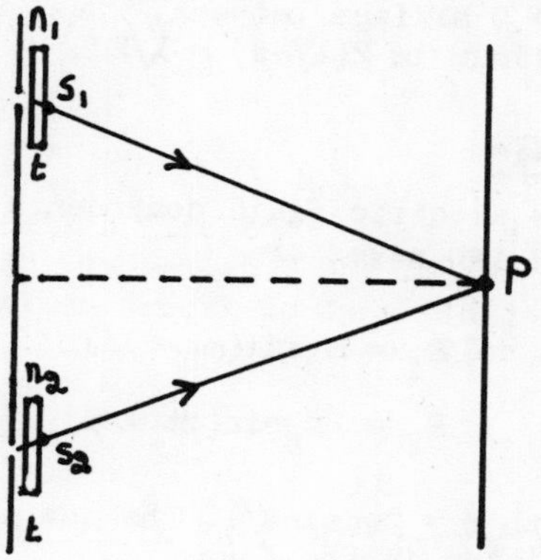

Before positioning the plates the central maximum fell at P. The geometrical path lengths from the points S_1 and S_2 at which the light emerges from the plates to P (with the plates in place) are equal. As these paths both are in air, i.e., in the same medium, no phase difference is introduced between the paths S_1P and S_2P. The wavelengths of light in the plates are

$$\lambda_1 = \lambda/n_1, \ \lambda_2 = \lambda/n_2.$$

The number of wavelengths of light in the plates are t/λ_1 and t/λ_2 and these must differ by five; therefore,

$$\frac{tn_1}{\lambda} - \frac{tn_2}{\lambda} = 5,$$

$$t = \frac{5\lambda}{n_1 - n_2} = \frac{5(480)}{1.7 - 1.4} = 8.0\ \mu\text{m} \quad \underline{\text{Ans.}}$$

44-16

The intensity I is given by

$$I = I_m\cos^2\beta,$$

with $\beta = \pi d\sin\theta/\lambda \approx \pi d\theta/\lambda$. At a point of half-maximum intensity,

$$\tfrac{1}{2}I_m = \tfrac{1}{2}I_m\cos^2\beta.$$

The smallest positive β satisfying this equation is $\beta = \pi/4$. Hence, the first half-intensity point occurs where θ has the value implied from

$$\pi/4 = \pi\theta d/\lambda,$$

$$\theta = \lambda/4d.$$

A symmetrical half-intensity point falls at $\theta = -\lambda/4d$, with the $\theta = 0$ maximum between. Thus, the half-width of this central maximum is $2(\lambda/4d) = \lambda/2d$.

<u>45-17</u>

The electric field components of the two waves are

$$E_1 = E_0\sin(\omega t + \phi),$$

$$E_2 = 2E_0\sin(\omega t + \phi)$$

with $\phi = 2\pi d\sin\theta/\lambda$. The sum is written in the form

$$E_1 + E_2 = E = E_\theta\sin(\omega t + \beta),$$

and the intensity I of this resultant wave is proportional to E_θ^2. Applying the law of cosines to the triangle formed by E_1, E_2, E in the phasor diagram gives

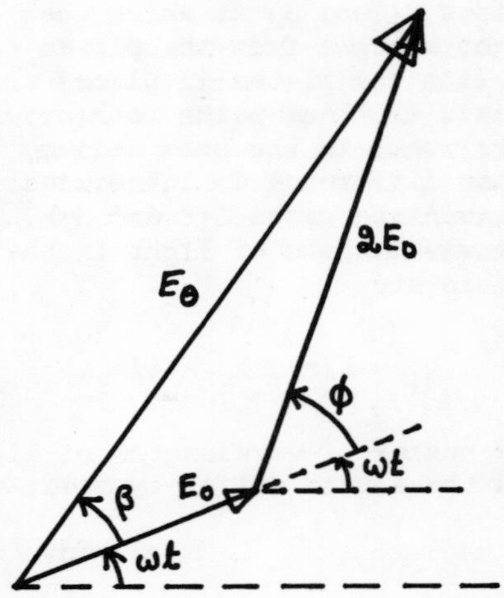

$$E_\theta^2 = E_0^2 + (2E_0)^2 - 2(E_0)(2E_0)\cos(\pi - \phi),$$

$$E_\theta^2 = E_0^2(5 + 4\cos\phi).$$

Hence,

$$I = AE_0^2(5 + 4\cos\phi),$$

with A the constant of proportionality. Now,

$$\cos\phi = \cos(\tfrac{1}{2}\phi + \tfrac{1}{2}\phi) = 2\cos^2(\tfrac{1}{2}\phi) - 1.$$

Substituting this gives

$$I = AE_0^2[1 + 8\cos^2(\tfrac{1}{2}\phi)].$$

If $\theta = 0$, then $\phi = 0$ so that $I(\theta = 0) = 9AE_0^2$. Define I_m by $I(\theta = 0) = I_m$; then $A = I_m/9E_0^2$ and therefore in terms of I_m, the intensity of the central maximum,

$$I = \frac{1}{9} I_m[1 + 8\cos^2(\pi d\sin\theta/\lambda)] \quad \underline{\text{Ans}}.$$

45-18

(a) If D is the detector, the optical path difference $S_2D - S_1D$ equals the geometrical path difference since the paths from both sources are in air. For maxima this path difference must equal an integral number of wavelengths:

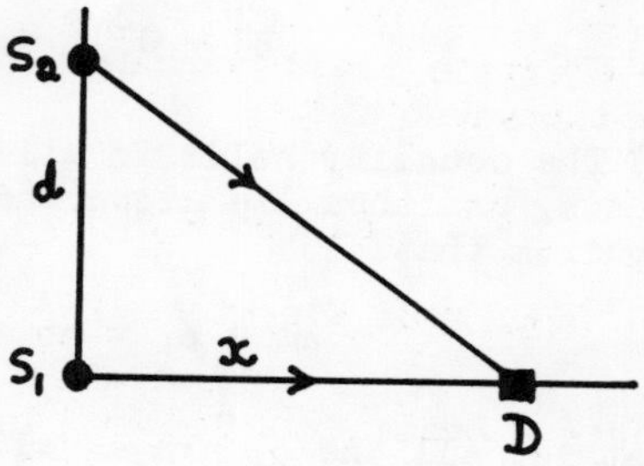

$$(d^2 + x^2)^{\frac{1}{2}} - x = n\lambda,$$

$n = 1, 2, \ldots$; $n = 0$ is not possible since $d \neq 0$. This gives

$$x = (d^2 - n^2\lambda^2)/2n.$$

Putting $n = 3, 2, 1$, together with $d = 4.0$ m, $\lambda = 1.0$ m yields

$$x_3 = \frac{7}{6} \text{ m}, \; x_2 = 3.0 \text{ m}, \; x_1 = \frac{15}{2} \text{ m} \quad \underline{\text{Ans}}.$$

(b) For completely destructive interference the waves must arrive at D exactly 180° out of phase and with equal amplitudes. The first requirement can be met but the second cannot since the waves, being spherical, experience a $1/r$ drop-off in amplitude with distance, and the distances r are different for the two waves.

45-21

(a) Let the sum in question be S:

$$S = A_1\sin(\omega t + \phi_1) + A_2\sin(\omega t + \phi_2) + \ldots. + A_n\sin(\omega t + \phi_n);$$

alternatively,

$$S = (\vec{A}_1 + \vec{A}_2 + \ldots. + \vec{A}_n)_y,$$

the y-component of the vector sum. This may also be written as

$$S = (A^*_{1x} + A^*_{2x} + \dots + A^*_{nx})\sin\omega t + (A^*_{1y} + A^*_{2y} + \dots + A^*_{ny})\cos\omega t,$$

$$S = B\sin\omega t + C\cos\omega t,$$

as shown in the figure.

(b) It can be seen from the sketch that $(B^2 + C^2)^{\frac{1}{2}}$ is the straight line distance OP. On the other hand, $A_1 + A_2 + \dots + A_n$ is the length of the zig-zag path along the A_i's from O to P. Thus,

$$(B^2 + C^2)^{\frac{1}{2}} \le A_1 + A_2 + \dots + A_n.$$

(c) The equality holds if all the A_i's make the same angle with the x-axis, and thus lie along the straight line from O to P. This requires that

$$\omega t + \phi_1 = \omega t + \phi_2 = \dots = \omega t + \phi_n,$$

that is, all the ϕ_i are equal.

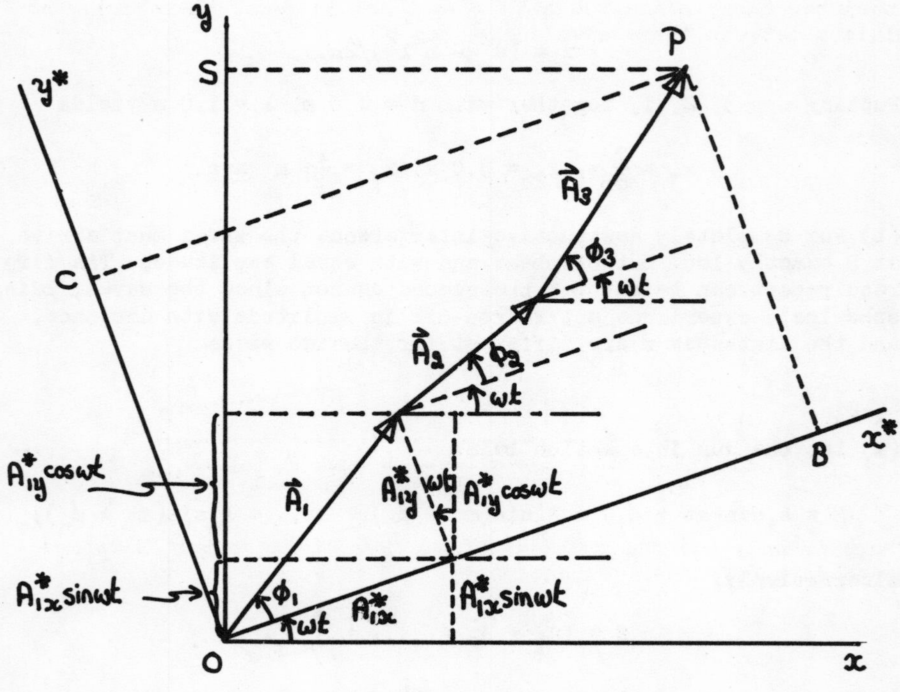

45-26

The phase changes on reflection for the two cases, $n < n_g$ and $n > n_g$, are shown in the figure. Thus the conditions for destructive interference in the two cases are

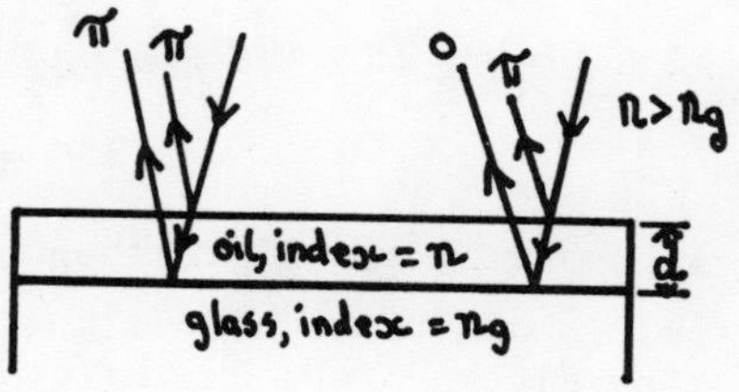

$$n > n_g: \quad 2nd = m\lambda, \quad m = 1, 2, \ldots;$$

$$n < n_g: 2nd = (m + \tfrac{1}{2})\lambda, \quad m = 0, 1, 2, \ldots .$$

Now, there is destructive interference at wavelengths 500 nm and 700 nm and therefore either

$$2nd = m_5(500), \quad 2nd = m_7(700), \; n > n_g,$$

or

$$2nd = (m_5 + \tfrac{1}{2})(500), \quad 2nd = (m_7 + \tfrac{1}{2})(700), \; n < n_g.$$

In either case, since $700 > 500$, $m_7 < m_5$. In fact, there being no minimum between these wavelengths,

$$m_7 = m_5 - 1;$$

hence,

$$m_7 = \frac{5}{2}, \; n > n_g; \quad m_7 = 2, \; n < n_g.$$

Since m_7 must be an integer, it must be that $n < n_g$.

45-31

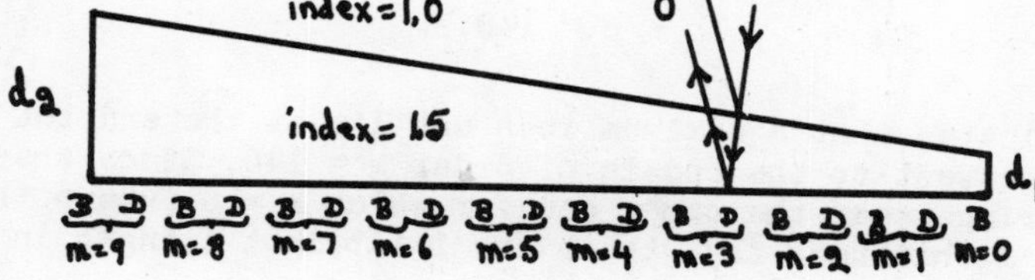

There is only one phase change of π; thus, for a bright B fringe and a dark D fringe, respectively,

$$2d = (m + \tfrac{1}{2})\frac{\lambda}{n} \text{ at B}; \qquad 2d = m\frac{\lambda}{n} \text{ at D}.$$

At the ends,

$$2d_1 = (0 + \tfrac{1}{2})(\frac{630}{1.5}),$$

$$d_1 = 105 \text{ nm};$$

$$2d_2 = (9 + \tfrac{1}{2})(\frac{630}{1.5}),$$

$$d_2 = 1995 \text{ nm}.$$

Thus the variation in thickness is

$$d_2 - d_1 = 1890 \text{ nm} = 1.89\ \mu\text{m} \quad \underline{\text{Ans}}.$$

45-32

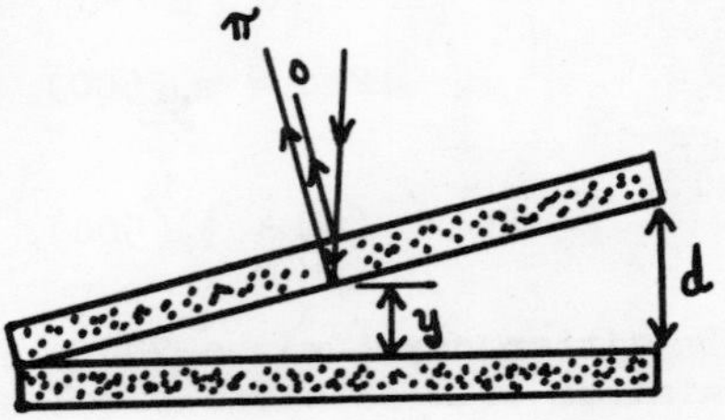

If the light waves are reflected from a region where the thickness of the film is y, the condition for a maximum is

$$2y = (m + \tfrac{1}{2})\lambda, \quad m = 0, 1, \ldots$$

with the $\frac{1}{2}$ compensating for the phase difference introduced in the reflection from the lower surface of the air film. If a maximum (bright fringe) appears at the end (y = d) it would correspond to an order number n given by

$$2d = (n + \tfrac{1}{2})\lambda,$$

$$2(48{,}000 \text{ nm}) = (n + \tfrac{1}{2})(680 \text{ nm}),$$

$$n = 140.7.$$

This indicates that a maximum does not lie at the end but that the maximum nearest to the end is of order m = 140. Since there is a bright fringe near the other end for which m = 0 (where the air film has a thickness $\frac{1}{4}\lambda$), there are 141 bright fringes in all.

45-33

There is a phase change of π at each reflection and therefore

$$2d = m\cdot\frac{\lambda}{n}, \text{ bright fringe}; \quad 2d = (m + \tfrac{1}{2})\cdot\frac{\lambda}{n}, \text{ dark fringe}.$$

(a) At the outer regions $d \approx 0$ and with $m \geq 0$ there can be only the bright fringe $m = 0$ whatever the wavelength of the light.

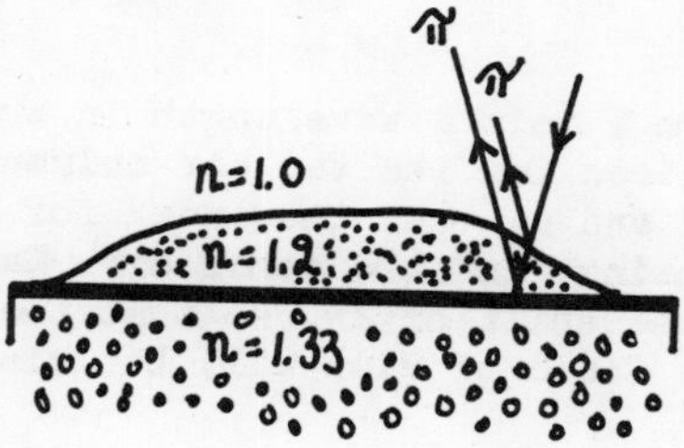

(b) For the third ($m = 3$) blue fringe,

$$2d = 3\,\frac{475\text{ nm}}{1.2},$$

$$d = 594\text{ nm} \quad \underline{\text{Ans}}.$$

(c) As the thicker regions of the film are examined, the fringes are seen to fall closer and closer together; eventually, they cannot be distinguished by the unaided eye.

45-36

The radius of ring m is

$$r_m = [(m + \tfrac{1}{2})\lambda R]^{\frac{1}{2}}.$$

Hence,

$$\Delta r = r_{m+1} - r_m = [(m + \tfrac{3}{2})\lambda R]^{\frac{1}{2}} - [(m + \tfrac{1}{2})\lambda R]^{\frac{1}{2}},$$

$$\Delta r = (\lambda Rm)^{\frac{1}{2}}[(1 + \frac{3}{2m})^{\frac{1}{2}} - (1 + \frac{1}{2m})^{\frac{1}{2}}].$$

Suppose $m >> 1$; then

$$\Delta r \approx (\lambda Rm)^{\frac{1}{2}}[(1 + \frac{3}{4m}) - (1 + \frac{1}{4m})] = (\lambda Rm)^{\frac{1}{2}}(\frac{1}{2m}) = \tfrac{1}{2}(\frac{\lambda R}{m})^{\frac{1}{2}} \quad \underline{\text{Ans}}.$$

45-41

Let L be the length of the chamber. The phase differences between the two rays arises from the unequal lengths of the interferometer arms and from the passage of light in one arm through a length 2L of air of density less (once the pumping starts) than the air "in" the other arm. If the lengths of the arms are not altered during the pumping process, then at any stage during the pumping the phase difference is introduced by the difference in air densities and is

$$\Delta\phi = 2\pi[\frac{2L}{\lambda_1} - \frac{2L}{\lambda_2}] = 2\pi[\frac{2L}{\lambda/n_1} - \frac{2L}{\lambda/n_2}],$$

$$\Delta\phi = \frac{4\pi L}{\lambda}(n_1 - n_2).$$

Here λ is the wavelength in a vacuum and n_1, n_2 the refractive indices for the two air columns. Assuming perfect pumping, put $n_2 = 1$ and $n_1 = n$, the index for air under atmospheric pressure that remains "in" the other arm. Each fringe shift corresponds to a phase shift of 2π, so that the number N of fringes that pass across the field of view will be, since 50 mm = 5×10^7 nm, 5000 A = 500nm,

$$N = \frac{\Delta\phi}{2\pi} = \frac{2L}{\lambda}(n - 1),$$

$$60 = \frac{2(5 \times 10^7)}{500}(n - 1)$$

$$n = 1.0003 \text{ \underline{Ans.}}$$

CHAPTER 46

46-1

From superposition, it is expected that the amplitude of the wave at P due to screen A added to the amplitude at P due to screen B will give the amplitude at P due to an unscreened wave. But, with $x \gg \lambda$, P lies in the geometrical shadow of the hole: thus the last quantity is zero. Hence,

$$E_{P,A} + E_{P,B} = 0,$$
$$E_{P,A} = -E_{P,B}.$$

But the intensity I is proportional to E_P^2 and therefore $I_{P,A} = I_{P,B}$.

46-6

Assuming that parallel "rays" strike the lens, the diffraction image is formed a distance f = 70 cm behind the lens. The angular distance from the center of the pattern to the m th minimum is given by

$$\sin\theta_m = m\lambda/a.$$

But the linear distance sought is

$$x_m = f\tan\theta_m.$$

Eliminating θ_m between these equations gives

$$x_m = \frac{f}{[(a/m\lambda)^2 - 1]^{\frac{1}{2}}}.$$

Now $a/\lambda \gg 1$ so that, approximately,

$$x_m = m\lambda f/a = (0.10325\text{ cm})m,$$

since f = 70 cm, a = 0.04 cm and $\lambda = 5.9 \times 10^{-5}$ cm.

(a) For m = 1, x = 1.0325 mm, and

(b) with m = 2, x = 2.065 mm.

46-10

(a) The intensity is given by

$$I = I_m \frac{\sin^2\alpha}{\alpha^2}.$$

To find the maxima and minima set $dI/d\alpha = 0$:

$$\frac{dI}{d\alpha} = 2I_m \frac{\sin\alpha}{\alpha}\left(\frac{\cos\alpha}{\alpha} - \frac{\sin\alpha}{\alpha^2}\right).$$

This will be zero if either
(i) $\sin\alpha/\alpha = 0$, giving $\alpha = m\pi$, m = 1, 2, ... These are minima since I = 0 for these values of α; or
(ii) $\cos\alpha/\alpha - \sin\alpha/\alpha^2 = 0$, which can be arranged into the form

$$\tan\alpha = \alpha$$

and are the maxima.

(b) Clearly $\alpha = 0$ is a solution of (ii). The next solution, as shown, is close to $\alpha = 3\pi/2$. Write this second solution as $\alpha = 3\pi/2 - x$. Then, in terms of x the equation $\tan\alpha = \alpha$ becomes

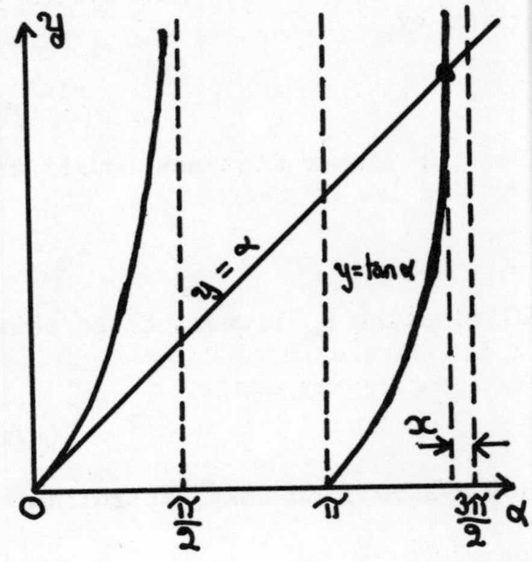

$$\frac{1}{\tan x} = \frac{3\pi}{2} - x.$$

Suppose $x \ll 1$; neglecting powers of x higher than two, this gives

$$\frac{1}{x}\left(1 - \frac{x^2}{3}\right) = \frac{3\pi}{2} - x,$$

which can be rearranged as

$$\frac{2}{3}x^2 - \frac{3\pi}{2}x + 1 = 0,$$

of which the solution is $x = 0.219$ rad, corresponding to $\alpha = 4.493$ rad, or $\alpha = 257^{\circ}$.

(c) For maxima, write

$$\alpha = (m + \tfrac{1}{2})\pi.$$

Then, the central maximum ($\alpha = 0$) has $m = -\tfrac{1}{2}$ and the next maximum, $\alpha = 4.493$, yields $m = 0.93$.

<u>46-12</u>

Consider a slit of given width divided into N strips of equal width so that there are N phasors. These are all parallel at the central maximum. If the slit width is doubled there will be 2N phasors, each having the same magnitude as for the original slit. Hence, the amplitude of the resultant field at the central maximum doubles and the intensity, proportional to the amplitude squared, increases by a factor of four. However, the diffraction pattern is narrower and the area under a curve of I vs. θ, equal to the rate at which energy passes through the slit, will only double.

<u>46-13</u>

Let d be the separation of the headlights a distance r away. The light rays make an angle θ with each other as they enter the eye and an angle θ^* after passing through the pupil. If n be the index of refraction of the fluid in the eye, then

$$\sin(\tfrac{1}{2}\theta) = n\sin(\tfrac{1}{2}\theta^*),$$

by the law of refraction. With the headlights just barely resolved,

$$\sin\theta^* = 1.22\,\frac{\lambda/n}{a},$$

λ the wavelength of the light in air and <u>a</u> the pupil diameter. If the angles are small,

$$r = \frac{d}{\theta} = \frac{d}{n\theta^*} = \frac{d}{1.22\,\lambda/a} = \frac{ad}{1.22\,\lambda} = 10.4 \text{ km} \quad \underline{\text{Ans}}.$$

<u>46-22</u>

(a) The sketches illustrate the formation of a halo as the moon shines through a cloud containing suspended water droplets. The ring will appear red if blue light is absent. Since the angle θ

for the first minimum is given by $\sin\theta = 1.22\lambda/d$, blue light, as it has the shortest wavelength in the visible spectrum, will have its first minimum closer to the moon (smallest θ gives the smallest ϕ) than any other color, giving the ring its reddish appearance.

(b) Since the rays MP and MO are virtually parallel, $\theta \approx \phi = (1.5)\cdot(\frac{1}{4}^{\circ}) = 22.5'$, the diameter of the moon being about $\frac{1}{2}^{\circ}$. Then, by (a)

$$\sin\phi \approx \sin\theta = 1.22\lambda_{blue}/d,$$

d the droplet diameter and $\lambda_{blue} = 400$ nm. These give d = 70 µm.

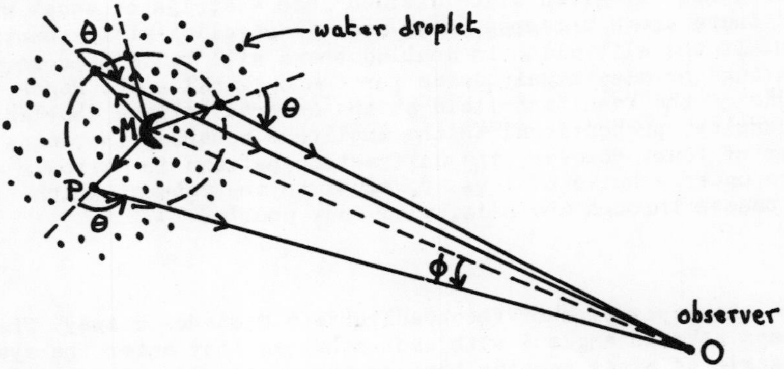

(c) A bluish ring will be seen at the scattering angle for the first minimum of red light. Since $\lambda(\text{red}) \approx 700$ nm $\approx 2\lambda(\text{blue})$, the radius of the blue ring will be about twice that of the red ring or three times the apparent lunar radius. (The intensity of the ring will be very low, however.) If water droplets of various sizes are present, various rings of these colors are present close together, giving a whitish appearance. This is, in fact, the most common form the rings assume.

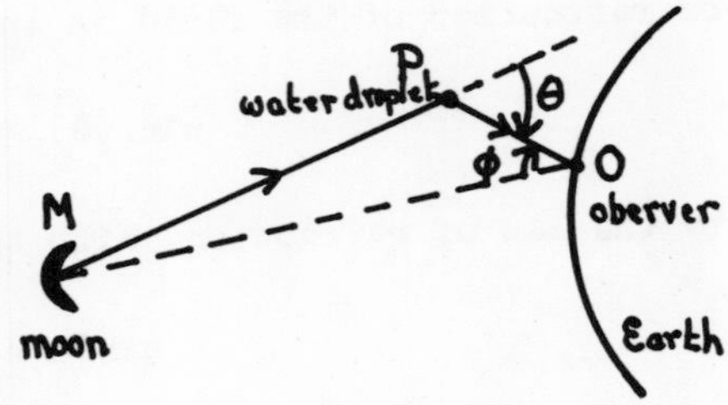

(d) These halos are a diffraction effect, a rainbow being formed by refraction.

46-27

(a) The location of the interference fringes is given by

$$d\sin\theta_1 = m\lambda, \quad m = 0, 1, 2, \ldots,$$

and the diffraction minima by

$$a\sin\theta_d = n\lambda, \quad n = 1, 2, \ldots .$$

Since d = 5a, the m = 5 interference fringe falls at the same position ($\theta_d = \theta_i$) as the n = 1 diffraction minimum and as a result is not seen. Inside the central diffraction envelope are the m = 1, 2, 3, 4 fringes on each side of the m = 0 fringe, giving nine fringes in all.

(b) The third fringe is located at an angle θ_i given from

$$\sin\theta_i = 3\lambda/d = 3\lambda/5a,$$

and has an intensity equal to that of the diffraction envelope at that location. Hence, the desired ratio is

$$\frac{I}{I_m} = \left(\frac{\sin\alpha}{\alpha}\right)^2$$

with

$$\alpha = \pi a\sin\theta_d/\lambda = \pi a\sin\theta_i/\lambda = 3\pi/5$$

so that $I/I_m = 0.255$ <u>Ans</u>.

47-6

With the slit width much less than the wavelength of light, the effects associated with a diffraction envelope can be ignored. The phase difference ϕ between rays from two adjacent slits is $\phi = 2\pi d\sin\theta/\lambda$ since the path difference for these rays is $d\sin\theta$, just as for the two-slit interference "grating". The electric field components of the three waves arriving at a given point on the screen can be written as

$$E_1 = E_0\sin\omega t,\ E_2 = E_0\sin(\omega t + \phi),\ E_3 = E_0\sin(\omega t + 2\phi).$$

It can be seen from the phasor diagram that the amplitude of the sum of these waves is

$$E_\theta = E_0\cos\phi + E_0 + E_0\cos\phi = E_0(1 + 2\cos\phi).$$

Since the intensity is proportional to the square of the amplitude,

$$I = AE_0^2(1 + 2\cos\phi)^2.$$

At the center of the pattern $\theta = \phi = 0$ and $I(\theta = 0) = 9AE_0^2 = I_m$, the central intensity. Thus, if $A = I_m/9E_0^2$,

$$I = \frac{1}{9}\, I_m(1 + 4\cos\phi + 4\cos^2\phi).$$

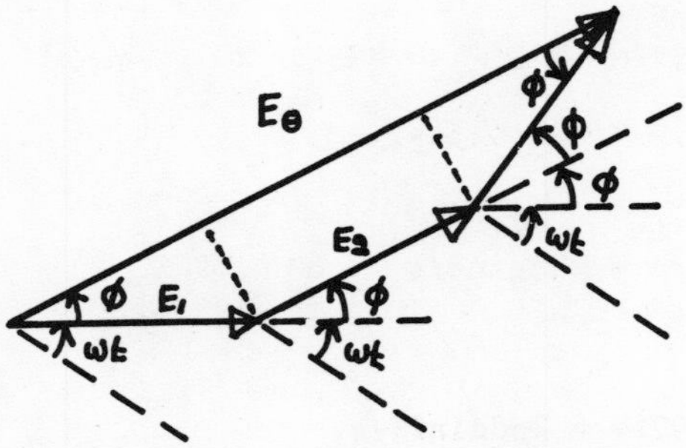

47-7

(a)

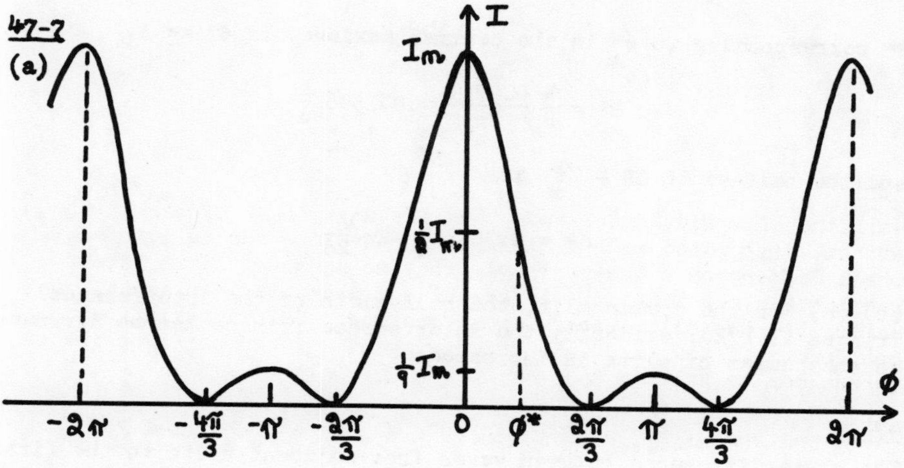

The maxima and minima of I vs. ϕ occur at those values of ϕ that satisfy $dI/d\phi = 0$:

$$\frac{dI}{d\phi} = 0 = -\frac{4}{9}I_m \sin\phi(1 + 2\cos\phi),$$

$$\sin\phi = 0; \ \phi = \pm n\pi, \ n = 0, 1, 2, \ldots ;$$

$$\cos\phi = -\tfrac{1}{2}; \ \phi = \pm\frac{2\pi}{3}, \pm\frac{4\pi}{3}, \pm\frac{8\pi}{3}, \ldots .$$

For $\phi = 0, \pm 2\pi$, etc, $I = I_m$: these are the principal maxima. At $\phi = \pm \pi$, etc., $I = I_m/9$ and these are the secondary maxima. All values of ϕ for which $\cos\phi = -\frac{1}{2}$ yield $I = 0$, the minima. To find the value of $\phi = \phi^*$ at which the intensity is half that of a principal maximum, set $I = \frac{1}{2}I_m$:

$$\tfrac{1}{2}I_m = \frac{1}{9}I_m(1 + 4\cos\phi^* + 4\cos^2\phi^*),$$

$$\cos\phi^* = 0.5607.$$

But,

$$\phi = 2\pi d\sin\theta/\lambda,$$

so that

$$0.9714 = 2\pi d\sin\theta^*/\lambda,$$

θ^* corresponding to ϕ^* in the central maximum. If $\theta^* \ll 1$,

$$\theta^* = \frac{\lambda}{d}\,\frac{0.9714}{2\pi} = 0.1546\,\frac{\lambda}{d},$$

and the half-width $\Delta\theta = 2\theta^*$ is

$$\Delta\theta = 0.3092\,\frac{\lambda}{d} = \frac{\lambda}{3.2d}.$$

(b),(c) For the double slit, the half-width of the interference fringes is $\lambda/2d$; evidently the interference fringes become narrower as the number of slits is increased.

47-10

The phase difference between waves from adjacent slits to the first minimum beyond the m th principal maximum is

$$\Delta\phi = 2\pi m + \frac{2\pi}{N},$$

which indicates a path difference ΔL given by

$$\Delta L = \frac{\Delta\phi}{2\pi}\cdot\lambda = m\lambda + \frac{\lambda}{N}.$$

But if θ_m is the position angle of the m th principal maximum, this path difference is also

$$\Delta L = d\sin(\theta_m + \Delta\theta).$$

Therefore,

$$d(\sin\theta_m\cos\Delta\theta + \cos\theta_m\sin\Delta\theta) = m\lambda + \frac{\lambda}{N}.$$

If $\Delta\theta \ll 1$, $\cos\Delta\theta \approx 1$ and $\sin\Delta\theta \approx \Delta\theta$; with $d\sin\theta_m = m\lambda$,

$$m\lambda + d\cos\theta_m\Delta\theta = m\lambda + \frac{\lambda}{N},$$

$$\Delta\theta = \frac{\lambda}{Nd\cos\theta_m}.$$

47-11

(a) For $N \gg 1$, the phasors form the arcs of circles. Let E = amplitude from one slit; the total length of the phasors then is NE. At the n th minimum the phasors form n closed circles, and at the k th maximum they form $(k + \frac{1}{2})$ circles, each of circumference πD_k. Thus,

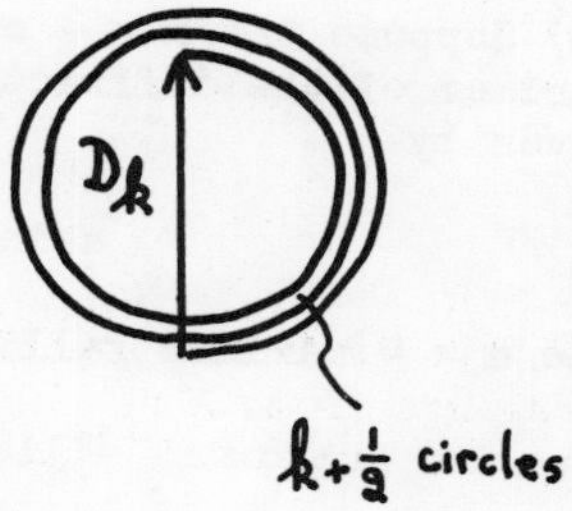

$$(k + \tfrac{1}{2})\pi D_k = NE,$$

$$D_k = \frac{NE}{(k + \frac{1}{2})\pi}.$$

Hence for the intensity $I = AD_k^2$, and $I_m = A(NE)^2$,

$$I_k = \frac{I_m}{(k + \frac{1}{2})^2\pi^2}.$$

For the single slit,

$$I = I_m\left(\frac{\sin\alpha}{\alpha}\right)^2.$$

At a maximum, $\alpha \approx (k + \frac{1}{2})\pi$, $\sin\alpha = \cos\pi k = \pm 1$, so that

$$I \approx \frac{I_m}{(k + \frac{1}{2})^2\pi^2}.$$

(b) Here $k \approx \frac{1}{2}N$ giving, for $N \gg 1$ so that $N + 1 \approx N$,

$$I_k \approx \frac{I_m}{N^2}(\tfrac{1}{2}\pi)^2 \approx I_m/N^2.$$

(c) In the single slit formula I_m corresponds to the phasors from all elements of the slit being lined up. In the grating case, I_m indicates that the phasors from all the slits (one phasor per slit since diffraction effects are neglected here) are lined up. From (a), the equations for the intensities are identical if the grating is considered one large slit of width Nd.

<u>47-18</u>

(a) Let the maxima in question be of order m and m + 1. Then,

$$d\sin\theta_m = m\lambda, \quad d\sin\theta_{m+1} = (m + 1)\lambda,$$

where $\sin\theta_m = 0.2$, $\sin\theta_{m+1} = 0.3$. Subtracting these equations gives

$$d(0.3 - 0.2) = \lambda = 600 \text{ nm},$$

$$d = 6.0\ \mu\text{m} \quad \underline{\text{Ans}}.$$

(b) Suppose the $m = 4$ maximum, which is missing, falls at the n th minimum of the diffraction envelope. These minima occur at angles ϕ given by

$$a\sin\phi = n\lambda, \quad n = 1, 2, \ldots .$$

The $m = 4$ maximum falls at an angle θ_4 satisfying

$$d\sin\theta_4 = 4\lambda.$$

If the two fall at the same place on the screen, $\phi = \theta_4$ so that

$$d\sin\theta_4 = 4\lambda, \quad a\sin\theta_4 = n\lambda,$$

$$a = \tfrac{1}{4}nd.$$

For the smallest <u>a</u> choose the smallest n, to wit, $n = 1$; then

$$a_{min} = \tfrac{1}{4}\ 6000 \text{ nm} = 1.5\ \mu\text{m} \quad \underline{\text{Ans}}.$$

(c) If the $m = 4$ maximum is missing, the $m = 8, 12, 16, \ldots$ will be missing also. Since $d = 10\lambda$, it is clear that the $m = 10$ maximum is deflected through 90° and will not appear on the screen. Hence, the visible orders are

$$m = 0, 1, 2, 3, 5, 6, 7, 9 \quad \underline{\text{Ans}}.$$

<u>47-22</u>

By "blazing" the grating (cutting the grooves at a specific angle), the deflection angle is fixed. From the grating equation

$$d\sin\theta = (1)(8000 \text{ nm}) = 8000 \text{ nm}.$$

For red light $\lambda = 700$ nm and therefore

$$d\sin\theta = 8000 = m\lambda = m(700),$$

$$m = 11.4;$$

for blue light λ = 400 nm, m = 20. Hence, the 11 th order of red and 20 th order of blue will appear at the same deflection angle, that for which the grating is blazed. Since the light of these wavelengths overlap on the screen, the spectrum will appear to be white.

47-26

If d is the distance between adjacent scattering centers, the path difference between parallel rays striking adjacent centers is

$$d_1 - d_2 = d\cos\gamma - d\cos 2\beta.$$

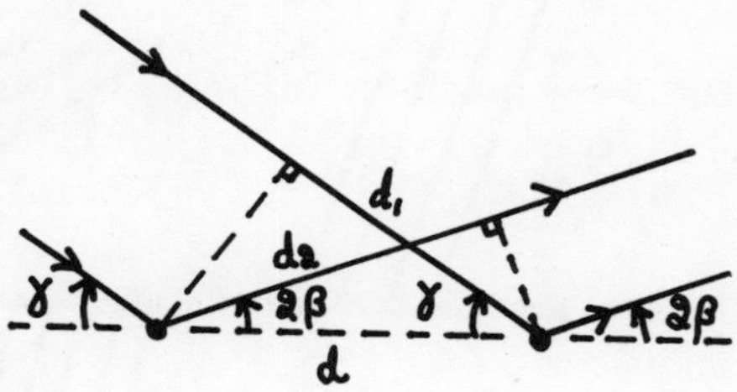

For maxima this must equal an integral number m of wavelengths:

$$d(\cos\gamma - \cos 2\beta) = m\lambda.$$

With γ small, $\cos\gamma \approx 1$; since d = 1500 nm, λ = 0.5 nm and m = 1 for the first order maximum,

$$1 - \cos 2\beta \approx \frac{1}{3000};$$

evidently, then, β is a small angle also. Setting $\cos 2\beta \approx 1 - 2\beta^2$ and $\cos\gamma \approx 1 - \frac{1}{2}\gamma^2$ gives, more precisely,

$$(1 - \tfrac{1}{2}\gamma^2) - (1 - 2\beta^2) = \frac{1}{3000},$$

$$\beta = (\frac{1}{6000} + \tfrac{1}{4}\gamma^2)^{\frac{1}{2}} \quad \underline{\text{Ans}}.$$

47-33

The spacings of the five sets of planes shown in the figure are

a: $a_0/(2)^{\frac{1}{2}}$;
b: $a_0/(5)^{\frac{1}{2}}$;
c: $a_0/(10)^{\frac{1}{2}}$;
d: $a_0/(13)^{\frac{1}{2}}$;
e: $a_0/(17)^{\frac{1}{2}}$.

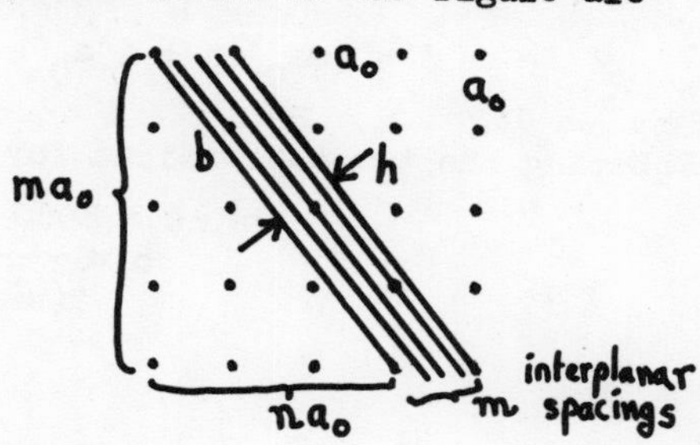

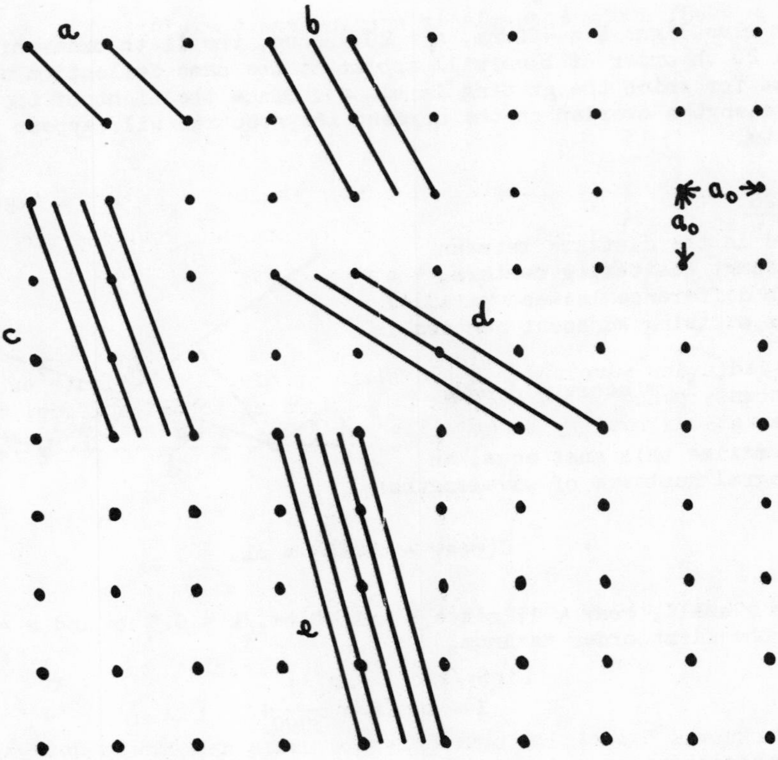

(b) For the general formula, see the sketch p. 379. The area of the enveloping parallelogram is

$$A = bh = [(ma_0)^2 + (na_0)^2]^{\frac{1}{2}}h.$$

It is also given by

$$A = (ma_0)(n + 1)a_0 - 2\cdot\tfrac{1}{2}(ma_0)(na_0) = ma_0^2.$$

Equating the two expressions for A gives

$$h = \frac{ma_0}{(m^2 + n^2)^{\frac{1}{2}}}.$$

But h = md, d the interplanar spacing and therefore

$$d = \frac{a_0}{(m^2 + n^2)^{\frac{1}{2}}}.$$

If the integers m, n contain a common factor, p say, the figure for planes with slope $\frac{m}{p}/\frac{n}{p}$ is being reproduced, yielding the same value of d: therefore, exclude these.

47-39

The radiation wavelength λ and plane spacing d constitute two unknowns; hence, observations at two orders should suffice; these observations satisfy Bragg's law

$$2d\sin\theta_1 = m_1\lambda,$$

$$2d\sin\theta_2 = m_2\lambda.$$

But upon dividing, these equations yield

$$\sin\theta_1/\sin\theta_2 = m_1/m_2,$$

the unknowns cancelling out. Only the ratio d/λ can be determined and this requires only one of these equations.

47-40

Let d be the thickness of a unit cell. For constructive interference between rays reflected from the top and bottom of the layer,

$$2d\sin\theta = \lambda$$

in first order. However, this indicates that

$$2\,\frac{d}{2}\sin\theta = d\sin\theta = \tfrac{1}{2}\lambda,$$

for interference between rays reflected from the top and middle of the same layer. But a shift of ½λ indicates destructive interference and hence the first order maximum is missing.

CHAPTER 48

48-5

Let I be the intensity of the polarized component and i the same for the unpolarized component in the incident beam; the intensity of this incident beam, then, is i + I. The transmitted intensity i_t of the unpolarized component is

$$i_t = \tfrac{1}{2}i,$$

independent of the angle θ of the polaroid sheet; the intensity I_t transmitted by the originally polarized component is

$$I_t = I\cos^2\theta.$$

Hence, the transmitted intensity is

$$i_t + I_t = \tfrac{1}{2}i + I\cos^2\theta.$$

As the polaroid sheet is rotated, $\cos^2\theta$ varies between 0 and 1, and if the transmitted intensity, as a consequence, varies by a factor of five, then

$$5(\tfrac{1}{2}i) = \tfrac{1}{2}i + I,$$

$$i = \tfrac{1}{2}I.$$

The fraction of the incident beam energy that is unpolarized, then, is

$$\frac{i}{i + I} = \frac{1}{3} = 33.3\%,$$

and the relative intensity of the polarized component is 100 - 33.3 = 66.7%.

48-6

(a) The electric field transmitted is

$$E_t = E_x\cos\theta \pm E_y\cos(\frac{\pi}{2} - \theta),$$

$$E_t = E_x\cos\theta \pm E_y\sin\theta,$$

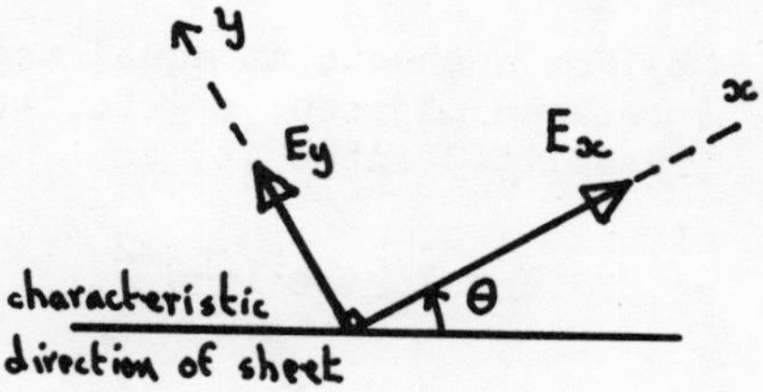

the $\pm$ sign suggesting the rapidly varying phase difference between E_x and E_y. Hence, with intensity proportional to the square of the electric field, on the average,

$$I_t = A(E_x^2\cos^2\theta + E_y^2\sin^2\theta \pm 2E_xE_y\sin\theta\cos\theta),$$

$$I_t = I\cos^2\theta + i\sin^2\theta,$$

the + occuring as often as the - in the first expression for I_t. Eliminating i in favor of p through

$$p = \frac{I - i}{I + i},$$

gives

$$I_t = I\cos^2\theta + I(\frac{1 - p}{1 + p})\sin^2\theta,$$

$$I_t = \frac{I}{1 + p}[(1 + p)\cos^2\theta + (1 - p)\sin^2\theta] = I\frac{1 + p\cos 2\theta}{1 + p}.$$

(b) If p = 1 (i.e., i = 0),

$$I_t = \frac{I}{2}(1 + \cos 2\theta) = I\cos^2\theta,$$

the law of Malus for plane-polarized light. With p = 0 (i = I),

$$I_t = I,$$

which is one-half the incident intensity, as expected since p = 0 describes completely unpolarized light.

48-7

(a) As illustrated in the sketch for two sheets, use a number of sheets with a total rotation angle of 90°.

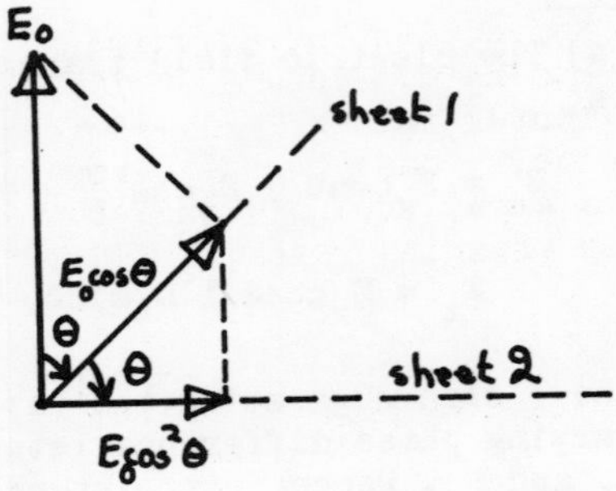

(b) For n sheets at equal angles θ between adjacent sheets, the transmitted intensity is

$$I = I_0[\cos^n(\frac{\pi}{2n})]^2,$$

since $n\theta = \pi/2$; I_0 is the intensity in the incident beam. For a loss of intensity of 5%, $I_0 - I = 0.05I_0$ so that $I = 0.95I_0$; thus,

$$0.95I_0 = I_0[\cos^n(\frac{\pi}{2n})]^2,$$

$$0.95 = \cos^{2n}(\frac{\pi}{2n}).$$

Let $x = \pi/2n$ so that

$$0.95 = (\cos x)^{\pi/x} = (1 - \tfrac{1}{2}x^2 + \ldots)^{\pi/x}.$$

Supposing x^2 to be small,

$$\ln(0.95) \approx \ln(1 - \tfrac{1}{2}x^2)^{\pi/x} = (\frac{\pi}{x})\ln(1 - \tfrac{1}{2}x^2) \approx (\frac{\pi}{x})(-\tfrac{1}{2}x^2),$$

$$x = -\frac{2}{\pi}\ln(0.95) = \frac{\pi}{2n},$$

$$n = -\frac{\pi^2}{4\ln(0.95)} = 48.104 \approx 48 \quad \underline{\text{Ans}}.$$

48-11

(a) For the deflection Δ of either beam, since $r\cos\theta = t$,

$$\Delta = r\sin(\theta_1 - \theta) = (\frac{t}{\cos\theta})\sin(\theta_1 - \theta),$$

with

$$\sin\theta_1 = n\sin\theta.$$

The deflection increases with increasing n. Since $n_o = 1.658$ is greater than $n_e = 1.486$ for calcite, y must be the o-ray and x the e-ray. The distance between these emerging rays is

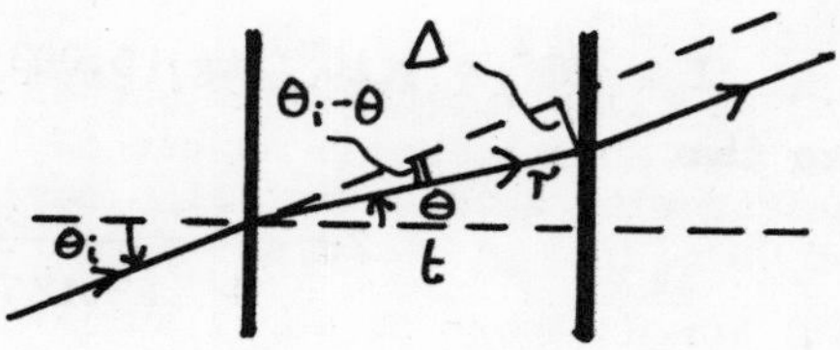

$$\Delta_o - \Delta_e = t\left[\frac{\sin(\theta_i - \theta_o)}{\cos\theta_o} - \frac{\sin(\theta_i - \theta_e)}{\cos\theta_e}\right].$$

Putting $\theta_i = 45°$ gives $\theta_o = 25°14.65'$, $\theta_e = 28°24.87'$, from Snell's law using the appropriate index of refraction. These give $\sin(\theta_i - \theta_o) = 0.33801$, $\sin(\theta_i - \theta_e) = 0.28545$, or

$$\Delta_o - \Delta_e = (1.0 \text{ cm})(0.049) = 0.49 \text{ mm} \quad \underline{\text{Ans}}.$$

(b) From Fig. 48-12 ray x, the e-ray, has $\vec{E}$ vibrating parallel to the optic axis and for ray y, the o-ray, $\vec{E}$ vibrates perpendicular to the optic axis.

(c) The o and e-rays are plane polarized at right-angles. Hence, upon rotating the polaroid in the two beams one or the other, alternatively, will be extinguished each 90° of rotation, for when the characteristic direction of the sheet is parallel to the plane of vibration of one ray, it is perpendicular to the plane of vibration of the other.

<u>46-18</u>

(a) Since the angular momentum delivered is $L = U/\omega$, the rate at which it is delivered is

$$\frac{dL}{dt} = \frac{P}{\omega} = \frac{\lambda P}{2\pi c}$$

$$\frac{dL}{dt} = \frac{(5 \times 10^{-7} \text{ m})(100 \text{ W})}{2\pi(3 \times 10^{8} \text{ m/s})} = 2.65 \times 10^{-14} \text{ J} \quad \underline{\text{Ans}}.$$

(b) Since $L = I\omega$, then assuming constant angular acceleration,

$$\omega = \frac{dL/dt}{I}\, t,$$

the initial angular velocity being zero. The rotational inertia is

$$I = \tfrac{1}{2}MR^2 = \tfrac{1}{2}(10^{-5}\ \text{kg})(0.0025\ \text{m})^2 = 3.125 \times 10^{-11}\ \text{kg}\cdot\text{m}^2,$$

so that

$$2\pi\ \text{s}^{-1} = \frac{2.65 \times 10^{-14}\ \text{J}}{3.125 \times 10^{-11}\ \text{kg}\cdot\text{m}^2}\, t,$$

$$t = 7.41 \times 10^3\ \text{s} = 2.1\ \text{h}\quad \underline{\text{Ans}}.$$

CHAPTER 49

49-4

(a) Let S = solar constant, R = radius of the earth. The rate at which energy from the sun impinges on the earth is $S\pi R^2$. Assuming a uniform distribution of this energy over the surface of the earth during a year, the average rate dE/dt at which solar energy is absorbed by a unit area of the earth's surface is given by

$$(4\pi R^2)\frac{dE}{dt} = S\pi R^2,$$

$$\frac{dE}{dt} = \tfrac{1}{4}S.$$

If the earth's average surface temperature is constant in time, the rate at which this energy is reradiated back into space must, on the average, equal this; hence, the rate of reradiation is

$$\frac{dE}{dt} = \tfrac{1}{4}S = \tfrac{1}{4}(1340) = 335\ \text{W/m}^2,$$

in agreement with the values given.

(b) For a cavity radiator, with a unit area $A = 1\ \text{m}^2$,

$$\frac{dE}{dt} = \sigma AT^4,$$

$$335 = (5.67 \times 10^{-8})(1)T^4,$$

$$T = 277\ \text{K} = 4^\circ\text{C} \quad \underline{\text{Ans}}.$$

49-6

Through the opening of area A the oven emits into the room energy at the rate $A\sigma T_0^4$, and receives energy from the room at the rate $A\sigma T_r^4$. Thus, converting temperatures to kelvins and the area A to square meters,

$$P = A\sigma(T_0^4 - T_r^4),$$

P the net power transferred. Numerically,

$$P = (5 \times 10^{-4})(5.67 \times 10^{-8})(500^4 - 300^4) = 1.54 \text{ W} \quad \underline{\text{Ans.}}$$

49-9

The integral in question is

$$I = \int_0^{\infty} \frac{2\pi hc^2}{\lambda^5} \frac{1}{e^{hc/\lambda kT} - 1} d\lambda.$$

Let $y = hc/\lambda kT$ so that $d\lambda = -(hc/kTy^2)dy$. Also, $y = 0$ corresponds to $\lambda = \infty$ and $y = \infty$ to $\lambda = 0$. Thus,

$$I = 2\pi hc^2 \left(\frac{kT}{hc}\right)^5 \left(\frac{hc}{kT}\right) \int_0^{\infty} y^5 \frac{1}{e^y - 1} \frac{dy}{y^2},$$

$$I = \left[\frac{2\pi k^4}{h^3 c^2} \int_0^{\infty} \frac{y^3 dy}{e^y - 1}\right] T^4 = AT^4,$$

the quantity in the square brackets being a constant, i.e., being independent of T.

49-10

(a) The rate at which energy escapes from a hole of area A is

$$P = A \int_b^r R_\lambda d\lambda,$$

considering only wavelengths between λ_b and λ_r (λ_b = 400 nm, λ_r = 700 nm). With T = 4000 K, it is found that

$$hc/\lambda_b kT = 9.01; \quad hc/\lambda_r kT = 5.15.$$

With these values, Wein's formula for the radiancy should be sufficiently accurate:

$$R_\lambda = \frac{2\pi c^2 h}{\lambda^5} e^{-hc/\lambda kT}.$$

Let $x = hc/\lambda kT$; then,

$$P = \frac{2\pi Ak^4T^4}{h^3c^2}\int_{5.15}^{9.01} x^3e^{-x}dx = \frac{2\pi Ak^4T^4}{h^3c^2}(x^3 + 3x^2 + 6x + 6)e^{-x}\Big|_{5.15}^{9.01}$$

$$P = -(2.98 \times 10^6 \text{ W/m}^2)A.$$

Dropping the minus sign and setting $A = \frac{1}{4}\pi(0.005 \text{ m})^2 = 1.963 \times 10^{-5}$ m^2 gives $P = 58.6$ W <u>Ans</u>.

(b) The total power radiated is $A\sigma T^4$ and therefore the fraction radiated in the visible range is

$$f = \frac{(2.98 \times 10^6)A}{A\sigma T^4} = 0.21 \quad \underline{\text{Ans}},$$

at T = 4000 K.

<u>49-14</u>

(a) Let an area A be a distance r from the lamp where the photon density is n. If A be normal to the beam of radiation, all photons within a distance ct of the area will cross it in time t, since the photons travel with speed c. Hence, with $h\nu$ the energy of each photon, the energy falling on the area in a time t will be

$$E = n(Act)h\nu,$$

considering only photons of frequency ν. The intensity is

$$I = \frac{E}{At} = \frac{nhc^2}{\lambda} = \frac{P}{4\pi r^2},$$

P the power of the lamp. Thus,

$$r^2 = \frac{P\lambda}{4\pi nhc^2}.$$

Putting P = 100 W, $\lambda = 5.89 \times 10^{-7}$ m, $n = 10^7$ /m^3 gives r = 88.6 m.

(b) From (a) it is evident that, at two locations 1 and 2,

$$r_1^2 n_1 = r_2^2 n_2.$$

For $r_1 = 88.6$ m, $r_2 = 2$ m this gives for the density at the closer point,

$$n_2 = (\frac{r_1}{r_2})^2 n_1 = (\frac{88.6}{2})^2(10) = 1.96 \times 10^4 \text{ /cm}^3 \quad \underline{\text{Ans}}.$$

49-22

(a) For the photoelectric effect,

$$h\nu = \tfrac{1}{2}mv^2 + \phi$$

The energy $h\nu$ of the incident photon is

$$h\nu = \frac{1240}{200} = 6.2 \text{ eV}.$$

Now $\phi = \phi_0 = 4.2$ eV for the most loosely bound electron and the greatest amount of kinetic energy that a liberated electron can possess is, therefore,

$$\tfrac{1}{2}mv^2 = 6.2 - 4.2 = 2.0 \text{ eV} \quad \underline{\text{Ans}}.$$

(b) The minimum kinetic energy will be carried by an electron that has a binding energy to the surface of 6.2 eV: i.e., $K_{min} = 0$.

(c) The potential required to stop the 2.0 eV electron in (a) is, since the electron charge is one quantum unit, 2.0 V.

(d) A photon barely able to remove the electron bound with an energy of 4.2 eV has a wavelength

$$\lambda = \frac{1240}{4.2} = 295 \text{ nm} \quad \underline{\text{Ans}}.$$

49-24

Under the assumption that the photon gives all its energy to the initially stationary electron which, after the collision, moves off with speed v, the conservation of momentum and mass-energy gives

$$\frac{h\nu}{c} + 0 = 0 + \frac{m_0 v}{(1 - v^2/c^2)^{\frac{1}{2}}};$$

$$h\nu + m_0 c^2 = 0 + \frac{m_0 c^2}{(1 - v^2/c^2)^{\frac{1}{2}}};$$

the electron, of necessity (conservation of momentum), moves away in the same direction in which the incident photon was headed. In the second equation, transposing the rest energy to the other side and dividing by c yields

$$\frac{h\nu}{c} = \frac{m_0 c}{(1 - v^2/c^2)^{\frac{1}{2}}}[1 - (1 - v^2/c^2)^{\frac{1}{2}}].$$

This agrees with the first equation if

$$c[1 - (1 - v^2/c^2)^{\frac{1}{2}}] = v,$$

which requires either $v/c = 0$ or $v/c = 1$. The former value, by the first equation, leads to $\nu = 0$ also; there is little point in further discussion of this case as the photon energy is zero. The latter "solution", $v/c = 1$, is prohibited by the Special Theory of Relativity.

49-28

(a) The Compton shift is given by

$$\Delta\lambda = \frac{h}{m_0 c}(1 - \cos\theta).$$

The Compton wavelength is $\lambda_C = h/m_0 c = 0.0242$ A. Also, since $\nu\lambda = c$, it follows that $\Delta\lambda/\lambda = -\Delta\nu/\nu$. Hence, ignoring the sign,

$$\frac{\Delta\nu}{\nu} = \frac{\lambda_C}{\lambda}(1 - \cos\theta),$$

$$10^{-4} = \frac{0.0242}{2}(1 - \cos\theta),$$

$$\theta = 7.37^\circ \quad \underline{\text{Ans}}.$$

(b) The energy lost by the photon is given to the electron as kinetic energy. Since the energy of a photon is $E = h\nu$, the energy imparted to the electron must be

$$E^* = h\Delta\nu = E\cdot\frac{\Delta\nu}{\nu} = (6200 \text{ eV})(10^{-4}) = 0.62 \text{ eV} \quad \underline{\text{Ans}}.$$

49-30

Treating the electron classically, the conservation of momentum and of mass-energy yield

$$\frac{h\nu}{c} + 0 = m_0(\beta c) - \frac{h\nu^*}{c},$$

$$h\nu + m_0c^2 = m_0c^2 + \tfrac{1}{2}m_0(\beta c)^2 + h\nu^*,$$

m_0 the rest mass of the electron. Let $K = \frac{1}{2}m_0(\beta c)^2$; then the latter equation becomes

$$h\nu = K + h\nu^*.$$

Eliminate $h\nu^*$ between this and the first equation to obtain

$$2h\nu = K\,\frac{2 + \beta}{\beta}.$$

Since $\beta \ll 1$,

$$2h\nu \approx \frac{2K}{\beta},$$

$$\frac{K}{h\nu} = \beta,$$

proving the assertion.

49-42

If kinetic energy is not conserved, some of the 6.0 eV of the neutron's initial kinetic energy must be used to excite the hydrogen atom. However, to reach the first excited level ($n = 2$) from the ground level ($n = 1$) the atom must be supplied with an amount of energy E, with

$$E = 13.6 - 13.6/2^2 = 10.2 \text{ eV}.$$

Since the neutron does not possess this energy the collision will be elastic, the hydrogen atom remaining in the ground state.

49-44

A neutral helium atom has two electrons. With one electron gone only one remains, so that the singly ionized helium atom is "hydrogen-like" in that it possesses one electron only. According to the Bohr theory, the ionization energy of a one-electron atom from the ground state is

$$E = \frac{Z^2 me^4}{8\epsilon_0^2 h^2} = (13.6 \text{ eV})Z^2.$$

For helium Z = 2 and therefore the ionization energy is 54.4 eV. A small additional correction due to the finite mass of the nucleus is ignored.

49-47

(a) For hydrogen the energy of the allowed quantized states is proportional to m, the mass of the electron. Strictly speaking, to take account of the motion of the proton and electron relative to the center of mass of the atom, m should be the reduced mass of the system:

$$m = \frac{m_e m_p}{m_e + m_p} \approx m_e$$

since $m_e \ll m_p$. In positronium the proton is replaced with another electron so that the reduced mass is

$$m = \frac{m_e^2}{2m_e} = \tfrac{1}{2}m_e,$$

of half the value for hydrogen. But the energies of the photons emitted in transitions is the difference between the energies of the allowed states between which the transitions take place, Hence, in positronium the frequencies of photons are one-half as great as those for the corresponding transitions in hydrogen. With $\lambda = c/\nu$ the wavelengths are $1/(\frac{1}{2})$ = twice as long as in hydrogen.

(b) The radius of the relative orbit is inversely proportional to m for any n; hence, the radius of the ground state orbit in positronium is twice that for hydrogen. In hydrogen the radius of the electron orbit relative to the proton almost equals the radius of the orbit about the center of mass, since the proton is located almost exactly at the center of mass. For positronium, however, the center of mass lies midway between the electrons, and it follows that the distance of either electron from the center of mass in any state virtually equals the radius of the corresponding Bohr orbit in hydrogen.

<u>49-49</u>

(a) By assumption $L = I\omega = nh/2\pi$. The rotational inertia I is

$$I = 2m(\tfrac{1}{2}d)^2 = \tfrac{1}{2}md^2$$

so that

$$\tfrac{1}{2}md^2\omega = nh/2\pi,$$

$$\omega = nh/\pi md^2 \quad \underline{\text{Ans}}.$$

(b) The rotational energy is

$$E = \tfrac{1}{2}I\omega^2 = \frac{L^2}{2I} = \frac{(nh/2\pi)^2}{md^2},$$

$$E = n^2h^2/4\pi^2md^2 \quad \underline{\text{Ans}}.$$

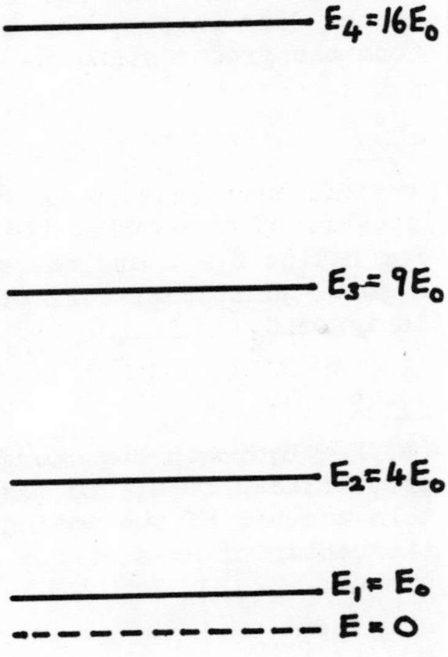

(c) The energies are proportional to n^2. If $E_0 = h^2/4\pi^2md^2$, the energy diagram for the first few states is as shown above.

<u>49-51</u>

The common factor $me^4/8\epsilon_0^2h^3$ cancels and the last column is easily seen to be just

$$100\,\frac{\nu - \nu_0}{\nu} = 100\,\frac{\dfrac{2n - 1}{(n - 1)^2 n^2} - \dfrac{2}{n^3}}{\dfrac{2n - 1}{(n - 1)^2\, n^2}} = \frac{100}{n}\left(\frac{3n - 2}{2n - 1}\right).$$

As n becomes very large, $3n \gg 2$ and $2n \gg 1$ so that

$$\frac{3n - 2}{2n - 1} \approx \frac{3n}{2n} = \frac{3}{2},$$

and therefore

$$100\,\frac{\nu - \nu_0}{\nu} \approx \frac{150}{n}.$$

49-52

Consider a transition between the n and $(n - p)$ levels, p an integer, of necessity. The frequency of the photon involved is

$$\nu = \frac{me^4}{8\epsilon_0^2 h^3}\left[\frac{1}{(n - p)^2} - \frac{1}{n^2}\right] = \frac{n^3\nu_0}{2}\left[\frac{1}{(n - p)^2} - \frac{1}{n^2}\right] = \frac{np\nu_0}{2}\,\frac{2n - p}{(n - p)^2}.$$

For $n \gg p > 1$, $(2n - p)/(n - p)^2 \approx 2/n$ so that

$$\nu \approx \frac{np}{2}\,\nu_0\,\frac{2}{n} = p\nu_0$$

consistent with the correspondence principle.

CHAPTER 50

50-11

The wavelength of the neutrons, which at this kinetic energy much less than their rest energy can be calculated without resorting to relativistic mechanics, is

$$\lambda = \frac{h}{p} = \frac{h}{(2mK)^{\frac{1}{2}}}.$$

With $m = 1.67 \times 10^{-27}$ kg and $K = 4(1.6 \times 10^{-19}$ J), this gives λ = 0.1434 A. In the Bragg relation with m = 1,

$$\lambda = 2d\sin\theta,$$

$$0.1434 = 2(0.7323)\sin\theta,$$

$$\theta = 5.6^\circ \quad \underline{\text{Ans}}.$$

50-13

The speed of the hydrogen atoms is found from

$$\tfrac{1}{2}mv^2 = \frac{3}{2}kT,$$

$$v = \left(\frac{3kT}{m}\right)^{\frac{1}{2}} = 2.7 \text{ km/s},$$

with T = 20 + 273 = 293 K. Hence, the particle momentum is p = mv = 4.49×10^{-24} kg·m/s and its wavelength is

$$\lambda = \frac{h}{p} = 1.5 \text{ A} \quad \underline{\text{Ans}}.$$

50-16

(a) Consider that confining an electron in a nucleus is roughly analogous to the problem of a particle in a one-dimensional box. The energy of the electron, approximately and by nonrelativistic mechanics, would be

$$E = \frac{h^2}{8mL^2} = 3.08 \times 10^{-10}\ J = 1900\ MeV \quad \underline{\text{Ans}}.$$

(b) The nucleus cannot, with a binding energy of 2-3 MeV per nucleon, hold a 2000 MeV particle. Thus, electrons are not expected to be long-term residents of nuclei.

50-18

The wavefunction of the particle is

$$\psi = C\sin\left(\frac{n\pi x}{\ell}\right),$$

where C is a constant. Since the particle must be found somewhere in the box, C can be determined by setting the probability of finding the particle in the whole box equal to one:

$$1 = \int_0^{\ell} \psi^2 dx = \int_0^{\ell} C^2 \sin^2\left(\frac{n\pi x}{\ell}\right) dx,$$

$$C^2 = 2/\ell,$$

so that

$$\psi = \left(\frac{2}{\ell}\right)^{\frac{1}{2}} \sin\left(\frac{n\pi x}{\ell}\right).$$

The desired probability P is

$$P = \int_0^{\ell/3} \psi^2 dx = \frac{2}{\ell}\int_0^{\ell/3} \sin^2\left(\frac{n\pi x}{\ell}\right) dx = \frac{1}{3} - \frac{1}{2n\pi}\sin(2n\pi/3).$$

Successively putting n = 1, 2, 3 gives (a) P = 0.20, (b) P = 0.40, (c) P = 0.33. (d) Classically, the probability of finding the particle in a region of length D is D/ℓ. Thus, with $D = \ell/3$, P = 0.333.

50-19

The motion of the electron in the ground state of the hydrogen atom is described by the wavefunction

$$\psi = (1/\pi a^3)^{\frac{1}{2}} e^{-R/a},$$

where a is the radius of the ground-state Bohr orbit and R is the distance to the nucleus. Consider two hypothetical spherical shells centered on the nucleus, with radii R and R + dR. The probability dP of finding the electron within the infinitesimal shell bounded by these spheres is

$$dP = \psi^2(R)dV,$$

where dV is the volume between the spheres. Hence, the probability of finding the electron within a sphere of radius r centered on the nucleus (a non-infinitesimal volume) is

$$P_r = \int \psi^2(R) \cdot dV = \int_0^r \psi^2(R) \cdot 4\pi R^2 dR = (1/\pi a^3)\int_0^r e^{-2R/a} \cdot 4\pi R^2 dR,$$

$$P_r = (4/a^3)\int_0^r e^{-2R/a} \cdot R^2 dR = (4/a^3)(\tfrac{1}{2}a)^3 \int_0^{2r/a} x^2 e^{-x} dx,$$

$$P_r = 1 - e^{-2r/a}\left(\frac{2r^2}{a^2} + \frac{2r}{a} + 1\right).$$

(a) For $r = 0$, $P_r = 0$ and this is expected since the volume in which the electron is being sought has reduced to zero.

(b) With $r = \infty$, $P_r = 1$; this also is anticipated since the electron must be somewhere, so that if all space is examined, it must be found.

(c) $\psi^2(r)dr$ gives only the probability of finding the electron within the distances r and r + dr, an infinitesimal region. The probability found above refers to a finite region.

50-20

Use the result of Problem 50-19, which refers to a sphere of any radius r, by setting r = a, the radius of the ground-state orbit in the Bohr theory. The result becomes

$$P_r(a) = 1 - 5e^{-2} = 0.323 \quad \underline{\text{Ans}}.$$

50-23

By the uncertainty principle, with $\Delta x = h/p = \lambda$, the deBroglie wavelength,

$$\Delta p \Delta x \geq h,$$

$$\Delta p \cdot \frac{h}{p} \geq h,$$

$$\Delta p \geq p.$$

But the momentum is p = mv, so that

$$\Delta p = m\Delta v,$$

and therefore

$$m\Delta v \geq mv,$$

$$\Delta v \geq v,$$

as asserted. This analysis assumes that the particle can be described classically. If, on the other hand, the motion is relativistic,

$$p = mv = \frac{m_0 v}{(1 - v^2/c^2)^{\frac{1}{2}}},$$

m_0 the rest mass of the particle, then

$$\Delta p = \frac{\partial p}{\partial v}(\Delta v) = \frac{m\Delta v}{(1 - v^2/c^2)}$$

and therefore

$$\Delta v \geq v(1 - v^2/c^2).$$

Hence the result quoted in the problem applies only in the classical domain.

50-25

Call the radius vector R. Then $\Delta x = R\Delta\phi$; also $L = Rp_x$ so that $\Delta L = R\Delta p_x$. Therefore

$$\Delta p_x \Delta x \geq h,$$

$$\left(\frac{\Delta L}{R}\right)(R\Delta\phi) \geq h,$$

$$\Delta L \Delta\phi \geq h.$$